Advanced Technologies for Smart Agriculture

RIVER PUBLISHERS SERIES IN COMPUTING AND INFORMATION SCIENCE AND TECHNOLOGY

The "River Publishers Series in Computing and Information Science and Technology" covers research which ushers the 21st Century into an Internet and multimedia era. Networking suggests transportation of such multimedia contents among nodes in communication and/or computer networks, to facilitate the ultimate Internet.

Theory, technologies, protocols and standards, applications/services, practice and implementation of wired/wireless

The "River Publishers Series in Computing and Information Science and Technology" covers research which ushers the 21st Century into an Internet and multimedia era. Networking suggests transportation of such multimedia contents among nodes in communication and/or computer networks, to facilitate the ultimate Internet.

Theory, technologies, protocols and standards, applications/services, practice and implementation of wired/wireless networking are all within the scope of this series. Based on network and communication science, we further extend the scope for 21st Century life through the knowledge in machine learning, embedded systems, cognitive science, pattern recognition, quantum/biological/molecular computation and information processing, user behaviors and interface, and applications across healthcare and society.

Books published in the series include research monographs, edited volumes, handbooks and textbooks. The books provide professionals, researchers, educators, and advanced students in the field with an invaluable insight into the latest research and developments.

Topics included in the series are as follows:-

- Artificial Intelligence
- Cognitive Science and Brian Science
- Communication/Computer Networking Technologies and Applications
- Computation and Information Processing
- Computer Architectures
- Computer Networks
- Computer Science
- Embedded Systems
- Evolutionary Computation
- Information Modelling
- Information Theory
- Machine Intelligence
- Neural Computing and Machine Learning
- Parallel and Distributed Systems
- Programming Languages
- Reconfigurable Computing
- Research Informatics
- Soft Computing Techniques
- Software Development
- Software Engineering
- Software Maintenance

For a list of other books in this series, visit www.riverpublishers.com

Advanced Technologies for Smart Agriculture

Editors

Kalaiselvi K.
Saveetha College of Liberal Arts and Sciences,SIMATS, Chennai, India

A. Jose Anand
KCG College of Technology, India

Poonam Tanwar
Manav Rachna International Institute of Research & Studies, India

Haider Raza
University of Essex, UK

River Publishers

Routledge
Taylor & Francis Group
NEW YORK AND LONDON

Published 2024 by River Publishers
River Publishers
Alsbjergvej 10, 9260 Gistrup, Denmark
www.riverpublishers.com

Distributed exclusively by Routledge
605 Third Avenue, New York, NY 10017, USA
4 Park Square, Milton Park, Abingdon, Oxon OX14 4RN

Advanced Technologies for Smart Agriculture / Kalaiselvi K., A. Jose Anand, Poonam Tanwar and Haider Raza.

Routledge is an imprint of the Taylor & Francis Group, an informa business

ISBN 978-87-7022-848-0 (hardback)
ISBN 978-87-7004-049-5 (paperback)
ISBN 978-10-0381-040-7 (online)
ISBN 978-1-032-62874-5 (ebook master)

Contents

Preface

Farming is an essential human activity and is the primary segment for all the developing nations. Even though the farming volume has prolonged with the rise in population, the necessity of resources such as farm land, water, and manure are aggregating exponentially. Smart farming can be benefited greatly from its increased use and application. Technology as derived partly from the Internet of Things and analyzed according to specific algorithms has a large and beneficial role to play in preventing, monitoring, growth, and in improving farming. This book focuses on how advanced technologies such as artificial intelligence, machine learning, and deep learning play vital roles in analyzing the crop betterment that will make the readers the latest concepts with practiced affluence. Digitization and crop management will be much helpful to understand the procedures in smart farming in a lined manner. It shows how technology can enhance, streamline, and improve services for smart agriculture. Each chapter in this book focuses on a specific area of agriculture and how technology can be helpful to it, with background and current state-of-the-art examples provided. The book has chapters with case study that provide the readers a better understanding of the technology usage in the field of smart agriculture.

List of Contributors

Ahire, Pritam R., *Department of Computer Engineering, Nutan Maharashtra Institute of Engineering Technology, Savitribai Phule Pune University (SPPU), India*

Anbarasi, G., *Department of Biotechnology, PSGR Krishnammal College for Women, India*

Arunachalam, A.S., *Department of Computer Science, Vels Institute of Science Technology & Advanced Studies (VISTAS), India*

Ashok, J., *School of Business and Management, CHRIST (Deemed to be University), India*

Augustine, C., *Department of ECE, GRT Institute of Engineering & Technology, India*

Bacanin, Nebojsa, *Singidunum University, Belgrade, Serbia*

Balaji, K., *GRT Institute of Engineering & Technology, India*

Balaji, S., *Department of Electrical and Electronics Engineering, SRM Valliammai Engineering College, India*

Balamurugan, R., *Department of Computer Science, Vels Institute of Science, Technology & Advanced Studies, India*

Chacko, Shiju C., *Department of Electrical & Electronics Engineering, Uxbridge College, London*

Charanarur, Panem, *School of Cyber Security and Digital Forensics, National Forensic Sciences University (NFSU), India*

Chowdary, Savana Sunil Kumar, *ECE, Hindustan Institute of Technology and Science Chennai, India*

Dhanasekar, S., *Sri Eshwar College of Engineering, India*

Dharanikumar, S.V., *GRT Institute of Engineering & Technology, India*

Duraiarasu, E., *Department of Electronics and Communication Engineering, Rajalakshmi Engineering College, India*

El Rhalibi, Abdennour, *Department of Computer Science, Faculty of Engineering and Technology, Liverpool John Mooris University, UK*

Ezhilazhagan, C., *Department of Electronics & Communication Engineering, Dr. N.G.P. Institute of Technology, India*

Ganesh Kumar, C., *Department of Electronics and Communication Engineering. Rajalakshmi Engineering College, India*

Gangwar, Veer P., *Department of General Marketing, Mittal School of Business, Lovely Professional University, India*

Gao, Xia-Zhi, *Machine Vision and Pattern, School of Engineering Science, Lappeenranta University of Technology, Finland*

Gobinath, R., *Department of Computer Science, Christ University, India*

Gomathy, M., *School of Computing Sciences, Vels Institute of Science, Technology and Advanced Studies (VISTAS), India*

Govindaraj, V., *Department of Electronics & Communication Engineering, Dr. N.G.P. Institute of Technology, India*

Hanchate, Rohini, *Department of Computer Engineering, Nutan Maharashtra Institute of Engineering Technology, Savitribai Phule Pune University (SPPU), India*

Harihara Sudhan, S., *Department of Electronics and Communication Engineering. Rajalakshmi Engineering College, India*

Jeevitha, S., *Department of Electronics and Communication Engineering. Rajalakshmi Engineering College, India*

Jose Anand, A., *Department of Electronics and Communication Engineering, KCG College of Technology, India*

Jubair Ahmed, L., *Department of Electronics & Communication, Akshaya College of Engineering and Technology, India*

Kalaiselvi, K., *Department of Computer Applications, Saveetha College of Liberal Arts and Sciences (SIMATS), India*

Kamesh, M., *Department of ECE, Velammal Institute of Technology, India*

Kanchana Devi, V., *Vellore Institute of Technology, India*

Karmel, A., *Vellore Institute of Technology, India*

Kartheeswari, K.R., *Department of Computer Science & Engineering, Hindustan Institute of Technology & Science, India*

Mahalakshmi, S., *Department of Computer Science and Engineering, Chennai Institute of Technology, India*

Manikandan, T., *Department of Electronics and Communication Engineering, Rajalakshmi Engineering College, India*

Mohammed Asif, Y.S., *ECE, Hindustan Institute of Technology and Science Chennai, India*

Nirmala Devi, K., *Department of ECE, KCG College of Technology, India*

Nisha, K., *Department of CSE, Chennai Institute of Technology, India*

Olatunde, Olabiyisi Stephen, *Computer Science and Engineering, Ladoke Akintola University of Technology, Nigeria*

Perumal, S., *Department of Computer Science, Vels Institute of Science, Technology & Advanced Studies, India*

Rao Gundu, Srinivasa, *Department of Computer Science, Government Degree College Sitaphalmandi, India*

Sakthivel, V., *Konkuk Aerospace Design-Airworthiness Institute, Konkuk University, South Korea*

Samanta, Debabrata, *Department of Computational Information Technology, RIT Kosovo (A.U.K.), Rochester Institute of Technology – RIT Global, Dr. Shpetim Rrobaj, Kosovo*

Selvaraju, P., *Department of Mathematics, Rajalakshmi Institute of Technology, India*

Sreenivas, R., *Vellore Institute of Technology, India*

Subba Rao, Nagulakonda Naga Sai, *ECE, Hindustan Institute of Technology and Science Chennai, India*

Sujarani, Pulla, *Department of Computer Science, Vels Institute of Science, Technology and Advanced Studies (Formerly Vels University), India*

Thirumurthi Raja, A., *Directorate of Distance Education (DDE), SRM Institute of Science and Technology (SRMIST), India*

Umamaheswari, E., *Center for Cyber Physical System, Vellore Institute of Technology, India*

Vallikannu, R., *ECE, Hindustan Institute of Technology and Science Chennai, India*

Varadarajan, Vijayakumar, *School of Computer Science and Engineering, The University of New South Wales, Sydney, Australia (UNSW Ranks 43rd in the 2022 QS World University Rankings, 1st in Australia for Research Excellence and Impact)*

Vijayakumar, S., *Department of ECE, Paavai Engineering College, India*

Vijaylaxmi, J., *PVKK Degree & PG College, India*

Vishnupriya, B., *Department of Biotechnology, Kongunadu Arts and Science College, India*

List of Figures

List of Tables

List of Abbreviations

ADAS	Airborne data acquisition system
ADC	Analogue-to-digital converter
ADT	Android development tools
AFSIS	Africa Soil Information Service
AgMIP	Agricultural Model Intercomparison and Improvement Project
AGRIBOT	Agricultural Robots
AGROVOC	(a multilingual controlled vocabulary covering all areas of interest of the Food and Agriculture Organization of the United Nations)
AI	Artificial intelligence
AMQP	Advanced Message Queuing Protocol
ANN	Artificial neural network
ANSI	American National Standards Institute
AR	Augmented reality
ARM	Advanced RISC Machines
AS	Autonomous system
BMU	Best matching unit
CAA	Crossbar adaptive array
CAD	Computer-aided design
CCD	Charged coupled device
CEA	Controlled environment agriculture
CGIAR	Consultative Group on International Agricultural Research
CGR	Crop growth rate
CNN	Convolutional neural networks
CoAP	Constrained application protocol
CP	Cropping pattern
CRUD	Create, read, update and delete
CSA	Climate smart agriculture
DAQ	Data acquisition
DAS	Direct attached storage
DAU	Direct active user

DBM	Deep Boltzmann machine
DBMS	Data base management system
DBN	Deep belief network
DCL	Data control language
DDL	Data definition language
DDS	Data distribution service
DHT	Digital humidity and temperature sensor
DIKW	Data insight expertise wisdom
DL	Deep learning
DML	Data manipulation language
DP	Dynamic programming
DPSF	Distributed and parallel simulation framework
DQL	Data query language
DRL	Deep reinforcement learning
DSS	Decision support system
EC	Environment controller
EFISCEN	European Forest Information Scenario Mode
EOS	Entrepreneurial operating system
ET	EvapoTranspiration
FAAR	Forum for African Agricultural Research
FAO	Food and Agriculture Organization
FC	Fully connected
FMS	Farm management software
FONTAGRO	Regional Fund for Agriculture Technology
GAN	Generative adversarial networks
GHG	Greenhouse gas
GIS	Geographic information systems
GIS	Geospatial information systems
GM	Genetically modified
GODAN	Global open data for agriculture and nutrition
GPGPU	General-purpose graphical processing unit
GPS	Global positioning system
GPU	Graphical processing unit
GSM	Global system for mobile
GSN	Global sensor network
GUI	Graphical user interface
HD	High-definition
HDPE	High-density polyethylene
HMM	Hidden Markov Models
HTTP	Hyper text transfer protocol

HYV	High yielding variety
I/O	Input/output
ICACCS	International Conference on Advanced Computing and Communication Systems
ICASA	The International Consortium for Agricultural Systems Applications
ICT	Information and communication technology
IDE	Integrated development environment
IIS	Internet Information Services
IJAERD	International Journal of Advance Engineering and Research Development
IK	Indigenous knowledge
IM	Information management
IoFT	Internet of food things
IoT	Internet of Things
IP	Internet protocol
IR	Infra Red
IRJET	International Research Journal of Engineering and Technology
ISCO	Intelligent Systems and Control
ISIMIP	The Inter-Sectoral Impact Model Intercomparison Project
ISO	International Standardization Organization
IT	Information technology
ITU	International Telecommunication Union
KAUST	King Abdullah University for Science and Technology
KDD	Knowledge data discovery
KNN	*k*-nearest neighbor
LAI	Leaf area index
LCC	Land cover classification
LED	Light emitting diode
LER	Land equivalent ratio
LOD	Linked open data
LPWA	Low power wide area network
LPWAN	Low power wide area network
LSTM	Long-short term memory
MCDM	Multi criteria decision making
MFAFF	Ministry for Food Department, Agriculture, Forestry and Fisheries
ML	Machine learning
MLP	Multi-layer perceptron
MODIS	Moderate resolution imaging spectroradiometer

MOX	Metal Oxide
MQTT	Message Queuing Telemetry Transport
MSP	Minimum support price
MTG	Manx Technology Group
NDVI	Normalised difference vegetation index
NIR	Near InfraRed
NLP	Natural language processing
NN	Neural network
NPK	Nitrogen, potassium, and phosphorus
NPP	Natural plant protection
NRICT	National Research Institute for Chemical Technology
OFDM	Orthogonal frequency division multiplexing
PALS	PanIIT Alumni Leadership Series
PCA	Principal component analysis
PIR	Passive InfraRed
PPD	Polarised phase differences
PV	Photovoltaic
PYP	Pitch your point
QPM	High-quality protein maize
RBFN	Radial basis function network
RBM	Restricted Boltzmann machine
RCT	Resource conserving technology
RDBMS	Relational data base management system
RDF	Resource description framework
RDSMS	Relational data stream management system
RF	Radio frequency
RFID	Radio frequency identification
RGB	Red, green and blue
RGM	Reduced gravity model
RMS	Rotronic monitoring system
RMSE	Root mean square error
RNN	Recurrent neural networks
RvNN	Recursive neural networks
RYT	Relative yield total
SAR	Synthetic aperture radar
SDMP	Soil data monitoring probe
SEM	Smart environment monitoring
SEQUEL	Structured English Query Language
SL	Supervised learning
SM	Soil moisture

SOM	Self-organizing map
SPARQL	SPARQL Protocol and RDF Query Language
SPP	Serial port profile
SQL	Structured query language
SRMIST	SRM Institute of Science and Technology
SSA	Sub Saharan Africa
SSN	Semantic sensor network
STI	Science, technology and innovation
STI	Sexually transmitted diseases
SUFIC	Smart urban farming integrated controller
SVM	Support Vector Machine
SVR	Support vector regression
TCN	Temporal convolutional networks
TCP	Transmission control protocol
TDR	Time domain reflectometry
TIVA	Texas Instruments Versatile ARM
TTA	Telecommunication Technology Association
UAV	Unmanned aerial vehicle
UDP	User datagram protocol
UV	Ultra Violet
UWB	Ultra Wide Band
VAO	Village administrative officer
VGG	Visual Geometry Group
VISTAS	Vels Institute of Science Technology & Advanced Studies
VLSI	Very large scale integration
VoID	Vocabulary of Interlinked Datasets
VRT	Variable rate technology
WPAN	Wireless personal area networks
WSN	Wireless sensor network
WTS	Wireless Telecommunications Symposium
WUR	Wageningen University and Research
XMPP	Extensible messaging and presence protocol

1

Introduction to Smart Agriculture

R. Balamurugan[1], K.R. Kartheeswari[2], and S. Perumal[1]

[1]Department of Computer Science, Vels Institute of Science, Technology & Advanced Studies, India
[2]Department of Computer Science & Engineering, Hindustan Institute of Technology & Science, India
Email: bmurugan.scs@velsuniv.ac.in; kr.kartheeswari@gmail.com; perumal.scs@velsuniv.ac.in

Abstract

A significant part of modern life is played by agriculture. This sector provides both foods for the people worldwide and animal feed. Additionally, it provides raw materials for businesses like tobacco, sugarcane, cotton, jute, and others. In India, one of the major industries that support many people's livelihood and sources of work is agriculture. Nearly 60% of people picked this field as their profession. According to the Economic Survey of India 2022 report, 303 million tons of agricultural products were produced overall. Our farmers have the option to plant a variety of crops because of the existence of vast agricultural lands and various climate conditions.

The farming industry has received a significant boost from the rapid expansion in population and the consistent development in urban–rural earning capability. Additionally, there is a sizable amount of demand for agriculture-related items including crops, fruits, vegetables, flowers, etc. in the export market. The farming sector needs to be updated in order to meet the aforementioned expectations. Today's farmers in India utilize high-yielding crops and several types of fertilizers to help them increase productivity. In addition, the availability of modern warehouses with cold storage and affordable prices aid farmers in securely storing their procedures.

The existing advancements are insufficient to accelerate the development of agriculture. Two significant challenges facing the entire planet today.

The first is population expansion, while the second is steadily rising urbanization. The farming industry is experiencing a labor storage as a result of urbanization. Problems will result from this in the upcoming years.

ICT-enabled technology was developed in the agricultural sector to address this issue. Traditional agricultural production involves a number of distinct phases as follows: (a) soil testing, (b) the choice of seeds, (c) irrigation control, (d) determine and use fertilizer, (e) tracking the expansion, (f) disease prevention and control, (g) harvesting, and (h) marketing.

All of the aforementioned agricultural tasks demand a combination of in-depth knowledge, quick decision-making, effective planning, and hard labor skills. It is challenging to complete all tasks in a routine manner. Consequently, farmers must switch from traditional agriculture to smart agriculture. Advanced techniques and modern algorithms combined with agriculture allow farmers to practice smart agriculture.

The use of smart agriculture helps to improve product quality, decrease harvesting time, and increase output. This chapter contains a detailed explanation of how various ICT-enabled technologies might be used with conventional methods to improve current agricultural activities.

1.1 Introduction to Agriculture Process

Food is a basic need of all living things. For food, we rely on both plants and animals. Ancient men started farming food in a small area and followed specific practices for managing and enhancing it. Agriculture is the practice of crop cultivation.

1.1.1 Soil preparation

The soil where a crop will be planted must first be prepared by ploughing, leveling, and manuring (Figure 1.1). The act of ploughing involves digging and loosening dirt with a plough. This aids in adequate soil aeration. Leveling is the process of distributing and leveling the soil after ploughing. After that, the soil is manured [1].

1.1.2 Sowing

The first step in sowing is choosing seeds from high-quality crop strains. This process of dispersing the seeds across the field after soil preparation is known as sowing. You can sow manually, by hand or with seed drilling equipment. Some crops, like paddy, are seedling-grown in a small area before being moved to the main field (Figure 1.2).

Figure 1.1 Soil preparation.

Figure 1.2 Sowing.

Figure 1.3 Manuring.

1.1.3 Manuring

Nutrients are necessary for crop growth and productivity. Therefore, a consistent supply of nutrients is required (Figure 1.3). Nutritional supplements are given throughout the manuring process, and they can be either natural (manure) or synthetic substances (fertilizers). Plant and animal wastes decompose into manure as a byproduct. Commercially produced fertilizers are chemical mixtures that contain plant nutrients. In addition to giving crops nutrition, manure restores soil fertility. Compost, crop rotation, and planting of leguminous plants are further techniques for replenishing soil [1].

1.1.4 Irrigation

Water is supplied by irrigation. Water sources include things like wells, ponds, lakes, canals, and dams. Waterlogging from excessive irrigation could harm the crop. Controlling this frequency and the time between successive irrigations is necessary (Figure 1.4).

1.1.5 Weeding

Unwanted plants known as weeds commonly grow among crops. They are eliminated using weedicides and hand pulling, and some are eliminated while the soil is being prepared (Figure 1.5).

Figure 1.4 Irrigation.

Figure 1.5 Weeding.

1.1.6 Harvesting

Once a crop has achieved maturity, harvesting involves cutting and gathering it. Following harvest, grains are separated from chaff either manually (winnowing) or by threshing in small-scale operations (Figure 1.6).

Figure 1.6 Harvesting.

1.1.7 Storage

At godowns, harvested grains are kept in granaries or bins for subsequent use or marketing. Therefore, crop protection techniques need to be improved. Prior to storage, grains are cleaned, dried, fumigated, and other insect and rodent protection measures are taken (Figure 1.7).

1.2 Role of Smart Agriculture in Soil Preparation

A soil monitoring system with the Internet of Things can improve production, reduce illness, and optimize resources by leveraging technology. IoT sensors can measure thermometry, NPK, volumetric water content, photosynthetic radiation, soil water potential, and soil oxygen levels. IoT sensors collect and transmit data to a central location for analysis, visualization, and trend analysis.

To increase crop yield and quality, agricultural practices can be optimized, trends can be detected and minor changes can be made to the environment. IoT is a key component of precision farming, and IoT is used in smart agriculture.

1.2.1 Activities of IoT sensors

As part of smart farming, soil, weather, and crop conditions are taken into account as well as smart environment (air quality) and smart water (pollution,

Figure 1.7 Storage.

turbidity, and nutrients). In the order of popularity, here is a list of the top IoT sensors:

1.2.1.1 Soil temperature

Moisture, conductivity, surface temperature, and soil temperature are all monitored by the smart agriculture node (Figure 1.8).

The activity of plants below ground is greatly influenced by soil temperature, which affects root development, respiration, nitrogen mineralization, and decomposition. A probe buried in the soil is the most precise method of determining soil temperature, but using IoT sensors to measure air temperature and other parameters may also be possible. There can be multiple probes inserted at varying depths based on the plant's root system. The temperature of the soil can be measured with IoT sensors that use IR technology.

1.2.1.2 Soil moisture

Buried probes with electrodes can also be used to measure the moisture content of the soil. In hydrology, soil chemistry, plant development, and groundwater recharge are all significantly influenced by agricultural moisture content and soil science. Several factors make the moisture content of the soil crucial: Water is a crucial ingredient for all plants and crops, and it is also a

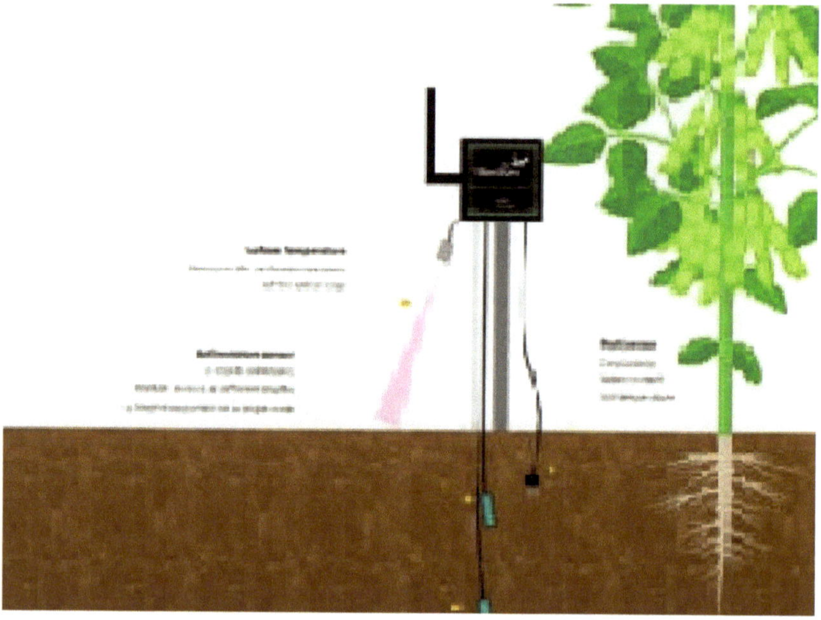

Figure 1.8 Soil temperature sensor usage.

necessary part of photosynthesis. The amount of water in the soil has a big impact on crop yield.

For the growth of plants, soil water is a crucial carrier of soluble food components. Water in the soil aids in controlling soil temperatures.

Given how complicated soil science is, it should go without saying that this essay cannot cover it. Libelium's IoT and smart agriculture technologies may gauge the following:

- Soil hydration (3 × depths).
- Conductivity.
- Water volume content.
- Potential water in the soil.

1.2.1.3 Weather

Different sun radiation ray types that are essential to photosynthesis can be measured by IoT sensors (Figure 1.9). Beyond the most fundamental light levels, IoT can measure the following:

Figure 1.9 Weather sensor.

- A light source that stimulates photosynthesis.
- Sunlight and UV light.
- Shortwave – solar.

Plant development can be significantly impacted by solar radiation, and IoT makes it possible to track sun levels and identify correlations and trends [2].

1.2.1.4 NPK soil sensors

Although they are relatively new on the market, nitrogen, phosphorus, and potassium (potash) sensors offer a way to measure these important soil nutrients with IoT sensors. NPK IoT sensors employ a variety of technologies, but TDR is one that they frequently employ. The following elements can be measured using NPK sensors:

- Potassium

- pH
- EC
- Temperature
- Moisture

The accuracy and sensitivity of NPK sensors differ among vendors according to their development stage. NPK sensors enable RS 485, enabling them to be integrated into IoT solutions, such as LoRaWAN and data recorders. MTG is supporting certain next-generation NPK sensor trials that might improve the accuracy of these sensors.

1.2.1.5 Other measurements

We provide a wide range of additional IoT sensors for agriculture and are useful in certain specialized situations. They improve any IoT implementation even though they are not soil monitoring. The following additional sensors are among them:

- Soil oxygen levels and vapor pressure.
- The temperature of the leaves and flower buds.
- Leaf moisture.
- Fruit, stem, and trunk dimensions.

1.2.1.6 Wireless communications for IoT and smart agriculture

The availability of a wide variety of wireless communication methods is a fundamental advantage of IoT solutions. The use of smart farming is not limited to urban, rural or extremely remote environments. It can be applied to 4G, LoRaWAN, Zigbee, Sigfox, Wi- Fi, and satellite technology as well. The coverage is not limited to urban areas. Sensors and nodes from IoT networks require minimal power, making them compatible with battery, solar or other renewable sources of energy (Figure 1.10).

1.2.1.7 Applications

IoT can be used to optimize the operations of almost any agriculture or farming enterprise. Particular interest has been shown below:

- Marijuana and hemp
- Soybeans

Figure 1.10 Wireless communication sensor.

- Potato
- Cherries with apples
- Almonds

Libelium IoT supports a wide variety of communication protocols (Figure 1.11), is of relatively low bandwidth, and can be buried or placed on the soil surface. Through satellite communications, data can be sent back to the cloud from localized networks of sensors using LoRaWAN (15KM).

Transmission of this information can either be real-time or batch-based. Satellite, IoT, and agriculture applications are particularly relevant to rural farming villages and settlements dispersed over a large area, possibly with inadequate or non-existent mobile or traditional broadband coverage.

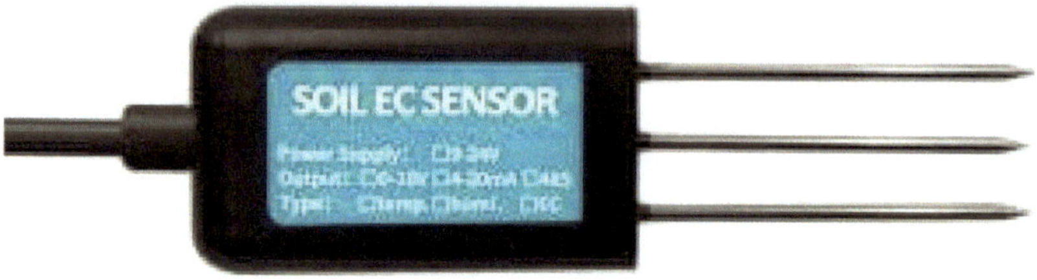

Figure 1.11 Soil sensor.

A precision farming or automation system's measurement component always comes first. Tens of thousands of data points are collected each day by smart agriculture solutions and IoT. As soon as the data has been processed, formatted and corroborated, the systems will be able to use it to optimize other aspects of your infrastructure or launch an intelligent response.

For instance, software can instruct your irrigation system to turn on when the soil moisture falls below a predetermined level. The technology can send notifications to operating or farming employees if moisture levels rise to unsatisfactory levels. The potential with IoT and agriculture is limitless.

Nitrates can be measured in water; however, phosphates are a little more challenging. You may therefore calculate fertilizer run-off after application using this kind of equipment. The majority of technology still uses reagents for phosphate detection in water, which is a little more specialized [2].

Your environment and application will actually determine how well you use data and insights.

Dashboards on the web can visualize data.

- Data visualization for mobile devices is possible using mobile apps and web sites.
- Analytical tools like Excel, PowerBI or Tableau can be used to retrieve or import data.
- If thresholds are surpassed, dashboard and monitoring systems can send out alerts (i.e. NPK levels, soil moisture, etc.)

1.3 Role of IoT Devices in Smart Agriculture

Different devices have a crucial role to play in the field of smart agriculture. The function of those gadgets is briefly outlined in this section.

Figure 1.12 Usage of robotics in agriculture.

1.3.1 Robotics in agriculture

Over the past century, automation has improved to handle complex tasks and increase productivity.

Farmers are starting to pay attention to ag-bots, sometimes called agriculture robots, as a result of the worldwide labor shortage and increased demand. Agriculture yield reduced approximately 2000 million tons per year due to labor shortage in the entire world. With the usage of modern equipment and cutting-edge technologies allowing systems to extract meaningful outcomes from their neighbors, ag-robotics are now more widely known. We are currently at the very beginning of an ag-robotics revolution (Figure 1.12).

1.3.1.1 Weeding robots

These cunning agricultural robots make use of advanced image processing methods to look for similarities between crops and weeds in photos in its database. They then use their robotic arms to remove the weeds or spray them directly. The growing pesticide resistance among plants is advantageous for both the environment and farmers who once applied pesticides liberally throughout their farms. Herbicides are administered at a cost of 1725 crores ($25B) per year, or about 13,000 kg (3 billion pounds), lowering their overall cost (Figure 1.13).

1.3.1.2 Machine navigation robots

Tractors and large pieces of ploughing equipment can be operated automatically from the comfort of your house using GPS, just like remote-controlled toy vehicles can. The labor-intensive duties are made simpler by these integrated automatic machines' excellent accuracy and ability to self-adjust when they notice variations in the terrain. Smartphones make it simple to monitor their whereabouts and work progress. These technology-driven motors are enabling advanced farming using IoT independently with features like

Figure 1.13 Wedding device.

Figure 1.14 Navigation robots.

automatic obstacle recognition thanks to improvements in IoT in agriculture and machine learning (Figure 1.14).

1.3.1.3 Harvesting robotics

Agribots are being used to pick crops, which is a solution to the manpower shortage issue. These cutting-edge machines can work around the clock, handling the delicate process of harvesting fruits and vegetables. These devices pick the fruits by using robotic arms and image processing to determine which ones to pick, thereby controlling the quality. Orchard fruits like apples are crops that are prioritized for agribot harvesting early on due to high operational expenses. These bots can also be used for the greenhouse harvesting of high-value crops like strawberries and tomatoes (Figure 1.15). These robots can accurately assess the growth stage of crops in greenhouses and harvest them when necessary [3].

1.3.1.4 Material handling robots

The hated physical labor jobs can be carried out by robots working alongside humans. They can lift big objects and space plants precisely, which improves the quality of the space and plant and lowers manufacturing costs.

Figure 1.15 Harvesting robot.

1.3.2 Drones in agriculture

Agriculture is one of the major businesses that uses drones. Drones equipped with sensors and cameras are used for farm imaging, mapping, and surveying. There are drones that are both airborne and ground-based. Mobile robots called ground drones are used to assess areas. Unmanned aerial vehicles often known as aerial drones or unmanned aircraft systems are flying robots. Thanks to flight plans that are software-controlled and coordinated with sensors and GPS in their embedded systems, drones can be remotely commanded or flown automatically.

Drones equipped with sensors and cameras are used for farm imaging, mapping, and surveying. The drone data may be used to learn about crop health, irrigation, spraying, planting, the soil and field, plant counts, yield forecast, and much more. Drone surveys may be prearranged or drones may be brought and kept close to farms where they be maintained and recharged. Drones must be brought to nearby labs for data analysis after the surveys, which will enhance the usage of IoT in agriculture (Figure 1.16).

1.3.3 Remote sensing sensors in agriculture

IoT-based remote sensing makes use of sensors installed next to farms, such as weather stations, to collect data that is then sent to analytical tools for examination. Sensors are tools that can detect irregularities. From the analytical dashboard, farmers can keep an eye on their crops and act based on their findings (Figure 1.17).

Crop monitoring sensors placed throughout the farms monitor for changes in light, humidity, temperature, shape, and size. The sensors check for irregularities and notify the farmer if any problem found. Therefore, remote sensing can help prevent the spread of illness and monitor the development

Figure 1.16 Usage of drones in agriculture.

Figure 1.17 Remote sensing sensor.

of crops. The data collected by sensors about humidity, temperature, moisture precipitation and dew detection helps in weather forecasting on farms so that suitable crops may be grown. A soil health research aids in identifying the nutrient value and drier farm areas, as well as the soil drainage capacity or acidity. This information is used to control the quantity of water needed for irrigation and to choose the most favorable type of cultivation [3].

1.3.4 Computer imaging in agriculture

Sensor cameras placed strategically around the farm or drones fitted with cameras are used in computer imaging to capture images that are then

Figure 1.18 Imaging sensor for agriculture.

processed digitally. The fundamental idea behind digital image processing is to use computer algorithms to modify an input image. Image processing analyses limiting factors and aids in improved farm management by seeing images in various spectral intensities, such as infrared, comparing photos gathered over time, and detecting anomalies (Figure 1.18).

1.4 Role of Irrigation in Smart Agriculture

Soil, weather, and plant IoT sensors can be employed in a smart irrigation solution, depending on the type of data that has to be collected.

1.4.1 Soil sensors

Sensors embedded in the soil collect pertinent information on factors including electrical conductivity, salinity, and volumetric water content. These sensors, which are strategically placed around the field, transmit data to a smart irrigation system to give farmers rapid insights into the condition of the soil and forecast irrigation requirements.

1.4.2 Weather sensors

Weather sensors, also known as evapotranspiration (ET) sensors, monitor extremely local environmental factors such as water evaporation from soil surfaces and plant transpiration. These sensors can aid in producing more precise water predictions when combined with the data supplied by a GIS-based solution [4].

1.4.3 Plant sensors

Although still in its infancy, plant-mounted sensors have considerable potential for monitoring a plant's water status. A sensor that is fastened to a plant's stalk or fruit can detect minute changes, such as swelling or shrinking, and warn farmers of low water content or yield-decreasing traits in the field.

1.4.4 Capabilities of a smart irrigation solution

Advanced software capabilities can be added to an IoT-based irrigation system to improve it even more. To get extra useful data and guide your decision-making on the watering of farmlands, you can integrate third-party APIs into your smart water irrigation solution and modify it.

1.4.4.1 Weather monitoring

Using satellite data and weather reports from weather stations to properly arrange your irrigation activities is one approach to achieve this. The technology may wait and automatically adjust the amount of water needed based on actual precipitation received because it is aware that rain is anticipated.

1.4.4.2 Location technology

By reducing expenses and collecting valuable information, precision mapping also assists agribusinesses in promoting smart farming. Farmers may monitor important land characteristics by using reusable maps that combine sensor data and satellite and drone imagery.

1.4.4.3 Artificial intelligence

The foundation of farming automation is evolving to include AI. In addition to helping to automate straightforward operations like data labeling, report writing and notice sending, AI also finds applications in unique yet infrequently used agricultural robotics systems. Such simple AI automation processes are getting simpler to deploy, but the effects on agribusinesses will be significant and noticeable in a matter of months.

1.4.5 Technology used in smart irrigation system

You can select from a variety of irrigation system types that can be improved with intelligent irrigation software depending on how water is dispersed around the field. Flood, sprinkler, center pivot, drip, and micro-irrigation systems are the most popular. Let us examine how each type's efficiency can be increased by technical advances in smart irrigation [4].

1.4.5.1 Sprinkler irrigation

A sprinkler irrigation system uses high-pressure overhead sprinklers to disperse water after it has been pumped through pipes. These sprinklers can be placed on a movable platform or in the middle of the field.

Function of software: After the rain ceases, thermal and auditory rain sensors identify the precipitation and gauge its intensity to plan the subsequent irrigation. A smart irrigation system analyses data and determines the monthly water budget. To avoid excessive water consumption and overwatering due to rain, sprinklers receive automated notifications.

1.4.5.2 Centre pivot irrigation

This type of sprinkler irrigation, sometimes referred to as water-wheel and circular irrigation, is the most common. In a conventional center pivot system, plants are irrigated by sprayers as a long irrigation conduit tied to a central tower gently circles the field.

Function of software: In-field sensors provide data insights that the system using circular irrigation sprinklers uses to modify the water stream or angle of flow. By doing so, you can reach plants that are far from the water source and prevent overwatering of nearby plants. The system plans irrigation and determines possible yield and harvest periods by examining weather information and soil moisture.

1.4.5.3 Drip irrigation

Through pipes with tiny perforations called drippers, water is delivered directly to the plant roots in this method of irrigation. This enables farmers to dramatically lower run-off and evaporation.

Function of software: The biggest difficulty with this kind of irrigation is making the watering process visible. The technology tells the user when watering begins and ends via an app. Additionally, it assesses soil characteristics both before and after irrigation.

1.4.5.4 Micro-irrigation

A low-pressure, low-volume technique called micro-irrigation provides precise control over watering. The technology delivers water straight to the roots of the plant, increasing distribution uniformity and increasing irrigation efficiency.

Software's function: Because the water is carefully managed, the system can calculate the precise dosage for each plant. Plant recognition and irrigation adjustments can be made using AI algorithms. Whatever kind of irrigation system you decide on, putting sophisticated IoT sensors in it will enable you to acquire data-driven, useful insights and keep track of your irrigation requirements [4].

1.5 Role of Harvesting in Smart Agriculture

Population growth has raised the need for food, which in turn has put more pressure on farmers to improve crop productivity and area farmed. Additionally, agriculture is practiced on a big scale in developing nations, which will in the near future greatly boost the demand for automation and smart harvesting technology.

Consumers are becoming more knowledgeable about nutrition and food safety. Today's farmer must meet consumer demands in addition to feeding the world's expanding population. Plant breeding is currently prioritizing crop enhancement to fulfill the avalanche of requests from numerous sectors as a result of the present pandemic's increased need for items made from plants. The secret to modern farming success is breeding crops with greater yields and consumer qualities like nutrition, shelf life, etc. Through the use of technology, our farmers have been able to increase crop yield by better understanding their soil, the nutrients it needs and the timing and pattern of planting. In a nutshell, farmers are experiencing benefits as a result of the integration of science and technology into farming.

Plant breeders are utilizing the cutting-edge science of gene editing to modify crops for increased yield, drought tolerance, disease resistance, high nutrition, and better taste. Two examples of the enormous impact gene editing can have on addressing significant farming challenges are citrus greening disease, which affect orange growers globally, and Panama disease of bananas, which struck the entire global and destroyed numerous acres of bananas in Asian countries, heavily affecting the life of small-scale farmers.

Researchers have precisely and successfully changed an organism's DNA to add desired qualities or remove unwanted ones by using gene editing. As a result, research takes only 2–3 years as opposed to the 7–25 years needed for conventional breeding.

The widespread application of gene-editing technology in agriculture will boost crop yields and as a result, worldwide food quality, benefiting all people [5].

1.6 Advantages of Smart Agriculture

- Smart agriculture sensors collect a terabyte of information, including data on weather, soil quality, crop development, and animal health.

This data can be used to track the general growth of agricultural products as well as worker efficiency, equipment performance, etc.

- Lessening of production hazards due to improved internal process control. If farmers predict the output of agriculture prediction, then they can set up a more efficient product distribution system.
- If the farmers are certain of the quantity of yield, then their items will not be left on the shelf unclaimed. Greater production control has resulted in cost control and waste reduction. If the agricultural people identify any irregularities in crop growth or livestock health, they can lessen the possibility of losing their output.
- Increased operational effectiveness brought on by automation of process. Smart devices may be used to automate a variety of processes throughout the entire production cycle, including irrigation, fertilization, best control, and distribution system.
- Increased production quantities and quality: A more automated production process allows higher standards to be maintained for crop quality and growth potential.
- This may lead to increased revenue in the long run.

1.7 Challenges in Smart Agriculture

The effectiveness of smart agriculture is influenced by a number of things. To achieve a better result, that must be addressed.

1.7.1 Hardware

These devices in agriculture must incorporate sensors so that an IoT solution can be constructed. Farmers are responsible for compiling and identifying the overall objective. As long as the data collected is accurate and reliable, the quality of the sensors is also crucial to the success of the product.

1.7.2 Brain

Analytics should be at the heart of every single smart agriculture solution. Data analysts are necessary to identify the insights in the collected information. For extracting hidden knowledge from the collected data, it is necessary to have strong data analytics, predictive algorithms and machine learning techniques.

1.7.3 Maintenance

The complexity of hardware maintenance is crucial for IoT solution in agriculture because the sensors are constantly used in the field and are easily damaged. It is necessary to choose devices that are reliable and flexible. Otherwise, physical supervision will be required in order to monitor the field.

1.7.4 Mobility

Applications for smart farming ought to be created specifically for field use. A farm manager or farm owner should have on-site or remote access to the data via desktop computer or smart phone. All connected devices must also be independent and have a large enough wireless coverage area to communicate with one another and transfer data to the main server.

1.7.5 Infrastructure

Smart farming applications require a solid internal infrastructure to function smoothly. Verifying the security of internal systems is equally important. Without it, hackers can easily hack into the system, steal data, and manipulate the results.

1.7.6 Connectivity

The demand for data transfer across various agricultural facilities continues to impact the implementation of smart farming. It should go without saying that the connection between these locations needs to be reliable enough to withstand bad weather and maintain service. IoT devices continue to employ a diversity of communication protocols despite the fact that there are currently efforts to create unified standards in this field.

1.7.7 Data gathering interval

Given the variety of data types in the agriculture industry, figuring out the appropriate frequency of data collection can be difficult. Data from field-based, aerial and environment sensors, applications, devices and equipment, as well as processed analytical data, may be subject to restrictions and legislation. The safe, prompt supply of this data and its sharing are two main issues with smart farming

1.7.8 Data security in the agriculture industry

Precision agriculture and IoT technologies necessitate working with enormous data volumes, which increase the likelihood of security holes that criminals might use for data theft and hacking attacks. Sadly, the idea of data security in agriculture is still in its infancy. For example, many farms use drones to transmit data to farm machinery. The Internet is connected to this equipment, but there are not many security precautions in place, including user passwords or remote access authentications.

The fundamental IoT security recommendations include monitoring data flow, encrypting sensitive data, using AI-based security technologies to immediately identify suspicious behavior, and storing data in the block chain to ensure its integrity. To effectively benefit from smart agriculture, farmers must familiarize themselves with the notion of data security, create internal security rules, and adhere to them [6].

1.8 Limitations of Smart Agriculture

- In order to operate and monitor farms, sensors, robots, drones, and cameras are required.
- In addition, expensive and sensitive equipment is needed for the deployment and management of bots and IoT.
- Hardware ongoing maintenance costs.
- Considerable upfront costs.
- Using cameras and drones operated manually, computer imaging is accomplished.
- There is no supply chain management and the system cannot be scaled since the data from each farm must be preserved independently.
- Issues with compatibility with already-installed devices and a lack of log information.
- Each piece of equipment performs a specific set of tasks, and hence none of them can provide all metrics.
- The infrastructure needs for smart agriculture systems with substantial upfront expenses involved in setting up sensors, drones, and robots.
- Proper and uninterrupted power connection is required to run and charge the drones and robots, as well as to hire qualified field employees for administration and operation.

- Hardware maintenance cost required for the entire year.
- Requirement of high-speed 24×7 Internet access facility.

1.9 Future Trends in Smart Agriculture

The greatest manufacturing sector is food production, which also makes it a top contender for technological advancement in contemporary business. Agriculture innovation may assist with problems including minimizing food waste, raising nutritional value, boosting output, and learning environmental impact across supply chain. Many new technologies and advancements will come to exist in the coming few years that will help to improve quantity and quality of agriculture product. In this section, we discuss about the future trends that are helpful in the field of smart agriculture.

1.9.1 The Internet of food things (IoFT)

In order to better understand how artificial intelligence, data analytics, and future technologies may advance the digitization of food supply chain, the Internet of Food Things brings together data and computer scientists, chemists, and economists. It has promoted the idea that a network of actuators, robots, sensors, cameras, drones, and other connected devices will enable agriculture to operate at a previously unheard-of level of automation and decision-making, enabling a long-lasting ecosystem of innovation in this oldest of industries. The outcome of IoFT provides the following advantages to the farming communities.

- Ensure the safety and hygienic of food items.
- Provide better distribution network of agricultural products.
- Regulate the supply chain cycle.
- Suggest modern storage techniques.
- Reduce the wastage occurred during production and distribution.

1.9.2 Advanced green revolution

The combination of IoT, artificial intelligence, and machine learning shows the way for third green revolution. This revolution makes use of big data analytics, unmanned aerial vehicle, robots, latest farming equipment, and advanced visualization and analytical tools. This helps to reduce the usage of water in the field and is used to optimize the amount of resources needed for each and every crop in the agricultural industry.

1.9.3 Biologically edited crops

It is expected to rise gene-edited crops by the year 2050 and it will help to produce more types of crops with better yield and size. It helps the scientist to produce different varieties of crops by modifying the existing crop's DNA structure. In addition to this, those varieties are capable of withstanding all climate conditions and diseases. It is possible to produce crop with balanced diet ingredients such as desirable amount of protein, fat and oil content. This will help to improve the quality of human being.

1.9.4 The robot farmers

In agriculture industry, for the employee time is a crucial factor to get quality and quantity of yield. Sometimes they are tired and get bored to involve in this type of repetitive work. That affects the outcome of the task. To overcome this issue, robots will be introduced to perform various tasks of farming. Tractor used with robots will save a lot of time and that can be effectively utilized for some other important activities. The usage of small robots allows us to gather geographical information and image of the plant individually for analytical purpose. Based on the report, the amount of resources needed for each crop in the field is fed by the robot. This will optimize the time, resource, and effort.

1.9.5 The automated greenhouse effect technology

In general, all over the world 95% of the plants are cultivated in the open field. Based on the natural climatic conditions, the farmers can get their yield. The heavy rain, cold, and drought situations affect the growth of the crop. To answer these uncertainties, automated greenhouse effect technology will be used in the near future. It is the technology of giving feasible climate conditions to the crops.

It is possible to design greenhouse based on shape, material, utility, and construction. The commonly used greenhouse based on shape (Figure 1.19) are:

- Lean-to type greenhouse.
- Even-span type greenhouse.

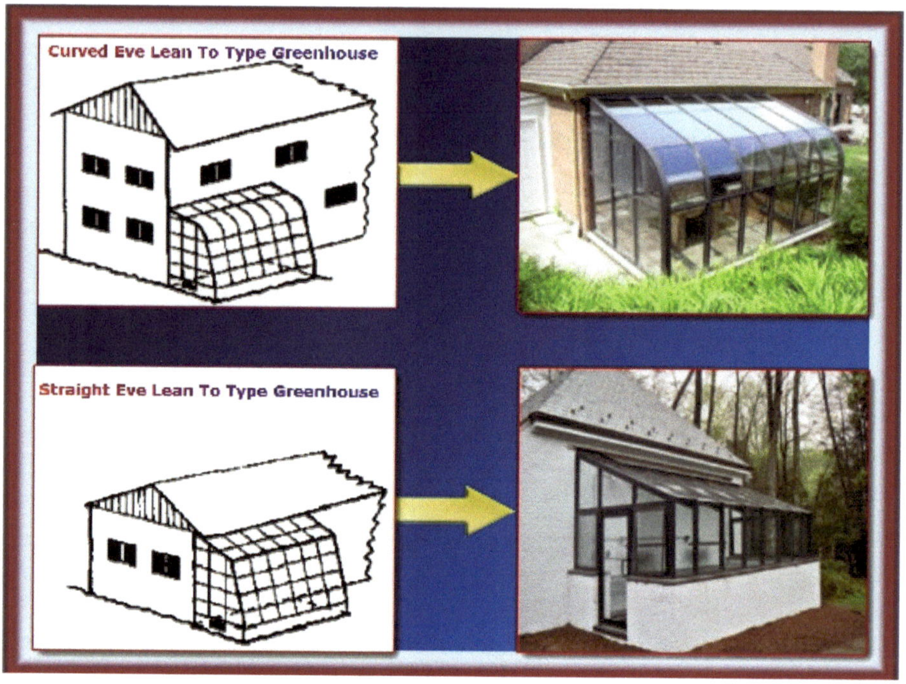

Figure 1.19 Greenhouse design-based on shape.

- Uneven-span type greenhouse.
- Ridge and furrow type.
- Sawtooth type.
- Quonset greenhouse.
- Interlocking ridges and furrow type Quonset greenhouse.
- Ground to ground greenhouse.

Based on utility, the design of greenhouse can be classified as follows:

- Greenhouses for active heating.
- Greenhouses for active cooling.

Based on construction (Figure 1.20), the greenhouse design can be classified as:

- Wooden framed structure.
- Pipe framed structure.
- Truss framed structure.

Figure 1.20 Greenhouse design-based on construction.

Based on covering material (Figure 1.21), the design of greenhouse can be classified as follows:

- Glass greenhouse
- Plastic film greenhouse
- Rigid panel greenhouse

Based on cost of construction (Figure 1.22), the greenhouse design can be classified as:

- High-cost greenhouse
- Medium-cost greenhouse
- Low-cost greenhouse

According to the survey conducted by WUR, the vegetable produced from closed environment with the help of greenhouse technology provides better quality than from the open environment. Since the plants are cultivated in closed manner, the product reaches the market at the estimated time.

Figure 1.21 Greenhouse design-based on covering material.

Figure 1.22 Greenhouse design-based on construction cost.

1.9.6 Climate smart agriculture

The way of changing and modifying agriculture process based on climate change is referred to as climate smart agriculture. The main objective of CSA is to achieve food security with higher productivity. In order to achieve that goal, three important factors are identified such as productivity, adaptation, and mitigation.

1.10 Conclusion

The increasing population, the need for high-yield crop, and the shortage of natural resources force the world to transmit from traditional agriculture to smart agriculture. The growth of information and communication technology help to implement modern agricultural process. It helps the farmers to cultivate plants in efficient, innovative and secured manner. Using this new technology, agriculture industry tries to resolve all the problems faced by the entire world.

References

[1] https://byjus.com/biology/agriculture-agricultural-practices/
[2] https://www.manxtechgroup.com/soil-monitoring-with-iot-smart-agriculture/.
[3] https://www.cropin.com/iot-in-agriculture
[4] https/intellias.com/smart-irrigation-in-agriculture
[5] https://www.apaari.org/web/smart-harvesting-a-must-for-better-crop-yield/
[6] https://easternpeak.com/blog/iot-in-agriculture-technology-use-cases-for-smart-farming-and-challenges-to-consider/
[7] https://www.iotforall.com/smart-farming-future-of-agriculture
[8] https://www.mckinsey.com/industries/agriculture/our-insights/agricultures-connected-future- how-technology-can-yield-new-growth
[9] https://www.worldbank.org/en/topic/climate-smart-agriculture
[10] https://www.syngenta-us.com/thrive/research/future-of-farming.html
[11] https://www.bbc.com/future/bespoke/follow-the-food/five-ways-we-can-feed-the-world-in-2050.html
[12] https://www.forbes.com/sites/jordanstrickler/2020/08/28/high-tech-greenhouses-could-be-the-future-of-agriculture
[13] https://www.hortidaily.com/article/9238914/market-potential-and-investment-opportunities-of-high-tech-greenhouse-vegetable-production-in-the-usa/
[14] https://www.lanner-america.com/blog/smart-farming-iot-5g-agriculture/
[15] https://www.iof2020.eu/about

2

Modern Agriculture Farming: Rack and Pinion Mechanism-based Remote Controlled Seed Sowing Robot

R. Vallikannu[1], Xia-Zhi Gao[2], Nagulakonda Naga Sai Subba Rao[1], Y.S. Mohammed Asif[1], and Savana Sunil Kumar Chowdary[1]

[1]ECE, Hindustan Institute of Technology and Science Chennai, India
[2]Machine Vision and Pattern, School of Engineering Science, Lappeenranta University of Technology, Finland
Email: vallikannu@hindustanuniv.ac.in; xiao.z.gao@gmai.com; 18121002@student.hindustanuniv.ac.in; 18121013@student.hindustanuniv.ac.in; 18121063@student.hindustanuniv.ac.in

Abstract

Nowadays the population is increasing faster and greater, and the demand for the food industry is enormous, leading to smart technology assistance for farmers to upgrade their cultivating process and increase the yield. The existing technology for agriculture focuses on monitoring. However, the initial process of seed sowing is still labor-intensive, and time-consuming leading to less productivity. Small-scale farmers find labor charges unaffordable. Therefore, there is a huge need for a smart seed sowing system to improve precise sowing and reduce labor costs. Hence, the project presented aims to design and develop a seed sowing robot or machine which is controlled via Bluetooth module at an affordable cost with a simple interface for farmers, which is user friendly.

The proposed model is controlled by a smartphone application. A Bluetooth module is used to receive the controlling signals and transmit them to the robot. A Rack and Pinion mechanism is considered for drilling the soil and a servo motor is connected to the seed tanker to disperse the seed. TIVA C series controller board controls all the operations in the robot. The distance between seeds and the considered depth of seeds planted are programmed for various crops. The designed robot monitors the path that has to be followed to sow the seeds.

The system structure is stable and a 3D design is developed to justify the stability of the system in motion. The robot is stable at various soil types at different humidity levels.

This system outperforms the existing model with high stability and precision in seed sowing distance as per the crop requirement.

2.1 Introduction

Robotics is a hopeful technology that subsidizes almost every aspect of the universal economy, from medicine to space learning. However, another sector is often lagging in agriculture. Today the environmental impact of agricultural production is extremely targeted and demand within the trade is increasing. Within the current context, most Indian cities do not have enough energy within the agricultural sector which disturbs the progress of emerging countries. Farmers ought to thus use advanced technology to try planting work (digging, sowing seeds, etc.) The recently developed prototype operates manually and there is no craftsmanship done by what is expected of seed sowing. The manual technique involves spreading the seeds by hand. Typically sowing is employed to create holes and drop the seeds by hand. Bulls are used to tilling and loosening the soil and dropping seeds. Thus it is time for the trade to mechanically overcome this downside. There is an urgent requirement to be innovative in the event of introducing agricultural machinery. The new thought of the purpose of the project makes it potential to dig and plant seed seeds and cover the soil mechanically so that human efforts are reduced by 90%. Agricultural Robots or AGRIBOT may be the right choice for agricultural functions.

Farmers defend the farm product from the time of sowing till the time it reaches the market. Their toil is the reason why most folks have food on our table daily. The proposed project has several benefits such as:

- Scale back the farmer's toil through AGRIBOT.
- Agribots do farmers' work, increase productivity and scale back trade reliance on handicrafts.
- One of the benefits of robots is that they will work 24/7, every year.

A number of the main applications for AI in agriculture include:

1. Gathering and selecting;

2. Self-cutting, sowing, spraying, and cutting;
3. Filtering and packaging;
4. Phenotyping.

As the population grows, farmers should use new technologies to stay in pace with the growing demand. By 2050, associate degrees calculable to nine billion individuals worldwide in square measure are expected to grow. The IEEE AI and Automation Society state that "agricultural production should be doubled if it is to satisfy the growing wants of food and nutrition." In the agricultural sector, several individuals blow over ancient sowing techniques with tractors, which have some disadvantages like price effectiveness and the need to have a lot of manpower to control.

2.1.1 Traditional sowing method

Old-fashioned strategies embody propagating manually, gap furrows through rustic cultivate besides dropping a few seeds via hand manually, referred to as "Kera," as well as dipping seeds within the holder over a bamboo or metal flue connected along rustic plough up. When it comes to sowing in subtle zones, i.e., creating pits or splits via stick or instrument besides dropping seeds manually by farmers is experienced. Ancient sowing strategies have several limitations such as manual seeding, labor, imprecision in location, planting at irregular depths and so on.

2.1.2 Problem statement

Many advanced strategies of farming provided an increase in productivity of yield and machinery usage. However, several farmers principally are not aware of advanced practices but follow strategies that result in less productivity and are not cost-effective, as this becomes a lot worse just in the case of tiny land (less than 30 ha) holdings.

Seed planting machineries and robots employed by farmers at the current stage come with several limitations like improper chassis design (mostly fabricated from thick plastic sheet mounted for support), robot wheels not supporting all types of fields, and also durability in case of wheel management.

2.1.3 Objectives

The basic objective is to develop a prototype model of a seed sowing robot, which can sense temperature, humidity, proximity, motion, object detection and seed sowing mechanism. The row-to-row arrangement, rate of seed sowing, seed beside seed arrangement, and depth of seed placement differ from various crops and depend completely on different weather conditions to attain maximum yield. Beneath concentrated cropping, the timeliness of processes is one of the foremost vital factors which might solely be achieved through acceptable usage of farming types of machinery. Manual methodology of seed sowing causes severe back aching for farmers and eventually leads to low productivity rate and low seed placement that limits the scale of ground that is supposed to be implanted.

A Bluetooth module is employed to regulate the robot during which it may be operated by a smartphone. A seed tanker is assembled to the servo motor which is mounted at the front facet of the robot. A driller is employed to drill the soil in dextrorotary and conjointly anti-clockwise to require to till the soil up. In most farms, the various strategies used for tilling and sowing are accustomed to be allotted consecutively and consumes lots of time. The labor work needed leads to a higher price. Tiny seeds usually got sowed too thickly and required to be diluted. The machinery used for sowing was not cheap for tiny landholders. The seed-based arrangement and also the specific depth of seed location differ from various crops that were not attainable accurately once done manually. Thus, there is a necessity for a machine that is capable to try and do these operations at the same time.

The factors that affect the propagation and development of seeds are:

- Consistency in the depth of the seed.
- Consistency of spreading of seeds on rows.
- Oblique displacement of seeds from the row.
- Stoppage underneath loose soil obtaining beneath the seed.
- Consistency of various soils covering the seed.
- After the sowing, soil can cowl the seed for these metal plates to hide the seed.

And factors thought about in coming up with a robot are:

To deposit the seed in a precise pattern and cover the soil to forestall the fast loss of moisture from the soil around the seed. To sow seeds with the correct distance is feasible by interfacing with TM4C123Gxl board with servo motor and DC motor to complete Rack and Pinion mechanism to sow and dispense seeds controlled by Bluetooth app.

2.1.4 Literature survey

In [1], the authors explicitly explained the far side of the engineering community and execute many roles additional with efficiency and success.

The robot was placed at the beginning location at the sphere by the worker. It generates its space by generous optimal arrangement among the sphere limitations and also lanes within the seeds square measure planted. The spacing between two repeated seeds was based on the values for the arrangement square measure as per the information provided by the government. The hole-digging mechanism had a roulette carried forward by the motor. The choppers on the helm square measure 10 varieties and have a pitch per tooth of 0.5 cm. The primary choppers include an altitude of 1.5 cm; also the last includes an altitude of vi.5 cm. The wheel rotates at a 60 rate and also the chopper square measure gives acceptable break, and up-to-date positions to reduce energy ingesting and interference to flaw flow. Once the opening is formed, the seed distribution system will be initiated.

The seed dispensing mechanism verified the availability and releases the seed upon an empty vessel when an alarm was triggered. The motor will be linked to a steam-power-driven battery.

In [2], the authors declared that the machine had a star panel-electrical device that provides environmentally friendly energy. The model was created from electrical phenomenon (PV) cells, absorbs daylight and alters this alternative energy to voltage. Seed Hopper contains a seed barrel created in two frustums. Tiny ends of these frustums area units are associated along plastic tubes. The massive split ends of solid area units associated with every different three holes area unit are created on the greater boundary to the hopper. The seed area unit was introduced within solids with the assistance of covered beginnings on the aspect of the frustum. Hopper can interchange concerning its essential-based axis. Seed arrangements are going to be sustained by holes that area units created on the border with equivalent reserve.

However, because of the quick rotation of the chopper, its times take away quantifiable measures from the external workpiece. A belt block system contains two additional lifters in communal to a belt. This enables machine-driven power, speed and torque to be conveyed through the shafts that encompass the winches. Henceforth, the claims adjuster must perform the entire procedure of the seed hopper. Seed should line by soil earlier irrigation method. Once casing seeds the soil, water is to bedspring.

In [3], the authors expressed that each one of the parts of this good seed disseminating instrument is designed to be made up within the house with the aim to possess active coaching of fabrication of varied parts of victimization 3D printing knowledge and also the use of IoT parts. A seed discharger unit and a lane hunter and follower were designed to discharge one of the seed hoopers to a hovel created within the ground. In addition, also the dropping distance between seed to seed is precise. The best disseminating speed will be preserved in the standardization with whole distance, remoteness, speed mechanism and choice of one seed beyond the hopper.

In [4], the authors stated that according to the calculations, the robot was not created properly and was automized utilizing various sensors and controls. They used mild steel to implement the chassis design. The main rationale for adopting mild steel is to raise the robot's weight, which reduces wheel sliding in the field and protects the robot from incorrect navigation. Inputs are specified through keypad and Arduino for the extent of field length, and the Arduino was programmed to consider the given input in feet, with being cast off as the entered control after input is prearranged. The field width is next requested, followed by the seeding point distance, and lastly, the amount of seeding points besides seed rows stay intended and shown on LCD after all of these inputs have been received. The Rack and Pinion mechanism was chosen to avoid causing the robot to lose control and move away from the seeding point. The rack's bottommost end is attached to a V-shaped sharp drill part, for easier drilling operations. The robot's movement is controlled by Arduino via the L298N motor driver, founded on the inputs and the pathway it must follow. The motors utilized have a 50 rpm speed and a strong torque, making it easy for the robot to pick up and manoeuver. The wheel movement has been given a delay for the robot to take the proper distance.

In [5], the system that offers speedy soil loosening, sowing seed and moving the mechanism forward in automatic mode was considered. In manual mode, the RF module is used to move the machine forward, backward, left and right. Main motor drivers are controlled using an Arduino Uno. The left and right motors are accustomed to driving the motor which flips left and switches right.

Seeds fall from the seed drum with every entire turn of the rotating wheel, and the seed planter process works flawlessly with no seed waste. Seed buckets took seeds from the chamber and, with the help of a plough, inseminate the bottom to the required depth. In addition, if an obstruction appears in front of the seed sowing machine, the IR detector detects it and sounds the buzzer.

In [6], seeding, sowing, ploughing and spraying pesticides are all multitasking working features with obstacle recognition. Furthermore, there are six motors in this AGRIBOT: two motors for forward and also backward motion,

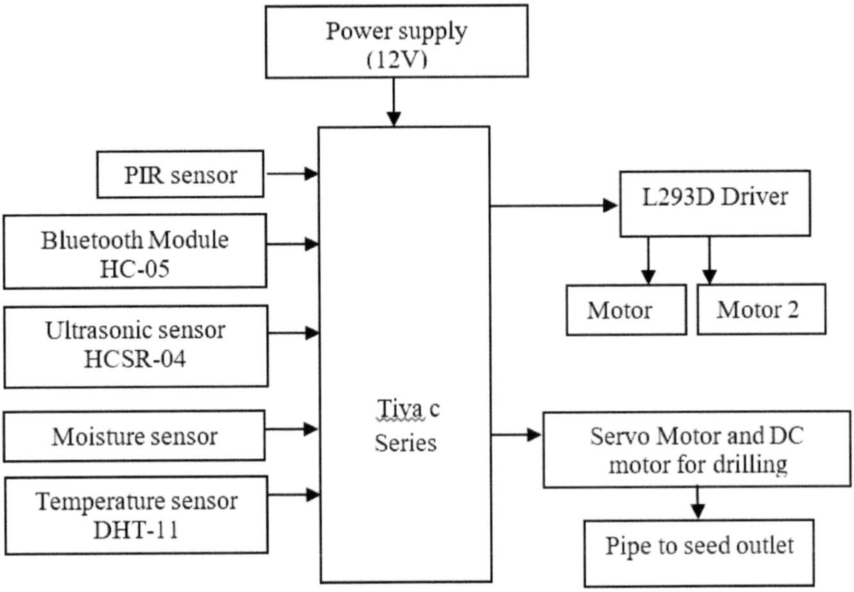

Figure 2.1 Block diagram.

two motors for the purpose of ploughing and sowing, then two motors for the purpose of grass cutting besides spraying. All these are connected to a motor-driven circuit [L29-3D IC] that controls the AGRIBOT's speed. Now robot's motions are controlled using a Bluetooth app, which sends commands such as frontward, spraying, cutting, backward, left, right, ploughing, sowing and also when an impediment is identified. As a result, operations such operation like grass cutting, seed sowing, pesticide spraying and ploughing are completed successfully. To prevent the robot from colliding with an unwelcome obstacle, Ultrasonic sensors have been utilized, which act as the primary requirement for each mobile robot. During the sowing process, the ultrasonic sensor sends a signal to any objects that fall in the robot's path. The device recorded precise gaps between the obstacle and the device. As a result, now controller can halt the robot's motion to sow more seeds.

2.2 Proposed Work

2.2.1 Block diagram

Figure 2.1 represents the proposed block diagram and it shows interfacing with the TIVA board.

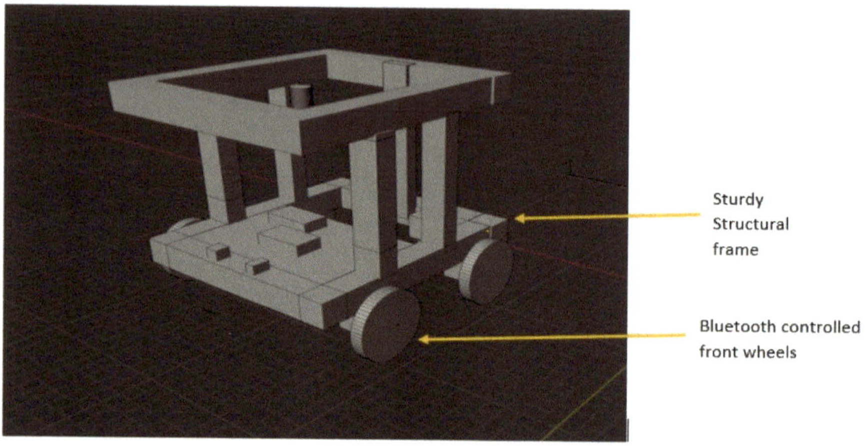

Figure 2.2 CAD model of robot.

2.2.2 Design of robot

The objective is to design a chassis model which can sustain most of the fields without having troubles and this mechanism should able to dig the soil and plant the seed as required [7–10].

Figures 2.2 and 2.3 show the CAD (computer-aided design) model of our seed sowing robot.

The major parts are given below:

1. Structural chassis of the robot
2. Bluetooth-controlled wheels
3. Seed storage container
4. Seed sowing mechanism (Rack and Pinion)
5. Servo motor
6. Bluetooth module HC-05
7. Sensors (temperature-humidity, Ultrasonic sensor HC – SR04, moisture sensor, PIR sensor)
8. TIVA TM4C123Gxl board

HC-05 Bluetooth module interfaced with TIVA TM4C123Gxl enables to monitor of sensor readings and using the Bluetooth Android app motion of the robot is controlled. A seed tanker is assembled to the servo motor which is fixed at the front side of the robot [11–14]. Then seeds are poured into the tanker so that the machine sows the seed by operating the servo motor.

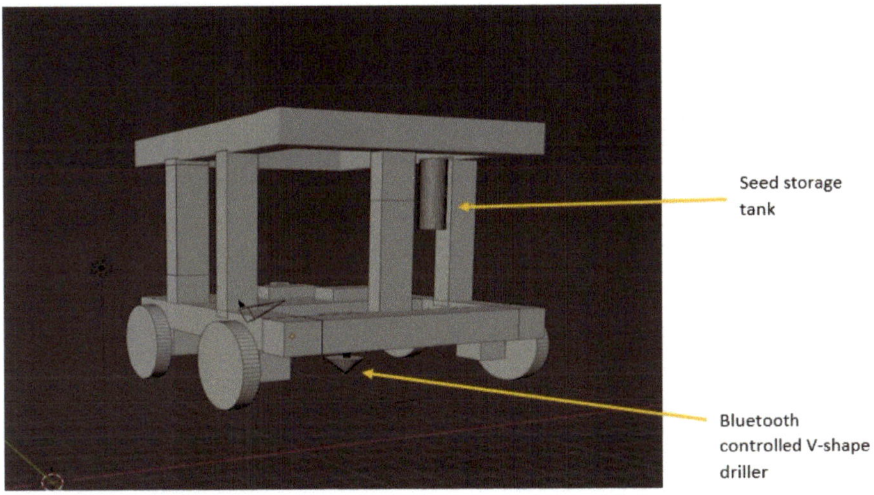

Figure 2.3 Another view of the CAD model of the robot.

A driller is used to drill the soil clockwise and also anti-clockwise to take the soil up. In most of the farms, the different methods used for ploughing, and sowing used to be carried out sequentially and required plenty of time. The drilling mechanism (rack and pinion) enables the robot to sow seeds in any kind of soil condition.

Precision agriculture comes at an affordable cost and it is compatible with multiple soil types [15–17].

2.2.3 Seed sowing mechanism

Rack and Pinion's mechanism is used for seed sowing. The Rack and Pinion linear actuator is made up of a circular gear that engages a linear gear for the conversion of rotational motion into linear. The rack is driven in a linear direction when the pinion is turned [18–20]. If the rack is driven linearly, the pinion will be pushed to rotate. Figure 2.4 resembles Rack and Pinion mechanism; arrows represent the linear movement of the rack and circular movement of the pinion.

2.2.4 Dispensing of seeds

A servo motor is coupled to the seed container to drop the seed. The mechanism of the servo motor is to travel at an angle of 30° in both the clockwise and anti-clockwise directions in which it drops the seed during sowing.

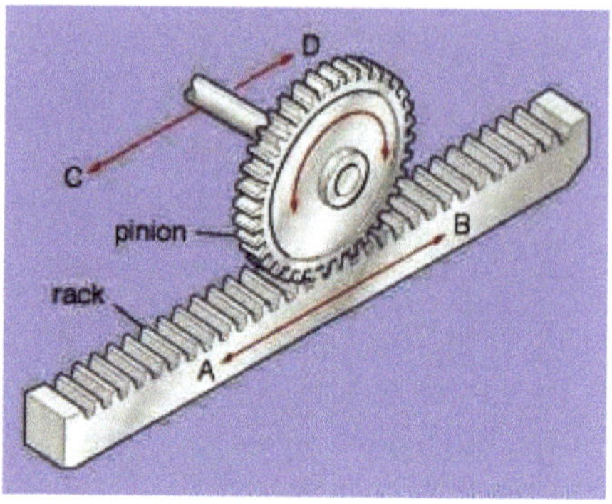

Figure 2.4 Rack and Pinion mechanism.

- Control of the seed depth and correct exploitation of seeds may be accomplished with minimal loss as a result of the seed sowing mechanism.
- Bluetooth module interfaces with the TIVA board through the serial port.
- Bluetooth module interfaces with TIVA board via serial port.
- HC-05 is a Bluetooth module intended for wireless communication, and it communicates with devices via a serial port.

2.2.5 TIVA C series

The TIVA TM4C Series TM4C123G pad analysis board (EK-TM4C123GXL) may serve as an affordable platform for ARM® CortexTM-M4F-based microcontrollers. For unique applications, the TIVA C Series pad has an RGB semiconductor diode and programmable user keys. The use of Booster Pack XL of the TIVA C Series TM4C123G pad is to extend the usefulness of the TIVA C Series pad by connecting various peripherals on numerous current Booster Pack add-on boards and upcoming goods.

Figure 2.5 shows the TIVA C Series pad TM4C123Gxl, it has programmable user keys and an RGB semiconductor diode.

2.2.6 BoosterPacks

The TIVA C Series pad makes it easy, and reasonable to design apps using a TM4C123GH6PM microcontroller. TIVA C Series BoosterPacks and MSP430 BoosterPacks extend the TIVA C Series pad's peripherals and

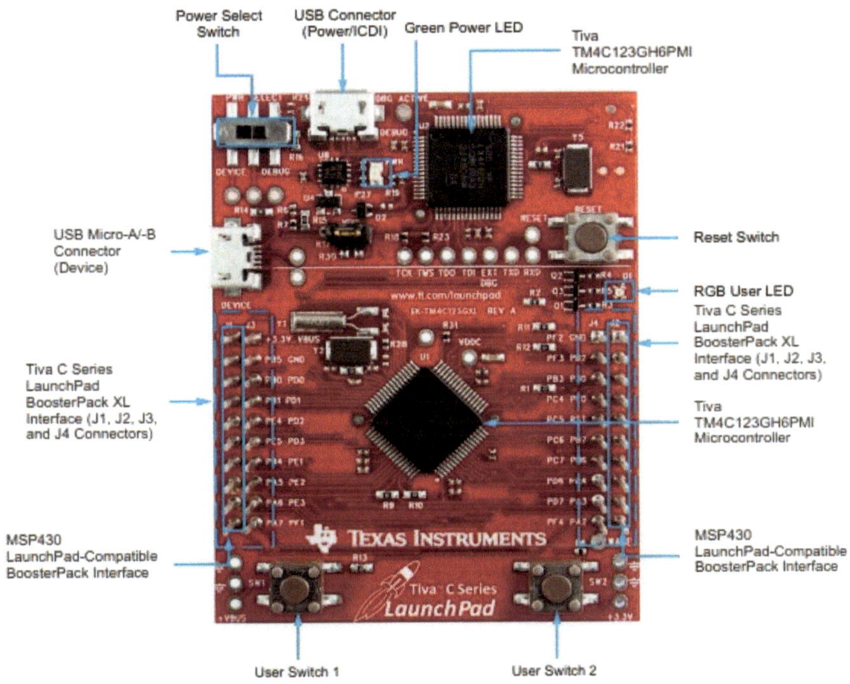

Figure 2.5 TIVA C Series TM4C123Gxl board.

possible uses. BoosterPacks will be utilized with the TIVA C Series pad; otherwise, the onboard TM4C123GH6PM microcontroller will be used as the CPU.

2.2.7 Software used: ENERGIA integrated development environment (IDE)

Energia platform is open source IDE. Energia has an enormous amount of libraries for programming the most used microcontrollers. Unlike many coding platforms, it provides versions of tools to many host platforms like Windows, Linux and Mac.

The main reason for this is that Energia IDE has been completely written in Java programming language. The main purpose of IDE is to build applications by combining common developer tools as a single GUI (graphical user interface). Figure 2.6 shows the user interface window of the Energia IDE platform.

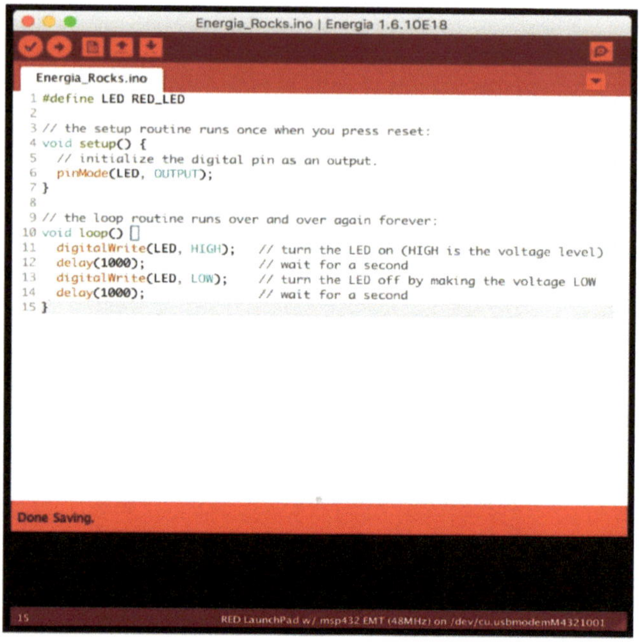

Figure 2.6 Energia IDE window.

2.2.8 Components used

2.2.8.1 Bluetooth module

HC-05 module (+5Vvcc) is a straightforward use Bluetooth SPP module. The term Bluetooth module (HC05) is employed for pairing between mobile devices and microcontrollers. It receives data the knowledge the data from a mobile device and sends information to the microcontroller within the style of binary bits.

Figure 2.7 shows HC-05 which is designed for wireless statements with transmit power of up to 4 dBm. The Bluetooth module is powered by 3 V. It has two working modes: order response and automatic association. Once the module has completed the automatic association job, it will use a defined method to communicate information mechanically. Once the module is in order-comeback mode, the user will send AT command to the segment interface to line up the management limitations and send management orders.

2.2.8.2 Motion PIR sensing element

PIR sensors distinguish motion and are frequently used, to determine whether or not someone was moved in or out of the sensor range. These sensors are

Figure 2.7 Bluetooth module (HC05).

small, low-cost, less-power and easy to operate. As a result, they are usually found in utilizations and devices used in homes (or) companies. This sensor is usually referred to as PIR, passive infrared or pyroelectric sensor.

PIRs are generally made up of pyroelectrical sensing element that detects the amount of infrared light radiation. The whole lot produces low-level radiation, the hotter something is, the more radiation it emits. An extreme motion detector's sensing element is divided into two halves. Since the goal is to identify motion rather than average IR levels. The two parts are connected in such a way they call off each out. When one observes further (or) less IR radiation than others, o/p might swing high (or) low.

Along with the pyroelectric sensing element, there may be a slew of supporting electrical components such as resistors and capacitors. The BISS0001 appears to be used by the majority of small amateur sensors ("Micro Power PIR Motion Detector IC"). This chip receives the sensing element's o/p and performs some little meting out to generate a digital o/p pulse from the analogue sensing element.

Basic statistics:

- Dimensions: Quadrilateral.
- Digital pulse high (3 V) when triggered, digital low while idling. Pulse lengths vary from sensing element to sensing element and are governed by resistors and capacitors on the PCB.
- Sensitivity varies: up to 20 feet (6 m) detecting range of 110° × 70°.
- Power supply: 5 V–12 V i/p voltage for many modules, although 5 V will be fine if the regulator has other specs.

Figure 2.8 DC mini motor (12,000 rpm, 12 V).

How PIRs Work

PIR gadget itself has two slots, each of which made unique IR-sensitive quantifiable. It is observed that two slots will see out past distance because the lens utilized here is not accomplishing a lot. The gadget is turned off, and each slot sees an equal amount of IR, which is transmitted from the area, walls and outside. When a heat body, such as a human or animal, passes, half of the PIR device is interrupted, causing a +ve difference adjustment between the two halves. When a lovely and comfy body departs the detecting space, the gadget creates a negative differential amendment, which causes the device to produce a negative difference amendment. The amendment pulses that are sensed.

IR gadget is protected by a window created of IR-transmissive physical that is hermetically sealed to improve noise/temperature/humidity.

2.2.8.3 DC motor

A direct current (DC) motor is a kind of rotational electrical equipment that converts voltage into energy. The furthermost common forms believe in magnetic field forces. Practically all DC motors have an internal mechanism, either mechanical or electronic, that changes the direction of the motor's current flow on an irregular basis.

A mini-DC motor is turned at an associate rev of 12,000 rpm for this mechanism the four DC motors square measure operated at 12 V and one is operated at 6 V voltage.

Figure 2.8 shows the DC mini motor which is given at bottom of the robot designed for moving in forward, backward, left turn also right turn. There is another motor used for drilling the soil upwards and downwards.

Figure 2.9 shows DC mini motor (6 V) which is active to ploy the shaft upwards in addition to downwards unto 5 cm.

Figure 2.9 DC mini motor (6 V).

2.2.8.4 L293D motor driver

L293 and L293D are high-powered fourfold half-H drivers. L293 is projected to produce duplex driving currents to nearly one amp at voltages ranging from 4.5 V to 36 V. The L293D is intended to generate 600 mA of duplex driving power at voltages ranging from 4.5 V to 36 V.

2.2.8.5 Servo motor

The rotary or linear actuator servo motor has the capability of controlling angular or linear position, and velocity besides acceleration precisely. It consists of a suitable motor and a location feedback device. It also necessitates the use of a complex controller, which is usually a separate module developed mostly for servomotors.

The word servomotor, while widely used to describe a motor suitable for use in a closed-loop control system, does not refer to an explicit type of motor.

2.2.8.6 DUAL power grid battery

The battery stores the power and emits power, thus it is used as a dual facility battery that contains 12 V DC power and 6 V DC power. Here 12 V is employed for rotating drive motors and 6 V is employed for servo motors and a few alternative motors. The dual facility battery will be a Li-ion battery or lead-acid accumulator. Lead-acid batteries are created from cells. Every cell is or so two volts, thus a 12 V battery has half-dozen individual cells. It seems

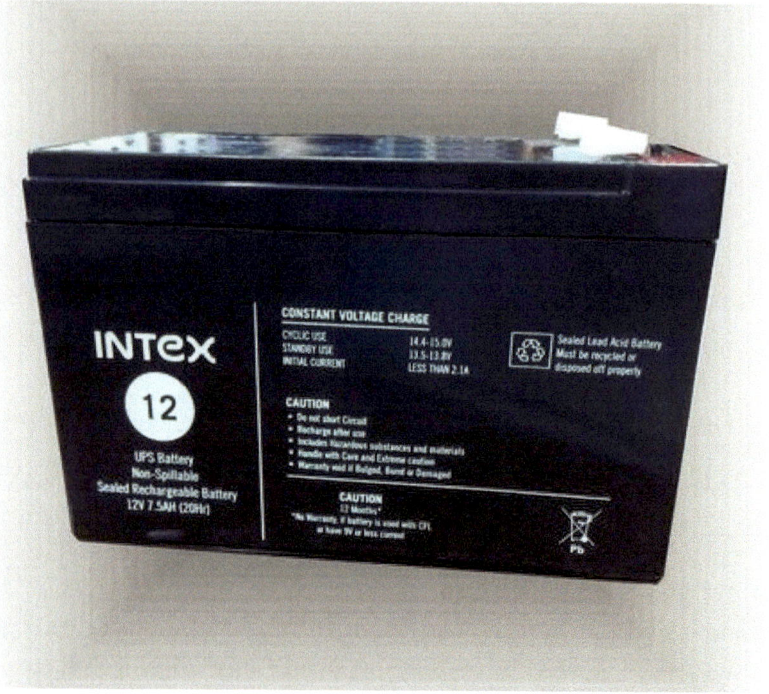

Figure 2.10 Battery (12 V and 6 V) DC power supply.

that a charged two-volt cell features a voltage of roughly 2.15 V. Interestingly enough, a discharged 2-V cell features a voltage of 1.9 V. Battery aims to store energy and convert this energy into power once the requirement arises. Figure 2.10 shows an image of a 12 V battery.

2.2.8.7 Humidity sensor

Measurement in addition to management of temperature and ratio finds many applications to varied extents. Recently devices square measure accessible that have each temperature, and humidity sensor through signal acquisition, ADC, activity also message edge interface all engineered within. The utilization of sensible sensors significantly modifies the look and lessens the general price. There is a tendency to be mentioned earlier concerning humidity besides temperature quantities along with Sensorium SHT1x/SHT7x named sensors. Sensirion sensors square measure accomplished of mensuration each temperature and ratio and supply mark digital based outputs, whereas SHT1x/SHT7x square measure correct sensor values. DHT11 device provides mark numerical outputs meant for temperature besides humidity, however, is comparatively cheap and inexpensive when compared to Sensirion sensors. DHT11 device practices a proprietary one-wire protocol that is used for exploring here besides applying with PIC16F628A microcontroller which will obtain the temperature in addition to humidity values through the device and show output on 16 × 2 character-based LCD.

2.2.8.8 Ultrasonic (UV) sensors

Bats are wondrous beings. Unsighted from the eyes and however an image thus accurate that could distinguish between a lepidopteran and a wrecked leaf even after flying by bursting speed. No hesitation the image is a swindler compared to ours besides is way on the distant side from the human abilities of the far sighting, however, there is no way taking place on the far side our considerate. Ultrasonic foldaway is a technique utilized by Nutty besides plenty of different individuals of the Animal kingdom for navigational purposes.

Soil moisture detector: A soil moisture detector measures the volumetric amount of liquid in soil and outputs the moisture level as one output. Because the detector has both analogue and digital outputs, it is frequently used in both analogue and digital modes.

Sensor working: Two probes are present in the soil moisture sensor that measures volumetric amounts (liquid). These two probes allow current to pass along the soil, and the resistance level is used for the calculation of moisture level.

If there is a lot of water then the soil can conduct a lot of electricity, which proposes that confrontation will be less. That resembles that the moisture level is high. Dry soil is meant to conduct electricity badly, thus once there will water, that soil conducts electricity in less amount, which suggests this fact there will present a lot of resistance. That resembles that the moisture level is low. The detector is often given connection as two modes: (a) analogue mode and (b) digital mode. Analogue mode connection goes first and later digital mode connection.

2.2.9 Performance analysis

In Table 2.1 whole work is divided into segments to clearly describe the work done and the outcome of the project.

2.2.9.1 Result analysis for sensing applications using Bluetooth module

By interfacing with the TIVA board, and Bluetooth module and making use of sensors for real-time monitoring of data, sensor readings are displayed in the Android app connected via Bluetooth.

Table 2.1 Work description in the form of segments.

S.no	Segment outcome	Description
1	Sensing applications using Bluetooth module	Interfacing tm4cgxl board with sensors: 1. PIR sensor 2. Ultrasonic sensor 3. Humidity sensor 4. Temperature sensor Sensor readings are noted via Bluetooth module HC-05
2	Seed sowing mechanism and dispensing of seeds	Interfacing tm4cgxl board with servo motor and dc motor to complete Rack and Pinion mechanism to sow and dispense seeds controlled by Bluetooth app.
3	Robot motion control using Bluetooth module	Interfacing the tm4cgxl board with dc motors for robot movement, which is initiated using Bluetooth app by delivering commands for operations like sowing and movement control (left/right, front/back).

Sensors interfaced with the TIVA tm4cgxl board are a PIR sensor, ultrasonic sensor, humidity sensor, and temperature sensor:

The various sensors that are mentioned above collect data and are displayed in Android phone which is connected using Bluetooth module.

Figure 2.11 shows the interface of the Android app. By clicking on the terminal module sensor readings are displayed on the Android app.

Figure 2.12 shows sensor readings which are displayed in an Android app connected via Bluetooth. Sensor readings will get updated every 30 m.

2.2.9.2 Result analysis for seed sowing mechanism and dispensing of seeds

A V-shaped driller is used to drill the soil clockwise and also anti-clockwise to take the soil up. The drilling mechanism used is Rack and Pinion which enables the robot to sow seeds in any kind of soil condition. The servo motor which is attached to the seed container drops the seed by moving at an angle of 30° in the direction of clockwise and anti-clockwise at the time of sowing. These actions are going to place simultaneously controlled by a Bluetooth module.

Figure 2.13 represents the drilling mechanism which is used for seed sowing as Rack and Pinion. Driller is used to drill the soil clockwise and also anti-clockwise to take the soil up.

Figure 2.14 represents dispensing of seeds from a container. The servo motor is attached to the seed container which moves at an angle of 30° in

Figure 2.11 Android app.

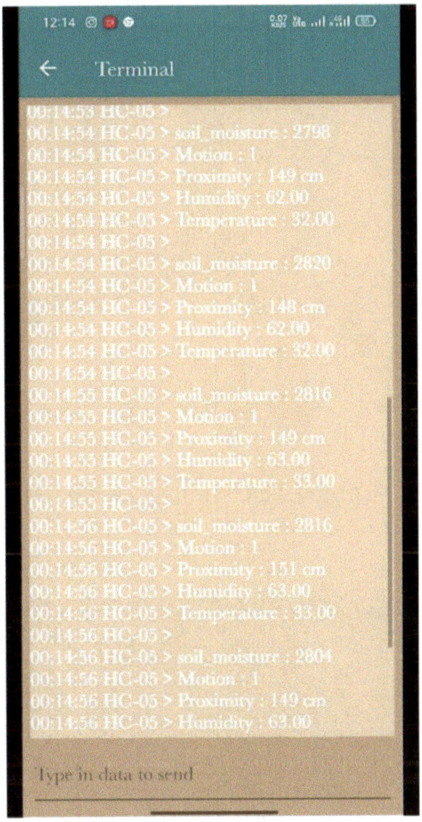

Figure 2.12 Readings of sensors.

the direction of clockwise and anti clockwise to drop the seed at the time of sowing.

2.2.9.3 Designing of CAD model

Figure 2.15 shows the window of CAD model designing and work done in CAD designing of the project.

Figure 2.16 shows the working of the project in the field and compares Rack and Pinion movement before drilling and after drilling.

Figure 2.17 shows the working of the project in the field performing operations.

Working of drilling mechanism: V-shaped driller rotates clockwise and anti-clockwise to dig the soil as shown in Figure 2.16, specific depth drilled by the V-shaped driller is set as 5 cm.

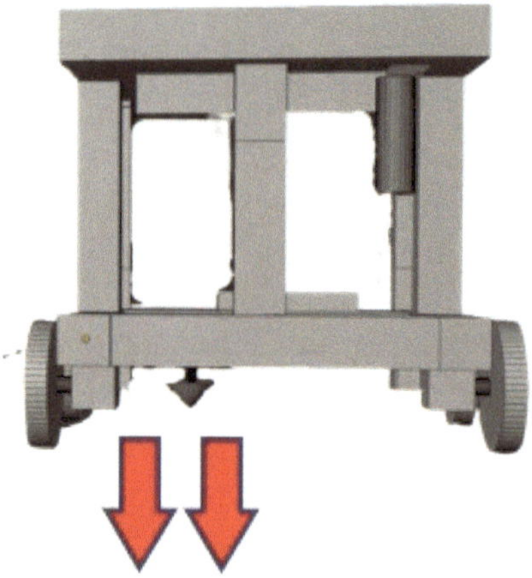

Figure 2.13 Drilling mechanism.

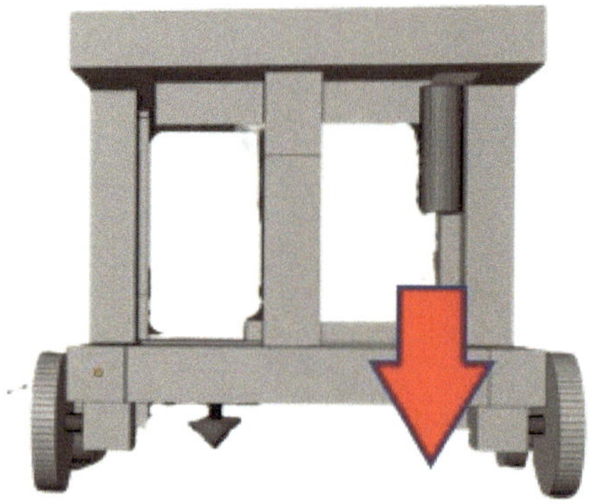

Figure 2.14 Dispensing of seeds.

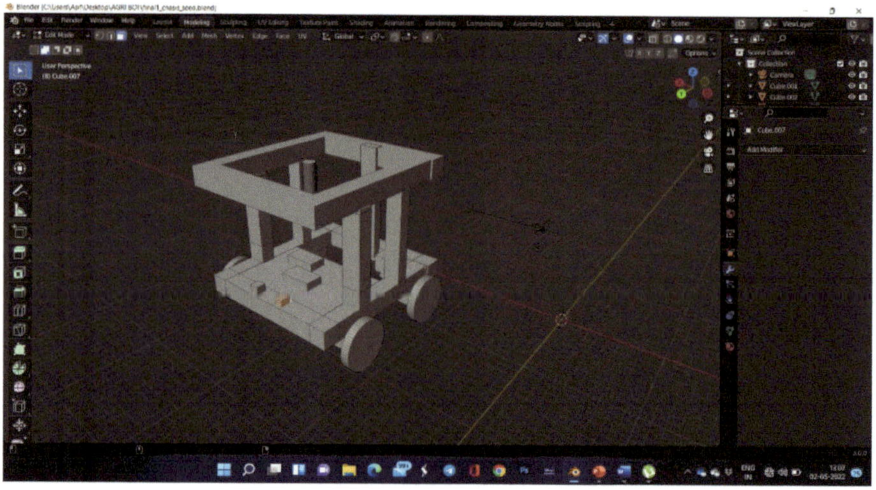

Figure 2.15 Window of CAD designing.

Depth can be modified in the Energia code according to the soil type amount of seeds dispensed.

Table 2.2 represents the pitch presentation of the project based on industry expectations. The project is designed in such a way that it can be a revenue model and a useful product for farmers.

Figure 2.16 Rack and Pinion movement before drilling and after drilling.

Figure 2.17 Working of the robot in the field.

2.3 Conclusion

The main focus of this program is its default method of sowing seeds. Seeds are sown in a sequence that should have the effect of proper germination. This method of sowing seeds used a robot that reduces the need for employing workers. Here seed loss is also greatly reduced. This robot is designed to sow seeds automatically. Here with the help of a robot, the seeds are spread in the ground in the right order thus reducing seed wastage. This robot will help farmers to make good farming practices.

Table 2.2 Shows the PYP of the project.

PROBLEM Manual sowing lacks maintaining acceptable spacing between row to row and plant to plant. The proposed model focuses on uniform and precise seed sowing processes and overcome the problem. 1	**SOLUTION** • Remote monitoring of Sensor readings with HC-05 • Various Seed Deposit in acceptable pattern. 4	**UNIQUE VALUE PROPOSITION** • Precision agriculture at an affordable cost • compatible multiple soil-types • Reduce wastage and increase efficiency 3	9 **UNIQUE ADVANTAGE** Precision in Agriculture Remote monitoring Scalable Reduce Wastage Multi-crop adaptability	**CUSTOMER SEGMENTS** • Small and Marginal farmers 2
EXISTING ALTERNATIVES • Fixed x-y axis chasis model	**KEY METRICS** Outreach to farmers (Nos) Acreage served Efficiency parameters vs. cost saved. 8		5 **CHANNELS** Direct Sales B2C (Digital) B2B with Big Basket etc. Farmer Cooperatives NGOs/MFIs Exhibitions in Farm Expos	**EARLY ADOPTERS** • New Age Farmers • Farmer Cooperatives • NGOs/MFIs
COST STRUCTURE MANUFACTURING COST PRICE in Bulk: Rs.6000 PROTOTYPE COST : Rs. 7000 6		**REVENUE STREAMS** • Asset Sale • Subscription model: Rentals/Leasing/Pay per use • Maintenance and Replacement • Licensing for manufacture 7		

After going through the limitations of the seed sowing machine, it is concluded that the automatic seed sowing machine can do the following:

- The model can maintain desired space in fields, using the HC-05 Bluetooth module.
- Code was given a certain depth for digging the soil using the TIVA board and the servo motor helps to place the seed in it, servo motor is open for certain milliseconds, so the robot can place some accurate amount of seeds every time it helps to prevent seed loss.
- The robot can do various operations simultaneously and it helps to save lots of money for farmers like labor requirements, and cost of labor the main thing is this robot can be affordable to farmers.

Acknowledgements

The authors thank PALS for evaluating the model and awarding incubation-worthy projects in PALS INNOWAH 2022. The authors thank the marginal farmers for the demo process in their agricultural land with a varsity of crops and fields. The authors thank the Texas Instruments Innovation Lab at the Hindustan Institute of Technology & Science for facilitating the processor and supporting software required to complete the project.

References

[1] ICCEMME 2019 IOP Conf. Series: Materials Science and Engineering 691 (2019) 012023 IOP Publishing doi:10.1088/1757-899X/691/1/012023

[2] International Journal of Applied Engineering Research ISSN 0973-4562 Volume 13, Number 6 (2018) pp. 259–262

[3] P. Kumar and G. Ashok, Design and fabrication of smart seed sowing robot, Materials Today: Proceedings, https://doi.org/10.1016/j.matpr(2020)

[4] Suibing Li, Shuai Li, Long Jin, "The Design and Physical Implementation of Seeding Robots in Deserts", Chinese Control Conference (CCC) 2020 39th, pp. 3892–3897, 2020.

[5] International Research Journal of Engineering and Technology (IRJET) e-ISSN: 2395-0056 Volume: 06 Issue: 04 | Apr 2019

[6] Proceedings of the Fourth International Conference on Trends in Electronics and Informatics (ICOEI 2020) IEEE Xplore Part Number: CFP20J32-ART; ISBN: 978-1-7281-5518-0

[7] Proceedings of the Second International Conference on Inventive Research in Computing Applications (ICIRCA-2020) IEEE Xplore Part Number: CFP20N67-ART; ISBN: 978-1-7281-5374-2

[8] International Journal of Advance Engineering and Research Development (IJAERD) Volume 6, Issue 02, February 2019, e-ISSN: 2348-4470, print-ISSN: 2348-6406.

[9] Kavitha. B.C. and Valikannu. R, "IoT Based Intelligent Industry Monitoring System using Raspberry-Pi", Proceedings of 6th IEEE International Conference on Signal Processing and Integrated Networks SPIN 2019. (Scopus Indexed).

[10] Kavitha, B.C. and Vallikannu, R. (2020), "Delay-aware Concurrent Data Management Method for IoT Collaborative Mobile Edge Computing Environment", Microprocessors and Microsystems. Elsevier., 74(1): 842–852. (SCI Journal – WoS and Scopus Indexed).

[11] Kavitha, BC. and Vallikannu, R., "IoT Assisted Predictive Maintenance & Worker Safety: An Initiative", 5th International Conference on Information and Communication Technology for Competitive Strategies (ICTCS 2020), Jaipur, India, 2020. (Scopus indexed Springer)

[12] IoT Based Agriculture Using AGRIBOT 2018 3rd IEEE International Conference on Recent Trends in Electronics, Information & Communication Technology (RTEICT-2018), MAY 18th & 19th 2018.

[13] Solar Powered Autonomous Multipurpose Agricultural Robot Using Bluetooth/Android App, Proceedings of the Third International Conference on Electronics Communication and Aerospace Technology [ICECA 2019] IEEE Conference Record #45616; IEEE Xplore ISBN: 978-1-7281-0167-5.

[14] Smart Agriculture with AI Sensor by Using Agrobot, 2020 Fourth International Conference on Computing Methodologies and Communication (ICCMC) 978-1-7281-4889-2/20.

[15] The Automation in Seed Sowing by using Smart Agri-Robot, International Journal for Research in Applied Science & Engineering Technology (IJRASET), ISSN: 2321-9653; IC Value: 45.98; SJ Impact Factor: 6.887, Volume 7 Issue IV, Apr 2019.

[16] Aditya R. Rao, Ajay H., Balavanan M., Lalit R., and J. Anand, "A Novel Cardiac Arrest Alerting System using IoT", International Journal of Science Technology & Engineering, Vol. 3, Issue 10, pp. 78–83, April 2017.

[17] Dhanalakshmi R., Jose Anand, Arun Kumar Sivaraman, and Sita Rani, "IoT-based Water Quality Monitoring System using Cloud for Agriculture Use", Cloud and Fog Computing Platforms for Internet of Things, Edited by Pankaj Bhambri, Sita Rani, Gaurav Gupta, Alex Khang, Routledge Taylor & Francis Group, May 2022.

[18] J. Anand, and J. Raja Paul Perinbam, "Automatic Irrigation System using Fuzzy Logic", AE International Journal of Multidisciplinary Research, Vol. 2, Issue 8, pp. 1–9, August 2014.

[19] K. Nirmala Devi, J. Anand, R. Kothai, J. M. Ajai Krishna, and R. Muthurampandian, "Sensor based Posture Detection System", Materials Today Proceedings, Vol 55, Part 2, pp. 359–364, 2022.

[20] S. Prema, J. Anand, P. Vanitha, Nirmala Devi K., M. Mohamed Yaseen, and Rajeswari C., "Smart Stick using Ultrasonic Sensors for Vissually Impaired", Advances in Parallel Computing Algorithms, Tools and Paradigms, Vol. 41, pp. 43–441, Nov 2022.

3

Crop Management System

L. Jubair Ahmed[1], S. Dhanasekar[2], V. Govindaraj[3], and C. Ezhilazhagan[3]

[1]Department of Electronics & Communication, Akshaya College of Engineering and Technology, India
[2]Sri Eshwar College of Engineering, India
[3]Department of Electronics & Communication Engineering, Dr. N.G.P. Institute of Technology, India
Email: jubaircbe@gmail.com; dhanasekar.sm@gmail.com; see1govind@gmail.com; ezhisang20@gmail.com

Abstract

Smart agriculture is emerging at a significant rate in recent times. It is a part of the information and communication technologies (ICT) movement that is ushering in agriculture in many ways, referred to as the Third Green Revolution. By enriching the crop yield process, smart farming enhances its manufacturing of high-quality food. This article focuses on crop cultivation techniques as well as applies deep learning (DL) and machine learning (ML) algorithms so as to offer responses to a number of challenges that emerge all through the cultivation process. Machine learning is a leading-edge technological innovation that assists farmers in reducing crop losses by providing specific crop recommendations and keen insight. Soil and water management, crop cultivation, crop disease detection, weed control, crop distribution, robust fruit counting and yield prediction are all examples of smart agricultural applications that employ deep learning. Farmers have experienced lots of new challenges recently, along with crop failure due to scarcity of rain, soil fertility issues, etc. As a result of the changing environment, this proposed study will assist by finding the most effective way to handle crops and harvest them.

3.1 Introduction

During the past two decades, extensive research has concentrated towards future farming techniques on agronomic crops. Cropping system changes will have a big impact on insect population trends and administration [1]. Crop production has been one of agriculture's most important branches. Crop production is essential for providing feed for livestock as well as food for the common people. Improving the economic efficiency of agricultural activities has been a primary goal over human agrarian history. Agricultural production sites must be examined regularly to obtain high-quality products, and they can also adopt every necessary crop production measures. Farmers increase the crop's cost by spending time as well as resources to each visit. Due to intensive monitoring and evaluation of crops that farmers do, smart agriculture has risen to become a necessity. Digitalization will encompass the majority of engineering fields, with a bigger impact over wide-area communication network involving fast data transmission [2–4].

Smart farming would be a novel model that uses advanced information technologies to make agriculture highly effective and efficient [5]. Farmers can better monitor various procedures as well as apply specific treatments chosen through machines of superhuman efficiency due to recent improvements in artificial intelligence, automation and connectivity. Engineers, data scientists and farmers are all working on ways to reduce the amount of human labor needed in agriculture. The crop products have been used for plant-based raw materials together in a wide range of sectors, which includes pharmaceuticals, food as well as fuel. Crop production would be a subset of agriculture that involves field crop cultivation, vegetable production and fruit production, among other things [6].

Artificial intelligence was introduced initially to solve simple mathematical or logical rules in computer programming tasks intellectually. Machine learning, a subset in the artificial intelligence, employs a self-learning method for extracting the data. Machine learning employs learning rules like reinforced, supervised, hybrid and unsupervised learning, which obtain useful associations of data [7]. Artificial neural network concepts have been used in machine learning, which is termed deep learning. Deep networks distinguish from the neural networks for its depth. Because of these characteristics, deep learning networks may detect latent structures in unlabeled and unstructured data. The deep learning networks particularly perform feature extraction while requiring human interaction, providing a substantial advantage over earlier techniques. Due to the rise of broadband wireless transmission networks, customer demands in high-speed wireless communication have risen significantly [8–10].

The deep anomaly method is more effective than region-based convolution neural networks at detecting humans at distances of 45–90 m (RCNN) [11]. This algorithm generates homogeneous field characteristics and detects anomaly. The deep learning categorization for land cover as well as crop types employing remote sensing data was described in this article [12]. CNN was compared to a traditional fully linked MLP as well as random forest. The use of self-learning convolutional neural networks to recognize individual plant classes based on visual sensor data is discussed [13]. In UAV images of line crops, automatic weed detection employing deep learning using unsupervised data labeling is presented [14]. Therefore, the fully automatic weed detection will conduct when CNNs are used to unsupervised training datasets. A crop disease classification system based on a mobile capture device due to deep residual neural network was introduced. Extensive testing improved the balancing accuracy between 0.78 and 0.8 [15].

The deep neural network including transfer learning is used to detect the mildew illness on millet crop images in order to identify the mildew disease [16]. The experimental result accuracy is 95%, precision is 90.0%, recall got 94.50% as well as the F1-score has been 91.75%. NDVI and RGB data collected through UAVs were used to estimate crop yields using deep convolutional neural network [17]. The RGB images outperform NDVI images while using CNN architecture. Rice grain productivity, CNN architecture as well as low-altitude remote sensing-based images were discussed as essential features [18]. Deep CNN performed much better and was more stable during the ripening stage. A multi-temporal crop classification based on deep learning has been studied [19]. XGBoost, SVM and RF parameters were compared to LTSM and Conv1D deep learning models. Another crop vision collection using deep learning categorization as well as precise farming recognition has been developed [20]. This suggested methodology outperformed VGG, DenseNet, ResNet, SqueeztNet and Inception on crop datasets, with the precision of 99.81%. The deep learning technology is used to properly recognize as well as distinguish crops in soil. As a data source, data from a high-resolution digital surface model is used. The transfer learning methodologies with convolutional neural networks are employed with automatic feature extraction for crop pest categorization [21]. Xie1, NBAUR and Xie2 datasets have the highest accuracy of 94.47%, 96.75% and 95.9%, respectively.

We evaluate the specific crop management system using deep learning technique and its model as well as frameworks used, nature, pre-processing and data sources as well as the performance attained using the metrics are studied. Since most agricultural firms have switched completely to crop production, the problem of crop production management has become increasingly relevant. The original study goal is to develop suggestions for enhancing crop production planning and management with modern business situations using deep learning models. The purpose of the study will be whether the deep learning technique has been utilized to improve crop management systems and rank the areas to encourage new researchers, to guide future research in agriculture production.

This article is structured such that Section 3.2 introduces the deep learning architecture using CNN. In Section 3.3, the role of deep learning in crop management system is discussed. The scope and challenges of artificial intelligence and deep learning in the crop management system have been described through Section 3.4. In the end, Section 3.5 concludes with remarks and future work for crop management system.

3.2 Insights of Deep Learning

Deep structured learning or hierarchical learning is a subset of the machine learning which focuses upon artificial neural network (ANN) algorithms that are inspired by the structure as well as activity of the human brain [22]. It performs excellently on a variety of complex cognitive tasks, approximating and even outperforming human performance. The ability to learn huge quantities of data is the advantages in deep learning technique. Deep learning popularity has been increased over the last decade, and it is effectively used for various traditional applications. In several sectors, including natural language processing, cyber security, bioinformatics, medical information processing and robotics and control, deep learning has excelled over well-known machine learning techniques.

The three main deep learning techniques are partially supervised (semi-supervised), supervised and unsupervised. Additionally, deep reinforcement learning (DRL), commonly referred to Recursive Learning, would be a form of learning approach that is typically classified as occasionally unsupervised and partially supervised techniques. Recurrent neural networks (RNNs), convolutional neural networks (CNNs) and recursive neural networks (RvNNs) are among three most well-known types of deep learning networks used. The most extensively used algorithm CNN which has key advantage over its predecessors would be automatically detecting significant features without the need for human intervention. The features have been extracted by various approaches in traditional machine learning

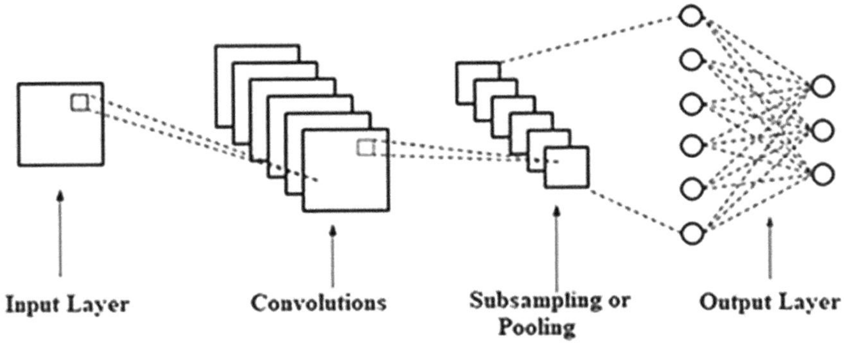

Figure 3.1 Convolutional neural network architecture.

algorithms; however, the convolutional neural network trains the filters for itself, indicating that the network extracts features autonomously. CNNs have now employed in various applications, such as computer vision, audio processing and facial recognition. CNNs are closely related neural networks in their structure which is inspired from neurons in animal and human brains. Convolutional neural networks were extensively deployed in the deep learning network type, which discusses the evolution for CNN architectures as well as their main characteristics. The model of a basic convolutional neural network architecture is shown in Figure 3.1.

A convolution layer is typically used in CNN, and it employs convolutional filter collections to recover numerous local features from each input region, resulting in a large number of feature maps. The layer has been represented mathematically as [23]

$$(cn)_{ij} = (Xn * q)_{ij} + dk, \tag{3.1}$$

where $(cn)_{ij}$ indicates the nth output feature map of (i, j) element.

The input feature maps are represented by q. The nth filter and bias are indicated by Xn and dk, respectively. A 2D spatial convolution operation is represented by the * symbol.

CNN architecture is divided into two components. Feature extraction is a convolution method which extracts and detects the image distinct characteristics of an image during evaluation. The fully connected network employs its convolution result for predicting image classification based on characteristics obtained earlier in the process. A CNN is made up of three layers namely pooling layers, fully connected (FC) layers and convolutional layers. These layers are combined to form the CNN architecture. The convolutional layer is being used to extract input image information. The mathematical operation of convolution has been carried out among the input images as well as filter of M × M size. A pooling or sub-sampling layer has been commonly used after our convolutional layer. The pooling layer reduces convolved feature map size with the reduced computational costs. While connecting both weights and biases in neurons, the fully connected (FC) layer will be connected with neurons between two layers. The CNN architecture's final few layers were usually placed prior to output layer. Finally, in the CNN model, an activation function employs important component which estimates any kind of continuous and complex link across network variables.

The CNN includes three distinguishing characteristics which make a potentially valuable tool in additional crop yield modeling. (i) A CNN model has been developed which represents environmental elements as well as the genetic transformation in seeds through time despite knowing their genotype. (ii) The model has proven the capacity that extends yield prediction across untested conditions while maintaining high accuracy. (iii) When combined with the back propagation approach, the model was able to demonstrate the level towards which weather prediction accuracy, weather conditions, management practices and soil characteristics may influence crop production variation. During 2008 and 2016, Khaki and Wang [24] developed the deep convolutional neural network technique which estimates the corn yield in 2247 locations. A deep learning architecture was developed by Kim et al. [25] during 2006 and 2015 on agricultural yield prediction. The soybean crop production in Argentina and Brazil, which uses deep learning and transfer learning approach, has been estimated by Wang et al. [26]. With remote sensed pictures, Yang et al. [27] evaluated effectiveness for yield of rice grain and yield prediction in ripening process by the CNN model.

3.3 Crop Management System: A Deep Learning Approach

As the world's population continues to grow, a significant increase in food production is required to ensure worldwide availability and good nutritional quality while also safeguarding natural ecosystems through sustainable agricultural practices [28]. The application of crop management methods guarantees the crop's productivity with higher yields and better quality. Crop management is a set of agricultural operations aimed at improving crop growth, development and yield. As indicated in Figure 3.2, it begins with land preparation, seed selection, seed sowing, irrigation, crop development, together with crop management and concludes with crop harvest, storage, along with commercialization.

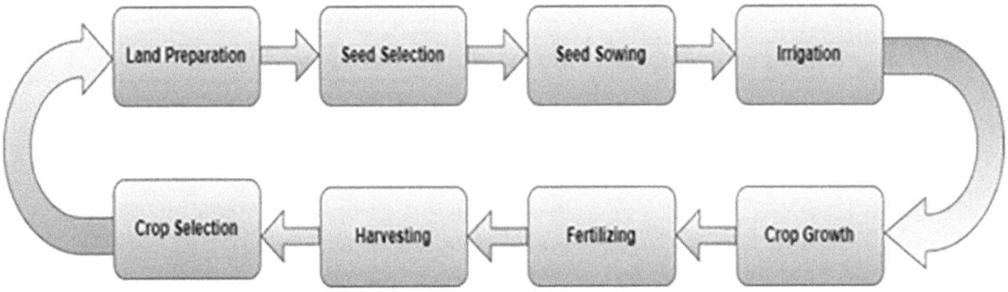

Figure 3.2 Crop management system.

Plant age, soil, climate and weather conditions all influence the timing and sequencing of agricultural processes. Broadcasting sowing method and row crops, winter or spring crops (harvested products such as grain, hay and silage), and weather conditions, plant age, soil and climate are all factors that influence the timing and sequencing of agricultural processes [29]. Computer vision applications in agricultural and popular techniques are shown in Table 3.1.

3.3.1 Soil health monitoring system

The foremost step involved in improving crop growth is preparing the seedbed. The ideal seedbed is constantly firm, wet to the surface and weed-free. "Good seed-to-soil contact essential" according to common comment held on seeding records. When seeds have sufficient soil contact, germination is improved. On the other side, a seedbed that is overly firm makes it harder to get the seed into the ground. [30]. Agriculture IoT may connect with sensors, communication protocols and microcontrollers in a current competitive environment to automate process executions and boost production. Deep learning performance yields acceptable findings and addresses a number of real-time challenges associated to agricultural developments. Sumathi et al. [28] discussed the design of a soil condition estimation system based on IoT network communication system. In modern agriculture, soil quality is a critical aspect in increasing production and regulating hydrological cycles. An improved deep learning model for IoT network-based automated evaluation of soil quality keeps track of the various soil properties and climatic elements that contribute to those issues. For soil quality prediction, a deep learning model was created with the capacity to fit huge data. Weight factors are calculated in order to precisely measure soil quality.

Table 3.1 Applications utilizing computer vision with agricultural with popular approaches.

S. No	Agriculture application	Data analysis techniques
1	Crop mapping/soil as well as vegetation	SVM, end-member extraction technique, linear polarizations (HH, VV, HV), image fusion, co-polarized phase differences (PPD), linear mixing models, logistic regression, distance-based classification, decision trees, ANN, NDVI
2	Crop canopy as well as the leaf area index	NDVI, linear regression analysis
3	Crop phenology	NDVI, Fourier transformations and wavelet-based filtering
4	Fertilizers' effect and biomass, crop height, estimation of yields	Vegetation index (NDVI, ICWSI), wavelet-based filtration, linear polarizations (VV), linear and exponential regression analysis
5	Crop monitoring	Step-wise discriminatory analysis (DISCRIM), regression with linearity, difference in co-polarized phasing (PPD), linear polarizations (HH, VV, HV, RR, along with RL) and classifier tree analysis are methods for feature extraction
6	Seed identification as well as realignment of species	Linear regression analysis, principal component analysis, feature extraction
7	Nitrogen content and treatment of soil and leaves, as well as salinity detection	Analysis of linear and exponential regression
8	Irrigation	Image classification methods (density slicing using thresholds, unsupervised clustering), NDVI and linear regression analysis are all used in decision trees
9	Pest detection and management	CEM nonlinear signal processing, Analysis of linear and exponential regression, NDVI, image processing using sample imagery, statistical analysis

3.3.2 Sorting of seeds based on deep learning

Seed sorting would be a mechanized process that aims to produce a high-purity, high-quality end product. Such processes are extremely difficult to predict and regulate. The mixed cropping seed classifier along with quality tester' is a deep learning-based system for precise seed classification based on quality tests on structure and color, together with texture. The dataset includes labeled pictures of healthy and damaged pearl millet and maize seeds. Its capacity to distinguish between infected and healthy maize seeds improves its food sector application. Chunlei Li et al. [30] proposed SeedSortNet, a compact CNN associated with visual attention that is quick and efficient. First, blocks of light weight feature extraction module with two branches are introduced. Shield block is detailed developed by conducting higher-dimensional spatial transformation, identity mapping and different receptive field modeling, and as a result, with fewer parameters and reduced computing complexity, it may reduce information loss while correctly defining the multi-scale feature.

3.3.3 Seed sowing using deep learning

Seed should be planted 1.5–2.0 inches deeper after the seedbed has been prepared to ensure enough moisture availability for effective seed germination. The seed requires certain conditions in order to germinate optimal moisture along with temperature conditions, thus constantly checking on the soil temperature and moisture requirements at several times [29]. Agriculture access is a starting point towards a better living, and agricultural tool development is the foundation for agricultural progress. Renuka Dhavale et al. [31] propose employing an agribot to design a system that saves operational costs, minimizes digging time and improves seed sowing performance. DC motors, moisture sensors, IR Sensors and ultrasonic sensors are employed inside this machine with the support of a Wi-Fi interface running android application on a field oeuvre robot. The seed-spreading and digging robot will walk through several rows of soil, digging up, sowing seeds and covering the soil with a cover.

Lukasz Gierz et al. [32] conducted survey at triticale seed samples of varying quality. The seeds were collected during tests that resemble actual planting circumstances. The experiments have been performed on such a specially developed testing facility by the respective authors. The air speed in the pneumatic tube conveying seeds was adjusted for sowing. The air speed in the pneumatic conduit conveying seeds (15, 20 and 25 m/s) was varied for sowing. The generated visual database allowed for the classification of six seed classes based on sowing speed and quality. The database seemed to be constructed in order to build training and validation, along with test sets. The neural models were created using statistical analysis and multi-layer perceptron networks.

3.3.4 Smart irrigation system using deep learning

Water plays a vital resource for crop production and is becoming increasingly limited, even as farmland continues to increase due to global population expansion. Farmers have been found to benefit from proper irrigation schedule in terms of crop quality and productivity, as well as reduced water usage. One of the most essential crop irrigation factors is soil moisture (SM). This represents the total amount of water from the soil. Estimating future soil moisture (forecasting) is a vital job for crop irrigation regarding water consumption optimization along with crop yields. The amount of moisture in the soil varies a lot depending on humidity, weather and time. As a possible solution, Paweena Suebsombut et al. [35] provide a novel long-short term memory (LSTM)-based approach to estimate future soil moisture levels based on information collected from multiple sensors. A dataset from the real world, including a collection of factors relevant to weather forecasts, soil moisture, along with other associated smart sensors, was used to gather parameters in a greenhouse in Chiang Mai province, Thailand, in order to train and verify this model.

Thomas Henry Colligan et al. [34] presented a novel approach for mapping irrigation, which they apply to Montana from 2000 to 2019. The approach totally depends on raw Landsat surface reflectance data, and it is built over an ensemble using convolutional neural networks. Without any supervision, the ensemble of network technique learns to mask clouds and disregard Landsat 7 scan-line failures, avoiding the requirement for data pre-processing or feature engineering. Jessica Kwok et al. [35] developed an irrigation system based on deep learning that can modify water quantities for each type of plant based on plant recognition. Software and hardware are the two primary components of the solutions. The former is linked to cameras for plant identification and utilizes a database to estimate the proper amount of water; the latter restricts the amount of water that may flow out.

3.3.5 Crop growth recognition using deep learning

Plant growth and yield forecasting are critical tasks for greenhouse growers and farmers in overall. Developing models that can accurately estimate growth and yield can aid producers in improving environmental management, matching supply and demand, and lowering prices. Machine learning, but particularly, deep learning, has recently advanced dramatically in which they can anticipate powerful new analytical tool. According to Yi Xie et al. [36], deep learning approaches were used to combine an crop growth methodology with a time series using remotely sensed data to improve the accuracy for local wheat-yield estimates in China's Henan Province. Models such as the LSTM, one-dimensional convolutional neural network as well as random forest were trained and evaluated using the leaf area index along with grain yield time series simulated aside crop environment resource to incorporate for wheat model. Finally, the regional LAI was calculated using a model of the exponential connection among field-measured LAI along with MODIS NDVI, which had been then fed to trained LSTM, 1D CNN, along with RF models to predict wheat yields within a certain wheat growing region.

The suggested work by Bashar Alhnaity et al. [37] employs machine learning as well as deep learning approaches to forecast yield and plant growth variance in controlled greenhouse conditions for two separate scenarios: tomato yield forecasting as well as Ficus benjamina stem development. While in prediction formulas a newly developed deep recurrent neural network was used, this utilizes long short-term memory neuron framework, by integrating a different leading-edge temporal sequence processing networks – temporal convolutional network as well as recurrent neural network. Liyun Gong et al. [38] established a novel greenhouse crop production prediction system. Different datasets are acquired from various genuine greenhouse sites, whereas tomato growth has been used to evaluate the proposed method in depth. Hafiza Ufaq Rehman et al. [39] provided a viewpoint planning method enabling harvest robots in order to enhance fruit detecting outcomes. Robotic systems can use viewpoint planning to consider active sensing techniques rather than simply one standard viewpoint. Thus, planner uses the recent environment for input as well as produces the optimum viewpoint from a list on pre-defined options: moves right, left or stay still. The perspective planning problem is formulated as a categorization issue, and it is implemented with a deep neural network. Using a fruit detector, they initially extract local fruit areas as well as nearby regions around the fruit from every current scenario. Following fruit-wise segmentation, they then employ the classifier's labels in the viewpoint planner to choose which would be the best perspective.

3.3.6 Fertilizer estimation using deep learning

Crop management may include fertilization. Before adding fertilizer to every crop, its soil should be tested for plant nutrients. The use of suitable fertilizers depending on the soil and/or plant analysis helps ensure that the nutritional requirements of the planted crop are met. Rice, wheat, barley, corn, oats, rye, sorghum and millet are all grain crops that are widely farmed all over the world. A crucial responsibility for sustainable agriculture is to monitor the health of grain plants and to detect diseases early. Early detection of various diseases can help with disease control through the adoption of appropriate pest control technologies to boost grain output. Manual diagnosis of grain plant problems might result in erroneous pesticide measurements.

Using neural networks, JuhiReshma et al. [40] proposed predicting the quantity of fertilizers needed for a specific banana crop and regression approaches for future plantations. One of the most important variables that affect food production and reduce losses will be plant diseases. Deep learning methods have significantly increased their usefulness in plant disease diagnosis, resulting in a complete tool with exceptionally accurate results. Andre Abade et al. [41] suggested the current cutting-edge technology on the use of convolutional neural networks (CNN) in the diagnosis and classification of plant diseases, as well as to identify trends and gaps.

3.3.7 Crop harvesting using deep learning

Abozar Nasirahmadi et al. [37] suggested using a digitized two-dimensional imaging system integrated with convolution approaches to identify apparent sugar beet mechanical damages while harvest using a harvester machine. Hafiza Ufaq Rehman et al. [39] expressed a point or view design enabling harvest robots in order to increase fruit detecting outcomes. Robots can use viewpoint planning to recommend active sensing techniques rather than just one standard viewpoint. As a result, the difficulties with viewpoint planning are reformulated as a classification issue, which is then implemented making use of a deep neural network. Using a fruit detector, we extract local fruit areas as well as nearby regions surrounding the fruit from each current scenario.

Then the labels supplied through the classifiers are used in the perspective planner to determine the optimum point of view after conducting fruit-wise classification (Tanmay U. Sane et al.) [42]. Various robotic harvesting systems were examined in his work, including those that have adopted or aim to apply such strategies to recognize a crop, travel to it and harvest it effectively and reliably. This explores the criteria for selecting an AI/DL approach as well as the challenges and benefits encountered in its field application. Based on its durability and performance, convoluted neural networks are a common choice of DL approach for such applications.

3.3.8 Crop recommendation system using deep learning

A crop recommendation system can be proposed by a digital farming solution to help farmers determine which crop to sow in their field depending on weather conditions, moisture and season. Deep learning and machine learning approaches offer a practical foundation for making data-driven decisions. Using the deep learning/ML recommendation engine, this application also

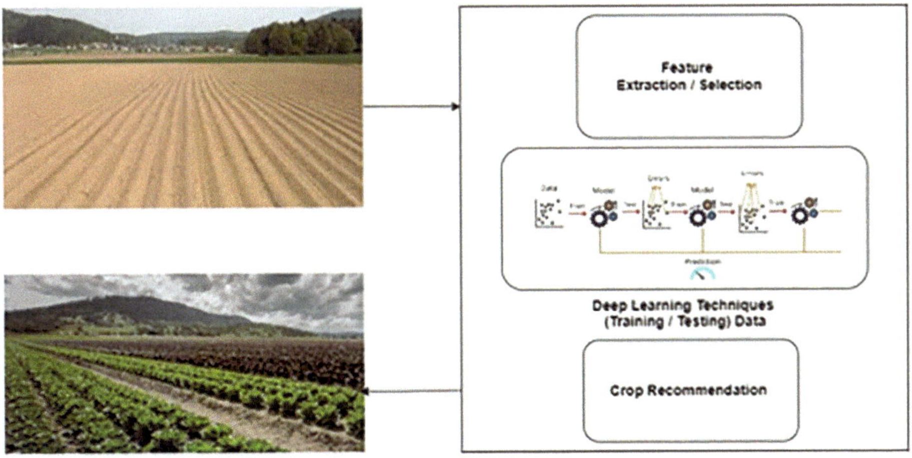

Figure 3.3 Crop recommendation system.

assists in identifying the optimal pesticide, seed spacing and seed as illustrated in Figure 3.3.

In the agricultural sector, seed quality is a crucial element [43]. Because certain seeds are naturally tiny, it might be difficult to distinguish and classify changes across species. Experts use the conventional way to classify these changes based on morphological structure, texture and color. This approach requires an expensive, subjective and time-consuming classifying procedure, possibly precluding the development of a system that can detect the kind of seed automatically. Using CNN, one of the deep learning algorithms, a smartphone application, has been built that accurately recognizes and classifies seed images.

Crop production is an ever-evolving approach to agricultural advancements and farming methods [44]. Dealing with climatic changes due to soil erosion and industry emissions are two of the challenges that farmers face. Nutritional deficiencies in the soil, caused by a lack of important minerals like potassium, nitrogen and phosphorus, can cause crop growth to be delayed. Farmers commit mistakes by farming the same crops every year without trying new types. Madhuri Shripathi Rao et al. recommend that farmers choose the best crop prediction model that can assist them pick what sort of crop to produce depending on the meteorological conditions and nutrients from the soil. To classify two distinct criteria Gini and entropy, popular approaches including decision tree, forest classifier and random K-nearest neighbor are employed. Fuzzy classifiers are used for health care and IoT-based system [47–52]. Table 3.2 shows the datasets available in public related to agriculture.

Table 3.2 Agriculture-related datasets that are publicly available.

S. No	Dataset/organization/link	Dataset characteristics
1.	Image Net Datasethttp://image-net.org/ explore?wnid=n07707451	Various plant images (trees, vegetables and flowers)
2.	Image Net Large Scale Visual Recognition Challenge (ILSVRC) http://image-net.org/challenges/LSVRC/2017/#det	Images which enable the location and identification of objects
3.	Plants Dataset from University – Arcansas https://plants. uaex.edu/herbicide http://www.uaex.edu/yard-garden/resource-library/diseases	Herbicide injury image database
4.	PFL, Plant Village Dataset https://www.plantvillage.org/en/crops	Different images of crops with related diseases
5.	Dataset for Leafsnap http://leafsnap.com/dataset	185 different tree species' leaves from the North-eastern US
6.	Dataset for LifeCLEF http://www.imageclef.org/2014/lifeclef/plant	Identification, geographic range, about the use of plants
7.	Visual Object Classes dataset for PASCAL http:// host.robots.ox.ac.uk/pascal/VCC	Various animal images (birds, cats, cows, dogs, horses, sheep, etc.)
8.	Africa Soil Information Service (AFSIS) dataset http://africasoils.net/services/data/	Continent-wide digital soil maps for sub-Saharan Africa
9.	UC Merced Land Use Dataset http://vision.ucmerced.edu/datasets/landuse.html	A collection with 21 classes of land use images
10.	MalayaKew Dataset http://web.fsktm.um.edu.my/~cschan/downloads_ MKLeaf_d ataset.html	Leaf scan-like photos representing 44 species classes
11.	Dataset of weeds as well as cropped images https://github.com/cwfid/dataset www.semanticscholar.org/pdfs/58a0/9b1351ddb447e6abd ede7233a4794d538155.pdf	Crop/weed plant group annotations, field pictures as well as vegetation segmentation masks

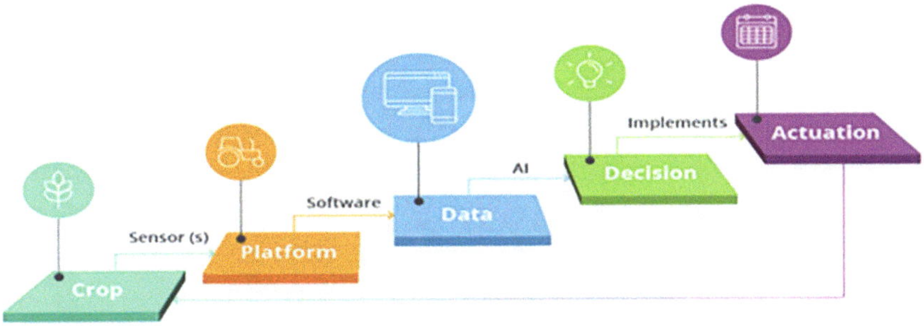

Figure 3.4 Artificial intelligence in smart farming.

3.4 Scope and Challenges in Crop Management System

This section presents a complete analysis of sophisticated strategies based on deep learning as well as machine learning that were used within smart farming. We specifically examine the scope, recent application and opportunities, problems, limits and future research avenues of machine learning – as well as deep learning-based advanced approaches used within smart farming.

3.4.1 Role of artificial intelligence in smart farming

Over the years, technology has reshaped farming, and technological breakthroughs have an influence on the agriculture industry by a variety of aspects. Agriculture is a major activity in several countries across the world, and as the world's population grows, so does agriculture, which, according to UN projections, will increase from 7.52 billion to 9.78 billion by 2050; there will also be additional land pressure, because just 4% of land will be farmed by that time [53–54]. As a result, farmers would have to do more with less. Also according to the survey, food production will need to increase by 60% in order to serve an additional 2 billion people. Traditional approaches are inadequate to meet this enormous need. This is causing farmers and agribusinesses to look for innovative ways to enhance output while reducing waste.

As a result, AI is progressively becoming a part of the agricultural industry's technical growth. Applications of computer vision in agriculture AI-powered solutions may not only help farmers increase efficiency, but they will also boost crop yield, quality and assure a faster time to market (Figure 3.4).

3.4.2 Farmers challenges in adopting new technologies

A number of challenges and factors influence the technology's acceptance and implementation. Farmers are the primary stakeholders in the technology

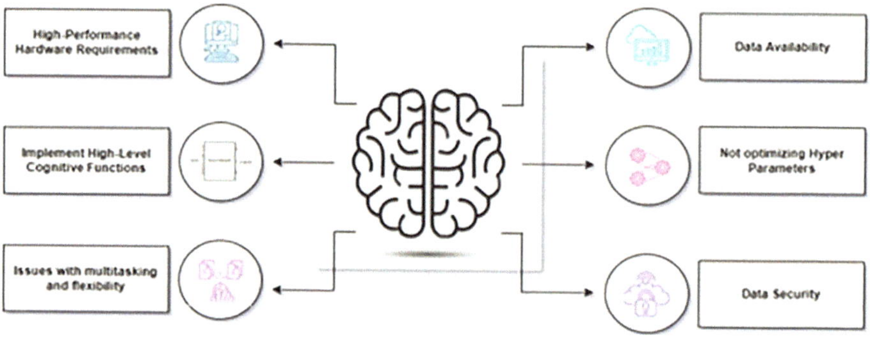

Figure 3.5 Deep learning system challenges.

since they will be using it to better the farming system. There is a lack of interest due to availability of minimal resources to build knowledge exchange platforms by accepting new technologies. The key reason for the lack of interest is a general lack of understanding and information about how technology may bring concrete progress both financially and environmentally to establish sustainable development architecture for all stakeholders. Although many research works are carried out in this sector, the practicality and application of these research works are sometimes ignored.

Even apart from applicability, there is a significant empty space in field work where qualified professionals can go about explaining the value of technology, which is the major obstacle to adopting the technology. The key difficulty is that security concerns might lead to agricultural harvesting disasters, resulting in food shortages, causing concerned parties to be hesitant to use the technology. Furthermore, issues and scopes are a continual process that must be addressed and altered often through check and balance in order to offer a sustainable development with real outcomes for the whole community [56].

3.4.3 Challenges in deep learning

Deep learning had already emerged among the most essential learning topics towards the development of intelligent systems. Deep learning techniques make use of artificial neural network (ANN) imitating human minds and progressively learning how to deal with a given particular situation. However, there are substantial challenges with deep learning systems that we must be aware of as shown in Figure 3.5.

3.4.3.1 Massive amount of data

Data is used to train deep learning algorithms to learn in a progressive manner. To ensure that the machine produces the appropriate results, large datasets are required. The equivalent artificial neural network needs a large quantity data, as the human brain need a large amount of experience to learn and interpret information. The greater the abstraction, hence more parameters that must be modified, and much more parameters demand more data.

3.4.3.2 High-performance hardware is needed

A huge volume of data is required to train dataset sufficient for deep learning solution. To tackle problem in the real world, the machine must have sufficient computing power. Data scientists use multi-core high-performance GPUs and equivalent processing units in order to improve efficiency and decrease time consumption. Such processing machines are costly and may consume more power. Deep learning systems at the industrial level require elevated data centers, whereas drones, robots and mobile devices are examples of smart technologies that require compact yet effective computational units. As a result, installing a deep learning system in the actual world is both expensive and energy intensive.

3.4.3.3 Neural network overfitting

There is sometimes a significant discrepancy between the errors encountered in the training dataset and also the issue identified in the newly declassified dataset. It happens in complicated framework if there are so many parameters in comparison with volumes of data. The effectiveness of a model is determined by its capacity to execute effectively for unknown dataset, rather than based on its effectiveness on training data provided to it. Typically, a model is developed by optimizing its efficiency on such a certain training dataset. As a result, the model remembers the training instances but it does not understand how to generalize current circumstances or datasets.

3.4.3.4 Hyperparameter optimization

The value of hyperparameters is determined before the learning process begins. A slight change in the value of such parameters might cause a significant difference which affects the performance of your model. Using its default configuration and without conducting hyperparameter optimization may possess substantial influence in terms of model effectiveness. Holding very few hyperparameters and adjusting those manually rather than employing optimization methods is also a type of performance-limiting factor.

3.4.3.5 Implementing high-level cognitive functions

If we know what our parameters of the model are, how do we feed a known data to neural networks and how are they assembled. However, seldom do we grasp how they arrive at a certain solution. Researchers struggle to grasp how neural networks draw conclusions since they are basically black boxes. The inability of neural networks to reason abstractly makes high-level cognitive processes challenging to accomplish. Furthermore, their functioning is mostly transparent to humans, making them inappropriate for areas where process verification is critical.

3.4.3.6 Lacking of multitasking and flexibility

Deep learning methods could provide extremely efficient and precise solutions (once trained) to specific problems. But, in today's world, neural network topologies were extremely customized for a particular purpose field. The majority of the systems are based on this concept, and they excel at handling a single issue. Even handling a related situation needs re-training and reassessment. Deep learning frameworks that could also multitask without having to redesign the entire architecture are being extensively investigated by researchers.

3.4.3.7 Data security

Deep learning has been identified in a lot of sectors and has delivered great performance in outcomes due to the advancement of algorithms in vast data and strong computing services. It is important in everyday applications and is gently affecting societal conventions, habits and behaviors. Data-based learning methodologies, on the other hand, are bound to create possible security and privacy risks, as well as raise public and government concerns regarding their use in the real world. In order to interrupt the training process or cause the model to deliver unexpected results, data security threats try to compromise the integrity or availability of data.

Black-box and white-box attackers are the two categories of attackers regarded in general. A DNN model's internal information, like optimization methods, parameter arrangement as well as training sets, is not accessible to a black-box attacker. In order to increase access to the search function, the white-box attacker gains a lot of information, such as the structure, DNN model specification as well as certain training data. White-box attackers were undoubtedly superior to black-box attackers in terms of effectiveness.

3.4.4 Recent deep learning algorithms in smart farming

With a flood of anticipation flowing into the DL sector, significant advances have been made in recent years. Image recognition is one of its agricultural applications, and it has overcome several difficulties that have slowed rapid progress with robotics and automation farming and agro-industry. Many parts of agriculture have improved, including weed control, counting of plants as well as plant disease identification. Researchers in agriculture are not necessarily skilled programmers. The researchers commonly use publicly available deep learning software frameworks before carefully analyzing their learning methods. Studying deep learning models will improve the analysis of data as well as agricultural research will be improved. Despite the availability of many software product architectures, deep learning algorithm has ideas, software restrictions, flow diagrams as well as sample codes that could aid agricultural researchers in quickly and efficiently learning important DL approaches.

A variety of techniques are used in deep learning models. Although networks are not perfect, several techniques were more suitable for specific tasks than others. It is critical to understand all of the essential algorithms in selecting the most suitable ones. During current history, deep learning techniques like recurrent neural networks (RNN), convolutional neural networks (CNN) and generative adversarial networks (GAN) had been intensively investigated as well as employed for a variety of industries, including agriculture. Table 3.3 summarizes the various deep learning algorithms. The focus of this research is to give agricultural researchers a comprehensive understanding about DL as well as to improve existing precision agricultural innovation.

3.5 Conclusion and Future Scope

The wide majority of recent agricultural developments produced through researchers were strongly associated with productivity and all other aspects of agriculture, to improve crop productivity, reduce and prepare for plant diseases, and promote mechanized and automated modernized agriculture and agro-industry. The principles, techniques, restrictions and algorithms of machine learning as well as deep learning are summarized elaborately in this paper. Deep learning applications in agriculture were discussed. Deep learning has been within the scope of smart applications for agriculture, including crop management systems. Crop cultivation, crop disease diagnostics, water along with soil management, crop distribution, weed control, rigorous fruit counting and yield prediction are all part of the process. The objective of this study is to inspire more scholars in order to explore with deep learning, utilizing it to address a wide range of agricultural issues involving not only prediction (or) classification but also computer vision along with image analysis, as well as generalized analysis of data. In the future, researchers may intend to extend the broad ideas and knowledge of deep learning, as indicated within this study, to broaden agricultural areas where all this contemporary

Table 3.3 Deep learning algorithms used in agriculture

Type	BP	CNN	RNN	GAN	BP	DBN
Different versions	RBF GRNN	LeNet, AlexNet VggNet	LSTM	DCGAN	MLP	DBM
Network infrastructure	Initial (I/P) Layer Final (O/P) Layer Hidden Layer	Initial (I/P) Layer Convolution layer Pooling layer	Initial (I/P) Layer Covert layer Final (O/P) Layer	Discrimination model Generation model	Initial (I/P) Layer Covert layer Final (O/P) Layer	Initial (I/P) Layer Covert layer 1 Covert layer 2 Covert layer 3 Output layer
Various application	Fitting data Pattern detection Classification	Image processing Speech signal Natural Language Processing	Analysis of time series Emotional examination Language of Nature Processing	Image creation Video creation	Image Recognition, Object detection, Route Adjustment	Motion capture Image generation Video recognition Image classification

approach has yet to be adopted. Deep learning's long-term benefits were promising because of its future application in to smarter, extra sustainable farming and then more secure food supply.

References

[1] M. Ngouajioa,, M.E. McGiffen Jrb, C.M. Hutchinsonc,"Effect of cover crop and management system on weed populations in lettuce", Crop Protection 22 (2003) 57–64.

[2] S. Dhanasekar, T. Jothy Stella, Mani Jayakumar, "Study of Polymer Matrix Composites for Electronics Applications", Journal of Nanomaterials, vol. 2022, Article ID 8605099, 7 pages, 2022. https://doi.org/10.1155/2022/8605099

[3] Dhanasekar S, Malin Bruntha P, Martin Sagayam K, "An Improved Area Efficient 16-QAM Transceiver Design using Vedic Multiplier for Wireless Applications", International Journal of Recent Technology and Engineering, vol.8,no. 3, pp.4419–4425, September 2019. doi: 10.35940/ijrte.C5535.098319

[4] Dhanasekar,S.Bruntha,P.M.Neebha,T.M,Arunkumar,N,Senathipathi,N, Priya, C, "An Area Effective OFDM Transceiver System with Multi-Radix FFT/IFFT Algorithm for Wireless Applications", In Proceedings of the 2021 7th International Conference on Advanced Computing and Communication Systems (ICACCS), Coimbatore, India, 19–20 March 2021, pp. 551–556. doi:10.1109/ICACCS51430.2021.9441694

[5] Zeynep Unal,Smart Farming Becomes Even Smarter With Deep Learning-A Bibliographical Analysis", IEEE Access, 8, 2020, doi:10.1109/ACCESS.2020.3000175

[6] Oksana Mamai, Velta Parsova, Natalya Lipatova, Julia Gazizyanova, and Igor Mamai, "The system of effective management of crop production in modern conditions", BIO Web of Conferences 17, 00027 (2020), doi:https://doi.org/10.1051/bioconf/20201700027

[7] C. Fyfe, "Articial neural networks and information theory," Ph.D. dissertation, Dept. Comput. Inf. Syst., Univ. Paisley, Paisley, U.K., 2000.

[8] Dhanasekar S, Ramesh J, "VLSI Implementation of Variable Bit Rate OFDM Transceiver System with Multi-Radix FFT/IFFT Processor for wireless applications", Journal of Electrical Engineering, vol. 18, 2018-Edition: 1 – Article 18.1.22. ISSN:1582-4594.

[9] Dhanasekar S, Ramesh J, "FPGA Implementation of Variable Bit Rate 16 QAM Transceiver System", International Journal of Applied Engineering Research, vol.10, pp. 26479–26507, 2015.

[10] Dhanasekar S, Suriavel Rao R S and Victor Du John H, "A Fast and Compact multiplier for Digital Signal Processors in sensor driven smart vehicles", International Journal of Mechanical Engineering and Technology, vol. 9,no.10, pp. 157–167, October 2018.

[11] P. Christiansen, L. N. Nielsen, K. A. Steen, R. N. Jørgensen, and H. Karstoft, "DeepAnomaly: Combining background subtraction and deep learning for detecting obstacles and anomalies in an agricultural field," Sensors, vol. 16, no. 11, pp. 2–21, 2016.

[12] N. Kussul, M. Lavreniuk, S. Skakun, and A. Shelestov, "Deep learning classification of land cover and crop types using remote sensing data," IEEE Geosci. Remote Sens. Lett., vol. 14, no. 5, pp. 778–782, May 2017.

[13] F. J. Knoll, V. Czymmek, S. Poczihoski, T. Holtorf, and S. Hussmann, "Improving efficiency of organic farming by using a deep learning classification approach," Comput. Electron. Agricult., vol. 153, pp. 347–356, Oct. 2018.

[14] M. Bah, A. Haane, and R. Canals, "Deep learning with unsupervised data labeling for weed detection in line crops in UAV images," Remote Sens., vol. 10, no. 11, p. 1690, Oct. 2018.

[15] A. Picon, A. Alvarez-Gila, M. Seitz, A. Ortiz-Barredo, J. Echazarra, and A. Johannes, "Deep convolutional neural networks for mobile capture device-based crop disease classication in the wild," Comput. Electron. Agricult., vol. 161, pp. 280290, Jun. 2019.

[16] S. Coulibaly, B. Kamsu-Foguem, D. Kamissoko, and D. Traore, "Deep Neural networks with transfer learning in millet crop images," Comput. Ind., vol. 108, pp. 115–120, Jun. 2019.

[17] P. Nevavuori, N. Narra, and T. Lipping, "Crop yield prediction with deep convolutional neural networks," Comput. Electron. Agricult., vol. 163, Aug. 2019, Art. no. 104859.

[18] Q. Yang, L. Shi, J. Han, Y. Zha, and P. Zhu, "Deep convolutional neural networks for Rice grain yield estimation at the ripening stage using UAV based remotely sensed images," Field Crops Res., vol. 235, pp. 142–153, Apr. 2019.

[19] L. Zhong, L. Hu, and H. Zhou, "Deep learning based multi-temporal crop classification," Remote Sens. Environ., vol. 221, pp. 430–443, Feb. 2019.

[20] S. Zhao, X. Liu, C. Ding, S. Liu, C. Wu, and L. Wu, "Mapping rice paddies in complex landscapes with convolutional neural networks and phenological metrics," GISci. Remote Sens., vol. 57, no. 1, pp. 37–48, 2019.

[21] K. Thenmozhi and U. Srinivasulu Reddy, "Crop pest classification based on deep convolutional neural network and transfer learning," Comput. Electron. Agricult., vol. 164, Sep. 2019, Art. no. 104906.

[22] Mayank Dixit, Abhishek Tiwari, Himanshu Pathak, Rani Astya, "An overview of deep learning architectures, libraries and its applications areas", International Conference on Advances in Computing, Communication Control and Networking (ICACCCN2018) 2018, pp: 293–297.

[23] V. Govindaraj, S. Dhanasekar, "Minimum Power Test Pattern Generator for Testing VLSI Circuits", 2022 6th International Conference on Devices, Circuits and Systems (ICDCS), pp.27–31. 2022.

[24] Khaki, S., and Wang, L. (2019). Crop yield prediction using deep neural networks. Front. In Plant Sci. 10, 621. doi: 10.3389/fpls.2019.00621

[25] Kim, N., Ha, K.-J., Park and Lee, Y.-W. (2019). A comparison between major artificial intelligence models for crop yield prediction: Case study of the midwestern united states, 2006–2015. ISPRS Int. J. Geo-Information 8, 240. doi: 10.3390/ijgi8050240

[26] Wang, A. X., Tran, C. and Ermon, S. (2018). "Deep transfer learning for crop yield prediction with remote sensing data," in Proceedings of the 1st ACM SIGCAS Conference on Computing and Sustainable Societies (San Jose CA, USA: ACM), 50

[27] Yang, Q., Shi, L., Han, J., Zha, Y., and Zhu,P. (2019). Deep convolutional neural networks for rice grain yield estimation at the ripening stage using uav-based remotely sensed images. Field Crops Res. 235, 142–153. doi: 10.1016/j.fcr.2019.02.022

[28] P. Sumathi, R. Subramanian, V. V. Karthikeyan, and S. Karthik, "Soil monitoring and evaluation system using EDL-ASQE: Enhanced deep learning model for IoT smart agriculture network," *Int. J. Commun. Syst.*, vol. 34, no. 11, pp. 1–12, 2021, doi: 10.1002/dac.4859.

[29] Y. Gulzar, Y. Hamid, A. B. Soomro, A. A. Alwan, and L. Journaux, "A convolution neural network-based seed classification system," *Symmetry (Basel).*, vol. 12, no. 12, pp. 1–18, 2020, doi: 10.3390/sym12122018.

[30] C. Li, H. Li, Z. Liu, B. Li, and Y. Huang, "SeedSortNet: a rapid and highly effificient lightweight CNN based on visual attention for seed sorting," *PeerJ Comput. Sci.*, vol. 7, pp. 1–21, 2021, doi: 10.7717/peerj-cs.639.

[31] S. K. B. P. P. C. A. P. S. Renuka Dhavale, "IRJET- Digging and Seed Sowing Agribot," *Irjet*, vol. 8, no. 7, pp. 3314–3319, 2021.

[32] Ł. Gierz, K. Przybył, K. Koszela, A. Duda, and W. Ostrowicz, "The use of image analysis to detect seed contamination—a case study of

triticale," *Sensors (Switzerland)*, vol. 21, no. 1, pp. 1–14, 2021, doi: 10.3390/s21010151.

[33] P. Suebsombut, A. Sekhari, P. Sureephong, A. Belhi, and A. Bouras, "Field data forecasting using lstm and bi-lstm approaches," *Appl. Sci.*, vol. 11, no. 24, 2021, doi: 10.3390/app112411820.

[34] G. S. Theses and A. N. Lommen, "ScholarWorks at University of Montana Relactation : A Phenomenological Study Let us know how access to this document benefits you .," 2014.

[35] J. Kwok and Y. Sun, "A smart IoT-based irrigation system with automated plant recognition using deep learning," *ACM Int. Conf. Proceeding Ser.*, pp. 87–91, 2018, doi: 10.1145/3177457.3177506.

[36] Y. Xie and J. Huang, "Integration of a crop growth model and deep learning methods to improve satellite-based yield estimation of winter wheat in henan province, china," *Remote Sens.*, vol. 13, no. 21, 2021, doi: 10.3390/rs13214372.

[37] A. Nasirahmadi, U. Wilczek, and O. Hensel, "Sugar beet damage detection during harvesting using different convolutional neural network models," *Agric.*, vol. 11, no. 11, 2021, doi: 10.3390/agriculture11111111.

[38] L. Gong, M. Yu, S. Jiang, V. Cutsuridis, and S. Pearson, "Deep learning based prediction on greenhouse crop yield combined TCN and RNN," *Sensors*, vol. 21, no. 13, pp. 1–16, 2021, doi: 10.3390/s21134537.

[39] H. U. Rehman and J. Miura, "Viewpoint Planning for Automated Fruit Harvesting Using Deep Learning," *2021 IEEE/SICE Int. Symp. Syst. Integr. SII 2021*, pp. 409–414, 2021, doi: 10.1109/IEEECONF49454.2021.9382628.

[40] D. J. Aravindhar, "Fertilizer Estimation using Deep Learning Approach," vol. 8, no. 4, pp. 5745–5752, 2021.

[41] A. Abade, P. A. Ferreira, and F. de Barros Vidal, "Plant diseases recognition on images using convolutional neural networks: A systematic review," *Comput. Electron. Agric.*, vol. 185, no. March, 2021, doi: 10.1016/j.compag.2021.106125.

[42] T. U. Sane and T. U. Sane, "Artificial Intelligence and Deep Learning Applications in Crop Harvesting Robots-A Survey," *3rd Int. Conf. Electr. Commun. Comput. Eng. ICECCE 2021*, no. June, 2021, doi: 10.1109/ICECCE52056.2021.9514232.

[43] Y. BAŞOL and S. TOKLU, "A Deep Learning-Based Seed Classification with Mobile Application," *Turkish J. Math. Comput. Sci.*, vol. 13, no. 1, pp. 192–203, 2021, doi: 10.47000/tjmcs.897631.

[44] A. Kamilaris and F. X. Prenafeta-Boldú, "Deep learning in agriculture: A survey," *Comput. Electron. Agric.*, vol. 147, pp. 70–90, 2018, doi: 10.1016/j.compag.2018.02.016.

[45] L. J. Ahmed, "Discrete Shearlet Transform Based Speckle Noise Removal in Ultrasound Images," Natl. Acad. Sci. Lett., vol. 41, no. 2, 2018, doi: 10.1007/s40009-018-0620-7.

[46] L. Jubairahmed, S. Satheeskumaran, and C. Venkatesan, "Contourlet transform based adaptive nonlinear diffusion filtering for speckle noise removal in ultrasound images," Cluster Comput., vol. 22, 2019, doi: 10.1007/s10586-017-1370-x.

[47] Ahmed, L.J., Anishfathima, B., Gokulavasan, B., Mahaboob, M. (2022). Fog-Assisted Real-Time Coronary Heart Disease Risk Detection in IoT-Based Healthcare System. In: Haldorai, A., Ramu, A., Mohanram, S., Lu, J. (eds) 3rd EAI International Conference on Big Data Innovation for Sustainable Cognitive Computing. EAI/Springer Innovations in Communication and Computing. Springer, Cham. https://doi.org/10.1007/978-3-030-78750-9_8

[48] Jubair Ahmed, L., Anish Fathima, B., Dhanasekar, S., Martin Sagayam, K. (2022). Modeling of Fuzzy Logic-Based Classification System Using the Gravitational Search Algorithm. In: Kumar, R., Sharma, R., Pattnaik, P.K. (eds) Multimedia Technologies in the Internet of Things Environment, Volume 3. Studies in Big Data, vol 108. Springer, Singapore. https://doi.org/10.1007/978-981-19-0924-5_5

[49] J. Anand, and J. Raja Paul Perinbam, "Automatic Irrigation System using Fuzzy Logic", AE International Journal of Multidisciplinary Research, Vol. 2, Issue 8, pp. 1–9, August 2014.

[50] Dhanalakshmi R., J. Anand, Arun Kumar Sivaraman, and Sita Rani, "IoT-based Water Quality Monitoring System using Cloud for Agriculture Use", Cloud and Fog Computing Platforms for Internet of Things, Edited by Pankaj Bhambri, Sita Rani, Gaurav Gupta, Alex Khang, Routledge Taylor & Francis Group, May 2022.

[51] Govindaraj, V & Ramesh, J 2017, 'An Improved Low Transition Test Pattern Generator for Low power Applications', Design Automation for Embedded Systems, vol. 21, no. 3, December, pp. 247–263.

[52] Govindaraj, V., Dhanasekar, S., Martinsagayam, K. et al. Low-power test pattern generator using modified LFSR. AS (2023). https://doi.org/10.1007/s42401-022-00191-5

[53] Govindaraj, V & Ramesh, J 2018, 'Adaptive neuro fuzzy inference system-based power estimation method for CMOS VLSI circuits', International Journal of Electronics, vol. 105, no.3, November, pp. 398–411.

[54] Govindaraj, V & Dr Aruna Devi B, J 2021, 'Machine learning based power estimation method for CMOS VLSI circuits', Applied Artificial Intelligence, vol. 35, no. 13, October, pp. 1043–1055.

4

Autonomous Devices in Smart Farming

T. Manikandan[1], E. Duraiarasu[1], C. Ganesh Kumar[1], S. Jeevitha[1], S. Harihara Sudhan[1], and Olabiyisi Stephen Olatunde[2]

[1]Department of Electronics and Communication Engineering, Rajalakshmi Engineering College, India
[2]Computer Science and Engineering, Ladoke Akintola University of Technology, Nigeria
Email: manikandan.t@rajalakshmi.edu.in; 200801064@rajalakshmi.edu.in; 200801066@rajalakshmi.edu.in; 200801087@rajalakshmi.edu.in; 20080101074@rajalakshmi.edu.in; soolabiyisi@lautech.edu.ng

Abstract

Co-growing or intercropping means growing two or more crops in the same field. The main purpose of intercropping is to obtain more net results from an open farm by using resources, and those techniques will not be used to raise a single crop. Intercropping-based agriculture makes farming easier to cultivate various crops simultaneously. This will be more efficient and time-saving in the small portion of land. The possibility of getting yield in two contrasting types of crops needs two requirements; proper irrigation and nutrients. An alert system for the safety and the physical phenomenon is required, that is why there is a need for an autonomous system. The system is capable of irrigating the proper amount of water to the different crop variety present in the fields. For this consider the other features like climate monitoring and forecasting, smart pest management, smart greenhouse, remote crop monitoring, etc. Smart farming helps to improve agriculture with techniques to operate, monitor, automate and analyze the operations using advanced techniques such as large data, the virtual database and internet-based technologies. It is simply a management concept. Smart agriculture, in other words, is known as precision agriculture, which works by the coordination of many sensors.

The increased growth of the global population increases the demand for food and with it comes the need for efficient use of natural resources; with that said there is also an increase in the usage of sophisticated information and communication technologies, thereby resulting in an increasing need for a smart climate tracking device. The importance of smart agriculture in daily life is increasing exponentially. The plant monitoring system is helpful to water the plants and monitor some parameters related to plant growth. This system is widely accepted in a few areas such as in tree nurseries and agriculture. The system frame contains a potentiometric sensor and physical sensors fixed in the root area of plants, and a vibration of frequency sensor is installed in the tank to check the water level in the tank. Smart systems have built-in detection, action and control capabilities to describe and analyze the situation and make decisions based on data available in predictive or adaptive ways, thereby smart actions. To execute in most cases, the "smartness" of a system can be attributed to controlling the energy efficiency and autonomous operation based on networking capabilities. Smart monitoring systems using technologies with internet-connected devices are adaptable for monitoring the plants effectively and efficiently. It is also suitable for checking the condition of the soil. This can help to get better growth of the crops. It may help in controlling the usage of water efficiently.

An autonomous system (AS) is a set of internet-based connected objects and prefixes that belong to a network or group of networks, all managed, controlled and overseen by a single entity or organization. The AS uses a common routing policy controlled by the entity. This system is capable of monitoring more than two crop varieties at the same time and the physical constraints have been accurately defined for efficient management of the crop under various conditions. Thus, the requisites of this type of agriculture will be given with the data which is a specifically suitable parameter for good yield of the specific crop variety, and these will be used as the thresholds. The autonomous system will react accordingly to the different situations and the system will give a response which is favorable to safeguard the specific type of crops. Furthermore, this autonomous system will help the different varieties of crops to sustain the situation which otherwise is not suitable for that crop variety. Hence, this chapter explores the various techniques and methods involved in it and explains how this autonomous system will be capable of overcoming these problems which are faced during this type of farming.

4.1 Introduction

Intercropping is a common practice of growing more than two crops of different varieties simultaneously in the same field. In other words, intercropping is the practice of cultivating more than one crop simultaneously over the same particular field. The most beneficial advantage of intercropping is that it produces a greater outcome in terms of yield within a short duration in a single land. The crops grown on that particular land use up the available resources to their full potential, while a single crop cannot utilize all the resources. Hence, the intercropping method is seen as one of the powerful ways of growing crops and producing much greater yield within a shorter period in a small field. Some well-known implementations of these techniques are planting crops that have deep roots along with a crop that has shallow roots and also planting a crop that is taller with a crop that is shorter which in this case requires a minimum amount of shade. There are various types of intercropping systems which vary on factors like the spatial and temporal mixture to some extent. This degree of temporal and spatial interference in the crops may differ, but the main thing is that both conditions should satisfy for a proper intercropping system to exist. To name a few, there are mixed intercropping systems, relay cropping systems and row cropping systems. One point to be noted is that crop rotation is indeed related, but not the intercropping system. The reason is different varieties of crops are grown in different seasons instead of growing within a single season. While carrying out this method, one should keep the most important thing in mind, which is careful planning. This plan should be taking into account the soil quality, climatic conditions and crops and their varieties. Also, the crops should not compete for the area, minerals, H_2O and natural light sources such as sunlight. After checking these conditions, other related agronomic effects can also be achieved.

There are many advantages to an intercropping system such as intercropping of compatible and suitable plants helping to achieve biodiversity and soil diversity, or providing a proper healthy environment for a variety of insects and soil organisms that might not be present in the case of a single crop environment. Micro-organisms might provide the crops with suitable and valuable nutrients, such as via (Nitrogen) N_2 fixation. And other such benefits are farmers could expect a higher percentage of yields during harvesting within a short period. Additionally, it also increases our overall economical value by improving agricultural productivity and reducing the challenges and poverty faced by the farmers. Even though there are many advantages to this system, there are still some drawbacks that are important to consider. Some of the main drawbacks of this system are poor pest management, difficulty in managing the growth of crops, especially in the irrigation process, when one needs to deliver a suitable amount of water to the suitable kind of crop, monitoring the climate to track and control the behavior of the plants and also controlling and preventing the pests and insects from damaging the crops and obliterating their quality. And all those problems lead to the wastage of money, energy, man labor, water, time and other such resources. All the hard work and efforts made by the courageous farmers get affected only by the improper management of those previously mentioned factors. This leads to a decrease in the entire economical growth degradation of crops in the agricultural sectors and also decreases the overall yield and quality drastically. As a result, one country might face serious issues like global warming, poverty and famine and other such obliterating disasters.

With ongoing developments in the field of agriculture, smart agriculture can be of huge benefit to the agricultural sectors. This will indeed uplift the overall productivity, quality, resource management and usage in the overall agricultural sector. Precision farming/agriculture is a similar kind of smart agriculture, which differs in the way how precise the total agriculture process happens (such as delivering the exact amount of water to the crops). It is a type of agriculture, which includes management strategies that gather, process and analyze the spatial, temporal and individual data to combine it with other information to support management decisions according to the estimated changes and variability for improving the usage of resources effectively, efficiently and increasing the profitability by improving productivity, quality and sustainability. Precision agriculture is the most efficient way of agriculture because it presents the most innovative and expected results which emerge from various experiments and research conducted in the field of agriculture. Furthermore, it provides an effective platform for disseminating original and fundamental experiences and research in this rapidly developing sector. The goal of this type of agriculture is to establish a decision support system (DSS) for the overall farm management considering the maximum optimal returns on the inputs while preserving the available resources.

Along with the help of smart agriculture, precision agriculture can achieve tremendous heights in the agricultural sectors with the help of an automated system that incorporates various modern enhanced technologies. This system can be deployed with numerous sensors, according to the needs of an individual which differs based on the workplace. Sensors that are installed on that particular system can be used to collect and gather data on various aspects and parameters. Some of the parameters are soil moisture, its quality and monitoring climate, tracking the water and nutrient supply. Finally, these collected data can be processed into information with the help of portable data acquisition systems. With the help of the Internet of Things (IoT), this system can be made to use and access wirelessly by installing a simple internet connectivity network with the help of wireless Wi-Fi modules [1–3]. Through this network, the data collected with the help of sensors can be stored and procured in an online platform, which is an online cloud service such as Amazon cloud drive. Moreover, this procured data can be converted into information using various applications and software such as LabView. Once the information is processed, that whole collection of information can be fed to an artificial intelligence framework which indeed uses machine learning models and algorithms. With the help of this fed information, this machine learning model will train according to the datasets given to it. The training process will be repeated until those models find a new way to automate the various systems, such as irrigation systems and climate monitoring systems.

4.2 Related Works

Due to advancements in our modern world, the agricultural sector is facing a lot of changes like shortage of land and workers which lead to a shortage of yield. To overcome these problems, a cost-efficient and modern technique is required. For those challenges, one solution is found, which is crop online monitoring using IoT technology which helps to know about all the information like animal intrusion, crop health, growth of weeds and water level in a tank. The main objective is that the farmer can access this network anywhere and anytime at any place. Using wireless network, farm conditions can be accessed and microcontrollers can be used to know about the farm's process. Wireless cameras also have been used to capture images and videos of the farm. To detect the conditions of the environment like temperature, humidity sensors have been used to retrieve and store databases. ATMEGA8535 and IC8817BS, analogue to digital conversion and wireless transceiver module based on Zig Protocol, have been used in this system. WSN is used to determine the timely need for watering which results in efficient use of water. The use of smart technologies and their combination helps the farmers to do farming effectively [4].

Climatic changes had a great impact in all aspects, especially in the agricultural field which had a major impact on people's life leading to poverty in farmer's life and a shortage of food. This smart agriculture is a great solution; the main objective is to send the farmers an SMS about the exact weather condition of the farm using the Rotronic Monitoring System (RMS) so that farmers can easily take some precautionary measures to protect the crops. By using certain parameters, detection of local agricultural parameters can be achieved. WINGZ (Wireless IP Network Gateway) using Zigbee technology works as a coordinator for WPAN networks. This system helps in obtaining the current weather condition of the farm including the weather forecasting of the previous days and provides an environment where users can monitor the data. Some researchers have concentrated on crop monitoring, and they have collected some spatial data about temperature and rainfall to reduce crop loss and increase crop production [5]. To increase yield production, crop has to be monitored from time to time, by using IoT major factors of the farm like soil moisture and certain other factors, which help in better planning for future production of crops. This technique has a major advantage as an energy conservation technique and it can also be used in different types of modes like standby mode, sleep mode and power-saving mode. When the system is powered on the microcontroller unit, the system activates the peripherals for communication between the devices. Then the connection between sensors is checked and the information is shared via the sensors, then it is shared with the computer. If a deviation occurs, required actions can be sent establishing a warning to the farmer and automatic switching ON of motor can be done with the help of sensors if required. For storing information, wireless sensors are used. Arduino board with ATmega microcontroller can react to the changes in the environment [6].

Another new modern technique for enhancing agriculture is CupCarbon Simulator, which shows greater efficiency compared to other techniques. Soil moisture is an important factor for plant growth. Excess moisture content in the soil leads to swelling of the roots of the plant and leads to the death of the plant or low yield. This technique helps in increasing productivity by 70% and also water is saved by 20.9%, and the growth rate is also increased by 41.2%. The humidity and temperature sensor (DHT22) was used to measure the humidity and temperature and the solenoid valve helps to control water flow on the field; the valve currently helps the coil to open and close the fluid flow valve. When water flows through the rotor, the sensor will detect the output pulse. In this proposed system, NodeMCU ESP8266 is used as the main processing unit and all sensors work together to send and receive information with the help of cloud firebase and data storage that help to monitor the farmland to track and record factors such as height, flower and fruit for a certain time to know the difference [7].

As the previous research used IoT and its advancement, here introducing new technology using the cloud computing concept and the LEACH protocol to rectify the disadvantage in the previous cases. Hopping Zigbee and LEACH protocol is used with machine learning that helps in getting data from the system, and then it is cross-checked and analyzed and then send to farmers for monitoring the soil condition; NPK sensor is used for measuring the nutritional contents of the soil, and photovoltaic module converts light energy source to electrical energy that helps in powering the sensors. For operating the sensors (Arduino Mega 2650, SKU: SEN0114, DTH 11 sensor), a maximum of 5 V is sufficient enough for the photovoltaic and solar energy techniques to work effectively. This system helps to know about the major factors for crop growth, exact temperature, humidity and nutrients and if these can match the expectations fixed for high yield, then it will be more useful. The main advantage of this system is that this mainly focuses on the nutrient content of the soil and PH of the soil; this is important because deficiency and excessive nutrient content of the soil both had a major impact on plant growth. So this system is broadly accepted because it majorly satisfies all the expectations of farmers [8].

The Internet of Things (IoT) and its advancements in technology help in resolving all kinds of environmental problems like controlling pollution and solution for day-to-day problems in agriculture; likewise smart environment monitoring with the help of IoT, modern sensors and in addition machine learning helps in advancement in the agriculture field. SEM (smart environment monitoring) helps in monitoring the quality of air, water and radiation. In agriculture, this machine learning also has many advantages, especially in monitoring the environment. The different aspects of SEM are agriculture system technologies with smart monitoring, home technologies with needed and smart advancements, health monitoring systems with advancements to attain patient's maximum needs, ecological systems, marine environment monitoring with the management of IoT, and also in pollution monitoring [9].

The farmer knew the crop quality with the help of SAR data to understand the quality of plants. Machine learning technique and Gaussian process model result in the output of 89% with the most accurate rate. AI with Naive Bayes and machine learning helps in operating sensor data. This helps to rectify problems such as reducing pesticide usage at the maximum level and the maximum amount of water for irrigation of the crops without wastage. Various applications of SEM are farming in an advanced manner, pest control, monitoring and crop area needs [10].

Many other factors like environmental changes, temperature, pests and insects on farmland affect farmers. Even with the present advanced technology, there is no complete solution for this problem without side effects.

Pesticides is widely used to control pest but has an adverse effect on human health leading to many harmful diseases like cancer and even change in genetic material. The Graphical processing unit (GPU) connects all sensors in addition to pest control and monitoring system. Distributed and parallel simulation framework (DPSF) with IoT is used for pest control and monitoring. This helps in handling data without crashing reducing execution time. DSPF with a multithreading concept helps in sharing tasks of one core to another core and there are four layers needed to carry out the task. Layer 1 is for crop management. Layer 2 is for pest detection and control, and layers 3 and 4 are for input and output. Artificial intelligence helps in remote control and also helps in processing atmospheric data in real time [11]. Innovations in the agricultural field are increasing day by day to rectify new problems and satisfy our needs. Here a new technology is implemented by remote crop monitoring mechanism using LoRaWAN; the main motto here is to improve the old techniques of agriculture with a wireless sensor network using LoRaWAN protocol. Using the system, the temperature of the environment is notified with the help of an IoT-based sensor which helps in sensing the inside greenhouse gas for plants like brinjal under the consideration of food production and growth of plants [12].

The improvement in agriculture is useful for the development of the economic condition. By using NodeMCU ESP8266, sensors and cloud computing are used for monitoring soil moisture, temperature and water control. The effects of growth rate, production rate and water-saving rate of crop planting are studied and a timer is used to control soil moisture and the greenhouse automatically. With this technique, the growth rate is better by 41.2%, soil moisture by 23.1% and productivity by 70%. The humidity and temperature sensor (DHT22) helps in measuring the humidity and temperature surrounding the air of the farm. The solenoid valve works on a mechanism based on an electromechanical-driven valve that uses current to generate a magnetic field to open and close a fluid flow valve. The NodeMCU ESP8266 system is a major part of the system. All sensors interact, and it serves as a nearby server for receiving sensor data with the cloud firebase for present-time data. To monitor and control the system, the above system can send data to the cloud and receive data from a cloud user [13].

The development of IoT helps to know many factors of a plant, especially relative humidity, temperature and pH. By using IoT, this information is compared with the current time data to find the problem and to give a solution. The system used here is open to adding on new sensors that are currently present here. The Raspberry Pi card helps scan data from the sensors in the network according to the request of the data. It is then added to the sensors linked to the board and communication via Free Software. The entire project is a combination of a microcontroller and a server. The sensors send data to the Raspberry Pi 3 card, which stores the data in itself and the server is placed in the microcontroller. The various web pages that contain the information on the server read those details from the database and help in generating graphs that the user can use. The microcontroller delivers messages to users regarding the information. A wireless system was designed so that the sensors used have a low-consuming microcontroller. Transmitter has a maximum consuming capacity of 50 mA and a low consumption of 30 μA. Due to a different bandwidth consumption, different data is stored. It is not easy to extract data from hundreds of sensors with the help of four sensors. Because of this, the data packet detection algorithm is used to transfer data later. A cyclic redundancy code is for confirming the receipt and it will be implemented within the TCP protocol. The main advantage of the system is that any object can be monitored, automate amounts controlled, and thus helps in getting information in large amounts with all kinds of details [14].

The advancements in the modern world not only increase new techniques but also increase laziness in people so automation in each sector satisfies human needs. In agriculture, automation is needed. On the contrary, the agriculture sector in India is facing losses due to many factors including loss of manpower which has affected the production capacity of the ecosystem for which this automation is the solution. There is an emerging need to solve the problem to save the vibrancy and to improve the higher growth. For large-scale production, a lot of maintenance, knowledge and supervision are required. The developed system mainly focuses on maintenance, insecticide and pesticide control, water management and crop monitoring all of which will be automated. The main advantage here is connectivity to a diverse range of devices, minimizing human labor, minimizing human labor, time-efficient, efficient communication and analytics [15, 16].

Wireless technology with agronomy helps in improving traditional techniques. A new IoT technique named AGRO-TECH is used here to store and update the activities which can be accessed by farmers to know the details in terms of soil and crop monitoring. Different sensors help to know about the certain needs of the farm; those factors are soil moisture, level of pH, temperature, humidity, light-dependent resistance and water level and all these factors are monitored and observed. With the help of these sensors, we can increase the production of the field. The sensor relay starts either the motor or the irrigation spray but starts only based on the soil moisture threshold value, overcoming the impact of water shortage through water level monitoring and also using the temperature sensor to know about surrounding temperature.

If the surrounding is maintaining a moist condition, then a water supply is not needed so this sensor is used here. This information about the above-mentioned factors will be updated through our system methodology. This information is then sent to the farmer through an embedded hardware kit and software. With the use of IoT software called AGRO-TECH, the system uses the creation of numerous sensors in embedded kits to detect soil and crop growth to update the sensor. During water scarcity, the irrigation sprinklers are actuated based on the value of the soil moisture sensor. A mobile-based SMS system will be issued to the farmer in case of an emergency. In short, the process in the above-mentioned registration about farmer details and the connection between sensors are already updated in the system. The water level sensor in the well and irrigation sprinkler store data and then AGRO-TECH software helps in coordinating the sensors [17].

IoT has a great impact on all fields especially in agriculture to provide a solution for problems. The technology used is precision crop monitoring with IoT which has a great impact to create a revolution in farming. The affordable framework of IoT will be useful to all farmers and helps them for monitoring the plants during their different seasons and also helps them in improving the yield by 10% while using this technique [18]. The IoT is a worldwide data network that includes web-connected products, software, sensors and other instruments that are considered to be critical components of the future-based internet. This solution assists an autonomous agricultural field monitoring system as well as the present update of the server using a Raspberry Pi camera. Depending on environmental temperature, humidity and soil moisture, the agriculture fields are monitored. Then automatic irrigation will be performed based on the results of the soil moisture sensor. Data collected in the field is monitored by the IoT, and processed, and relevant information is provided to the owner. The Raspberry Pi used here helps in taking instant pictures of the farm. The DC motor is used for irrigation according to the necessity of water on the farm. The live video monitoring of the farm is viewed by authorized persons, in this case, as the farm owners are provided with passwords and user names. The coordination of programming languages like python, HTML, Raspberry OS along with other hardware such as Raspberry Pi camera, Analog-to-Digital Converter (ADC), Embedded Main Board Module, ARM Processor and Humidity/Temperature Sensor helps in working of the system [19, 20].

In India, agriculture is the backbone for major sectors in terms of economic as well as healthy food availability and food is the most required thing to survive. Agriculture is always our country's priority. Food plays an important role in it because it involves the requirement of a secure level of food and that might lead to shortage and wastage of food. Reports regarding this reveal that a maximum of 50% of the produced food across the world does not even reach the people's demands due to food wastage during the process while they are transported and storing. This is the main threat to the farmers. By reducing losses available for the production of crops and rectifying the problem of wastage, it provides a solution to increase the amount of food available. Our research takes into account this issue that affects a lot of people around the world and has implemented efficient hacks to bring out an easy way by building a module that helps in monitoring and controlling the depositories' team. To help the farmer, WHO planned an efficient way of protecting crops that are stored in a warehouse with the help of an IoT system and it has been planned to be implemented in remote areas, which helps to increase food safety [21].

The new growing technology in this modern world is AI (artificial intelligence) which plays a role in the software as well as hardware sectors. Now it is also in the agricultural field helping the farmers to increase productivity. Without agriculture, the Indian economy will not be good so technological developments mainly aim at this sector. Smart farm operating system is mainly composed of mobile auto-control, integrated control system and cultivation management system and is based on machine learning that is monitored with the help of plants that have been surrounded for cultivation considering the image and conditions of weather and the entire history of cultivation stored as the data. Many sensors which have been placed outside, especially the external sensors, help users to know external environmental activities and internal activities that enable users to know about internal information of soil such as nutrient content, CO_2 content and pH level. The main controller guides the Environment Controller (EC) node and based on the data obtained from nodes, EC nodes automatically turn electric switches (motor, electricity, etc.) on/off; the other works of EC nodes are irrigation replacement and control of temperature and CCTV instant monitoring. This model helps in obtaining data about the location of the field, time and weather conditions. This automatic irrigation system model reads the moisture content of the soil at regular intervals in a day. The demo helps in working with various types of soil and crop variants in different weather conditions and seasons and increases feasibility. The smart controller enables automation in the irrigation of fields and also at homes through mobile smartphones. With the help of sensors, data collection is done, processed and sent to users for future purposes [22].

Existing agriculture applications with IoT help to increase the productivity of the crops, but there are some disadvantages too. To reduce the impacts, many types of research are carried out to overcome the challenges using futuristic methods by implementing a Green IoT (G-IoT) and green nanotechnology which are innovative technologies having no impact on the environment or human health; nonetheless, they are not easily adopted by farmers. So, this concentrates on building a real-time, cost-effective precision agriculture monitoring system with less power consumption, less greenhouse gas (GHG) emissions, and also helps farmers to know about the weather, water, soil, pest detection, intrusion detection, fire detection from anywhere with the help of their smartphones. The existing systems such as Zigbee and GSM protocols provide a short range of communication to overcome it. An additional feature of LoRa SX1276 technology is added to rectify the problem because it helps in providing long-range communications up to 8–10 km in urban areas, and 12 km in a rural area with 600 bps. For designing six levels of a system, the necessities are requirement level, specification level, architecture level, component level, integration level and application level. Additionally, these modules help in fire detection and intrusion of animals or insects in the field and remote monitoring [23].

The farmers in India have done different ways and different methods of agricultural techniques in most places to increase production. Our farmers lack technical knowledge and due to such uncertainties in weather and environmental changes farmers are facing major losses in their fields. The properties of light intensity, humidity levels and temperature are all factors that help the growth of crops. Smart crop monitoring with IoT, cloud and machine learning helps with its stability and auto-monitoring. Agriculture with modern techniques helps plan and improvise their agricultural activities by handling this, and the growth of the plant can be predicted helping to reach new achievements and also be helpful to get more and more yield [24, 25].

The main factors in agriculture environmental factors are climatic and environmental. Changes in meteorological conditions might result in output losses. For these problems, a simplification is found with the combinational work of IoT and deep learning. The system collects data and codes the info through sensors and transmits it to the cloud to know about the condition of the soil, nutritional imbalance and crop diseases. The stages of production are divided into three major categories: panicle initiation, panicle initiation for flowering and flowering to maturity. To increase productivity, deep learning is used for disease prediction systems. The system aims to maintain moisture and water levels for healthy crop production and better yielding. Deep learning in IoT helps in categorizing one of the crop diseases like leaf smut, bacterial leaf blight and brown spot [26–28].

There are numerous problems in the field of agriculture: no proper maintenance and conductance to the crops, ignorance of soil quality and water shortage; sometimes water is given to the crops more than it needs. These are some of the troublesome factors, especially when one is new to the agricultural field. The technology-based IoT helps in attaining advanced solutions to all the commonly mentioned problems. It helps to monitor the soil conditions and atmospheric conditions and humidity and other necessities using NodeMCU and various types of sensors connected with it. This system also gives a notification about the conditions and situations of the environment in the form of SMS to the farmer's mobile phones using wireless Wi-Fi modules [29].

Precision agriculture irrigation systems have been developed with low complex circuitry. Cloud computing is known as a new type of computing in which dynamic scaling and virtual resources are provided as additional support with the help of the internet. Two sensors have been used here to control the irrigation process such that troubleshooting can be done at ease whenever it is required. The obtained correlated data which are algorithm-based minimize the hardware's complexity when compared to the other systems. Threshold voltages have been particularly set for the rectifying/calibrating purpose of the sensors by taking the previous values of temperature and soil moisture values into consideration. The values of the threshold might be varying according to the crop and plantation. So machine learning is introduced here to simplify the complex process and to automate various agricultural-related works [30].

4.3 Proposed Methodology

Based on the literature, it is found that the autonomous devices developed for agricultural plant monitoring systems concentrated on one type of crop variety. The proposed methodology shown in Figure 4.1 concentrates on two or more types of vegetation at the same time and thus more variations in the parameters and requirements can be monitored at the same time. Hence, different parameters will be initialized to monitor the agricultural fields to operate as per the requirement of the crop variety and to nourish the land. Therefore, here Arduino Mega microcontroller has been established due to its diverse options and ability to use the analogue pins available for the exchange of the input signal from the sensor to the microcontroller. These input signals are the input from the different sensors used for crop monitoring. This mega Wi-Fi R3 type variant is a unique microcontroller where ESP32 is also associated with it. Thus this monitoring system can be monitored from different parts of the world. This advantage satisfies the need if the monitoring should be done over a long distance and if the crop fields are

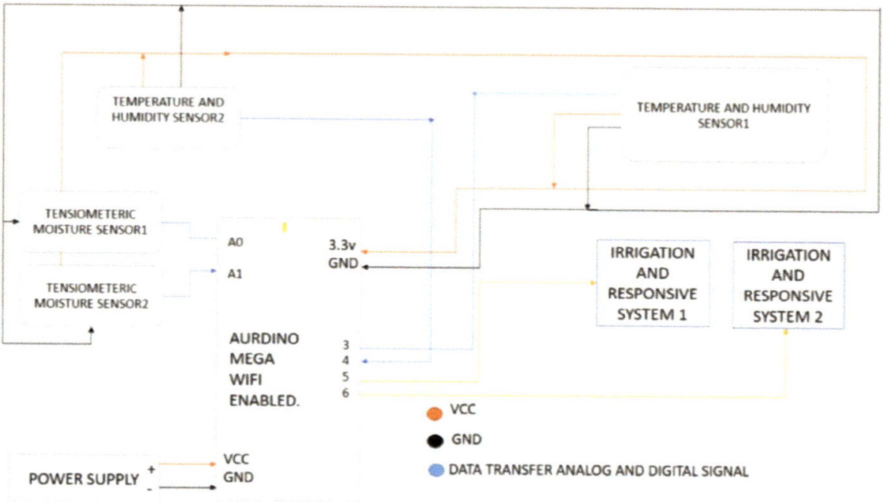

Figure 4.1 The circuit diagram of the proposed autonomous system for agricultural field monitoring.

far away from the residents or even can be used in farming land, horticulture-based farms and fields where the people in higher authorities can even go through the status of the farming land without relying on the persons who are responsible for the monitoring. This can be done through web-based applications and the real-time monitored data will be projected for time intervals in the situations like overloaded water during the rainy season to prevent the rotting of plants, and the emergency indication is also available to denote the situations.

In Figure 4.1, the first sensor is a tensiometry, moisture or water level indicator sensor. It is used to measure the moisture content of soil and estimate the amount of water stored at the ground level using the soil moisture sensor. This tensiometry sensor will work by measuring the soil moisture content by using a capacitive effect and will measure the water level in the soil horizon or the area where it is placed. Soil moisture sensors have two probes that measure the moisture content material throughout a vertical surface of the soil which usually spans several 30–120 cm. This soil sensor will intensively take care of monitoring the moisture content. These sorts of sensors should be placed or buried at the proper level from the top surface of the soil based on the crop and soil variety that is going to be sown in the agricultural field. To get a long duration of life and reduced reduction in the quality of the sensor, these are mostly preferred use which is built on plastic and PVC-based materials. To get more precision in the sensing, there are perforations in the sensor to establish a peripheral airflow through the probes of the sensor.

These sorts of sensors usually have a range of 0–100% of detection and external modules will be connected as an interconnect with soil in the probe and the microcontroller where it will convert the electrical signal which has been converted to both analogue and digital signals. These sorts of irrigation and monitoring systems will also require other parameters to make the crops grow much more efficiently and faster. Now that another sensor required for monitoring is simple, the DHT sensors with version DHT22 is used for improved precision of measuring the humidity and temperature where the range of these sensors is much larger than the DHT11 generation senor and that data retrieval is much faster and efficient in this module. The sensor is well known and capable of measuring and scaling the amount of humidity present in the surrounding or the fixed range of area and the temperature sensor will be integrated with the DHT22 sensor which will result in the data extracted in analogue and digital data formats with fewer data retrieval time. Using this sensor is much more beneficial as this provides the capability that it can sense or measure the temperature and humidity at the same time, whereas these two should be established separately to measure the data to monitor the crop in the agricultural land. This will significantly reduce the circuiting problem and complex connections and make it efficient to extract different transducer data in both analogue and digital signals. Then the sensor will be placed based on the position where it can give precise measurement of multiple inputs. Then these data will be sent to the Arduino Mega controller and then the data sent will be evaluated as per the algorithm that has been initialized within the microcontroller considering all the parameters. This parameter will be fixed as the threshold for different crop varieties based on the requirement of the crop variety which is decided to be grown. The motor pump and the irrigation systems are operated concerning the parameter initialized based on the physical change in the monitoring land. Growing different crop varieties will require a different moisture level in the soil. Then water will be released and moved to the selected area if it is initialized with a drip irrigation system or if it is normal irrigation land water will be released concerning the calculated amount of water required to increase the moisture level or the area of the land where the crop is grown.

The flowchart of the proposed autonomous device is shown in Figure 4.2. We could see from Figure 4.2 that whenever required the pump will be switched on and switched off using a relay switch which will turn on and off whenever required and when the constraints are met with the parameter which is given as input. Now the data will be sent to the platform as

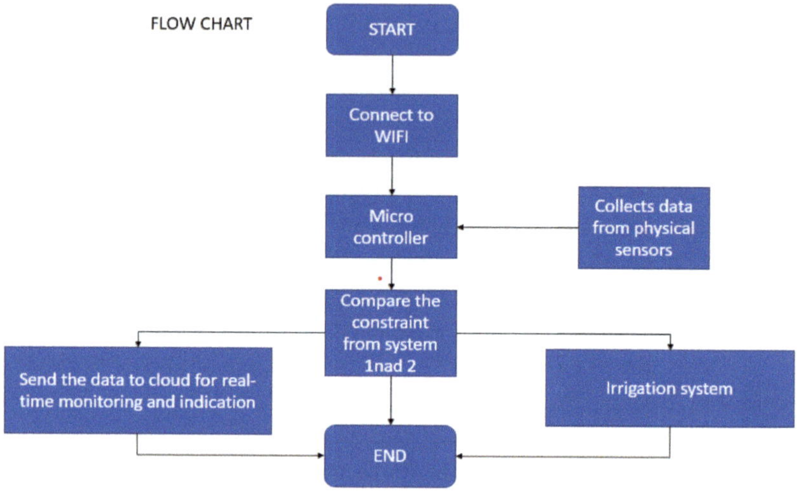

Figure 4.2 Flowchart of the proposed system

input for a continuous overview of these crop varieties and the time-compared analysis of the crops and the data on the variation of the physical and environmental changes observed in the deployed fields. Other plants and artifacts will be able to handle internet protocol (IP) and switch facts over the network. More and more comparators in more industries are using the IoT to work more efficiently, better recognize their consumers, give good buyers' support and improve solution making, and the cost to do an exchange and get profit is increasing. An environment with internet-based connected devices include web-enabled smart gadgets are used for broadcasting. This responds to the facts collected from the environment using embedded processors, sensors and some additional hardware. These IoT-based devices display the collected sensor data as a percentage by connecting to a dynamic IP generator or another sub-tool where the data is sent to the cloud database for analysis or local analysis. These gadgets may connect with various related gadgets and act on the information obtained from each other. This device performs various imagings with no human overview required, for example, interact with the device to set up the device, give commands and gain insight into the facts. Thus, the data is now established and the alerts are all established with an automated irrigation system and monitoring of the intercropping agricultural field; this is mainly possible only by initiating two separate operating versions of the output from the microcontroller which has sensors connected. The output decision will be based on the sensor where the input is observed and the range which it gives as we have a wide range of options from the microcontroller as it has

Figure 4.3 Status of the parameters of the agricultural field.

more numbers of input and output pins. Thus, the irrigation system will be established with the same constraints as the different irrigating pathways will be opened as per the command coming from the microcontroller, and the parameters will be fixed before initializing the intercropping monitoring system which will require different parameters for different crop varieties.

4.4 Results and Discussion

We have implemented the irrigation and monitoring system for two different crop varieties, based on the variety of crops chosen; the parameter or the physical constraints for the monitoring of the crops will differ, so while initializing the parameter for an overview of the requirements of the crop, two sets of different parameters are provided and according to that the actions will be made like there will be two types of crops. The system will have a different scenario to acquire and analyze the data.

From Figure 4.3, the data for the different crops are analyzed frequently whenever there is a requirement for the irrigation of the crops that will be activated accordingly as the parameters are important for the crop varieties' growth and to get more yield. Even the crop varieties could adopt drip irrigation using this methodology by increasing the tensiometer sensor and the crop varieties could be monitored more accurately; water requirement to irrigate the crop will reduce as the conservation of the resources is much important in this era.

The activation of the irrigation for the other set of varieties of crops is seen in the below Figure 4.4. This ensures that based on the parameters

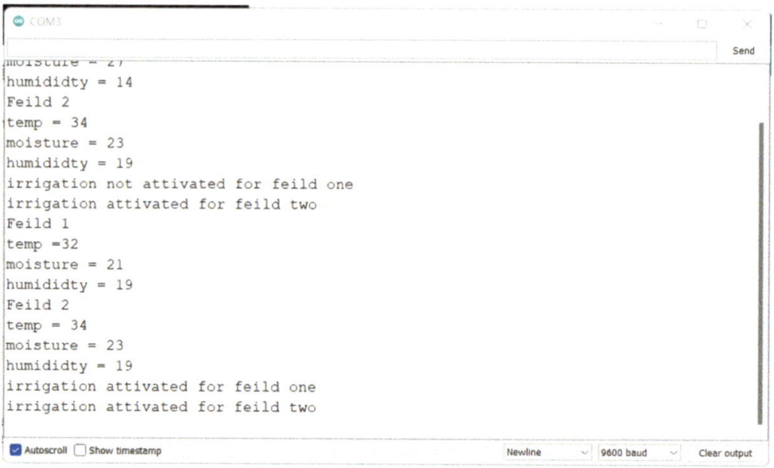

Figure 4.4 Status and parameters for irrigation of other crop varieties.

observed in the field for the separated partition of the crop variety here this avails us to monitor two varieties of the crop without the issues.

From Figure 4.4, we see that irrigation is established for the other type of crops when there is a requirement for the crops and when the cultivating land changes to unsuitable conditions for the crop which is chosen to be grown in the specific part of the land. If we also like to add the nutrients which are required for the plant to be delivered in the mean interval, then it is much more convenient to add them to the irrigation water tank so that it will be mixed with the water which is used for the irrigation. Thus directly the water will be placed to its surrounding and the crop would yield more efficiently.

In another case, both the partitioned land will be the basic level of requirement of the plant or crop. Both the irrigating system will be activated at the same time. Here it is not a rare condition because the same land will not become much differing in this physical parameter. Both the plants will not require the same amount of sunlight for its growth. Figure 4.5 depicts the activation of the irrigation field at the same time and both the different parameters will be counted on the minimum required level for the crop to stop the irrigation at the time it reaches the required threshold of the crop.

The controller used here is enabled with the Wi-Fi connecting facility; the data would be monitored from any device which you have with you like PC, laptop, mobile phone and smart tab. The indicators used in the irrigation system are used to indicate whether the system is turned on or turned off. The

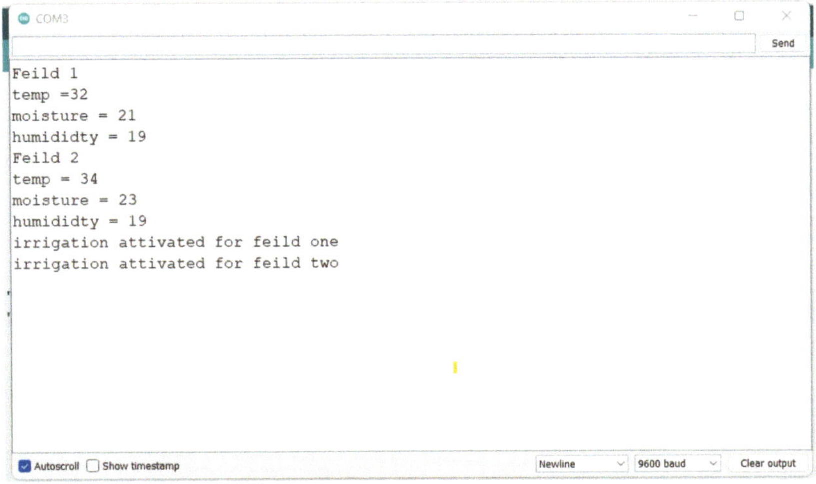

Figure 4.5 Activation of two irrigation systems.

indicators are also available for the time of irrigation; meanwhile, one can see the data collected in the different time intervals and the data will be stored in the cloud database to show the real-time graph to plot different parameter data.

From Figure 4.6, it is evident that the entire dashboard is composed of moisture data representation, indication or activation of the irrigation system. The temperature and humidity data are plotted concerning the data and the time of the data. If the monitoring person is interested to view the specific data point, the legend will appear and the data will be displayed with a specific date, time and parameter data. The data can be used to analyze the growth and the requirement of the crop to grow efficiently and the factors can be improvised by analyzing the data which are collected.

4.5 Conclusion

The intercropping methodology of the system will give a drastic change in the yield-efficient usage of the land resources and due to this the aim to increase in yield will also be achieved. The secondary crop variety provides more return even when the primary crop fails. While implementing this system the failure of the crop is prevented using an intercropping system. The ergonomic use of land is considered the main advantage of the efficient usage of land because even the dead space between the plants in the monocropping will also be used in this case. In this system, the

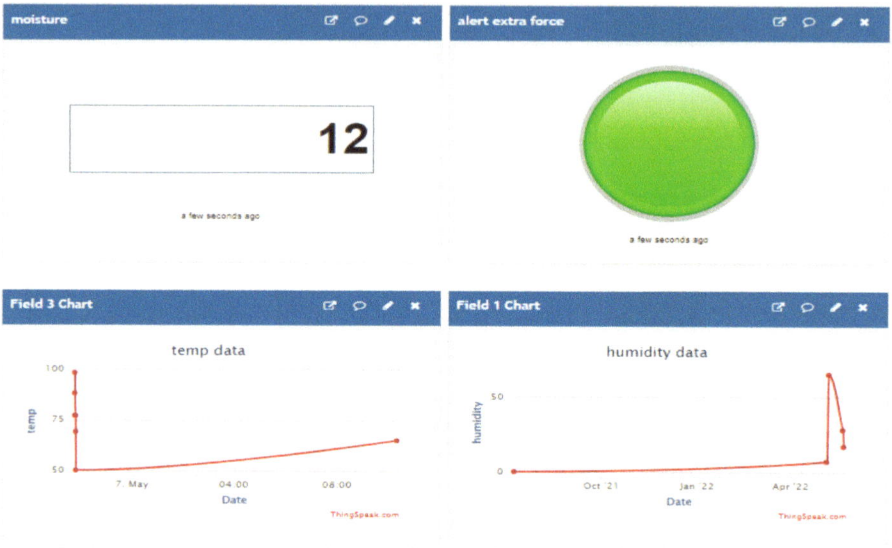

Figure 4.6 The entire dashboard.

intercropping methodology performs various functions and adds more value through its nature of implementation. Through this methodology the pest is trapped and pest management additional feature reduces chemical fertilizer application and saves money. This monitoring level of irrigation and the usage of water will reduce the erosion of the soil and the water is conserved; the resources are not used randomly but conservatively as per the requirement. Adopting this sort of agriculture pattern adds value to using this method because in intercropping methodology the leguminous plants are known for nitrogen fixation providing nitrogen for the neighboring crop varieties.

As intercropping adopts different crops to be grown on the same piece of land, it leads us to reduce the usage of synthesized fertilizers which are harmful to the crops and the health of the fertile land. The main advantage of using this autonomous device for intercropping is that usually it will reduce the usage of resources like solar energy and water. If this method of intercropping system is also established, it will again reduce the risk factor by 50% from that of the system its risk depends on. The system also provides improved weed management as the beneficial plants occupy the vacant space and therefore there is much less possibility to leave a space to grow a different variety of crops. Here more biodiversity and ecological stability are stabilized. The more the different species grow, the better the environment, the health and the nature of the agricultural field are preserved.

References

[1] R. Dhanalakshmi, J. Anand, Arun Kumar Sivaraman, Sita Rani, 'IoT-based Water Quality Monitoring System using Cloud for Agriculture Use', Cloud and Fog Computing Platforms for Internet of Things, Edited by Pankaj Bhambri, Sita Rani, Gaurav Gupta, Alex Khang, Routledge Taylor & Francis Group, 2022.

[2] J. Anand, J. Raja Paul Perinbam, 'Automatic Irrigation System using Fuzzy Logic', AE International Journal of Multidisciplinary Research, vol. 2(8), pp. 1–9, 2014.

[3] T. Manikandan, V. Nandalal, J.L. Mazher Iqbal, S. Joshua Kumaresan, S. Karthikeyan, 'Protection of the plantation from wild animals using wireless sensor network', Int. J. of psychosocial rehabilitation, vol. 24(5), pp. 2719–2723, 2020.

[4] D.K. Sreekantha. A. Kavya, 'Agricultural crop monitoring using IOT - a study', Proc. In 11th Int. Conf. on Intelligent Systems and Control (ISCO), pp. 134–139, 2017.

[5] K.A. KPatil, N.R. Kale. N.R, 'A model for smart agriculture using IoT', Proc. In Int. Conf. on Global Trends in Signal Processing, Info. Computing and Comm., pp. 543–545, 2016.

[6] Ram Krishna Jha, Santosh Kumar, Kireet Joshi, Rajneesh Pandey, 'Field monitoring using IoT in agriculture', Proc. In Int. Conf. on Intelligent Computing, Instrumentation and Control Technologies, pp. 1417–1420, 2017.

[7] Kamal Kumar Ghanshala, Rahul Chauhan, R.C Joshi, 'A Novel Framework for Smart Crop Monitoring Using Internet of Things (IOT)', Proc. In the first Int. Conf. on Secure Cyber Computing and Comm., pp.62–67, 2018.

[8] S.S. Ramaprasad, B.S. Sunil Kumar, Sivaprasad Lebaka, P. Rajendra Prasad, K.N. Sunil Kumar, G.N. Manohar, 'Intelligent Crop Monitoring and Protection System in Agricultural fields Using IoT', Proc. In the fourth Int. Conf. on Recent Trends on Electronics, Information, Communication & Technology, pp. 1527–1531, 2019.

[9] S.L. Ullo, G.R. Sinha, 'Advances in Smart Environment Monitoring Systems Using IoT and Sensors', Sensors, vol. 20(11), pp. 1–18, 2020.

[10] Tian Erlin, Li Zuhe, Huang Wei, Ma Haizhen, 'Distributed and Parallel simulation methods for pest control and crop monitoring with IoT

assistance', Acta Agriculturae Scandinavica, Section B-Soil & Plant Science', Taylor and Francis, vol. 71, pp. 884–898, 2021.

[11] N. Kaur, G. Deep, 'IoT-Based Brinjal Crop Monitoring System', Smart Sensors for Industrial Internet of Things, Springer Publications, 2021.

[12] Khampheth Boundary, Poutthasone Sibounnavong, Khampasith Chanthavong, Savath Saypadith, 'Smart Crop Cultivation Monitoring System by Using IoT', Proc. In fifth Int. Conf. on Engineering, Applied Sciences and Technology, pp. 100–110, 2019.

[13] Liqiang Zhao, Shouyi Yin, Leibo Liu, Zhen Zhang, Shaojun Wei, 'A Crop Monitoring System Based on Wireless Sensor Network', Procedia Environmental Sciences, Elsevier, vol. 11, pp. 558–565, 2011.

[14] Amaya Diaz Juan Carlos, Luzmila Rojas Estrada, Cardenas-Ruiz César Augusto, Ariza-Colpas Paola Patricia, Piñeres-Meloo Marlon Alberto, Ramón Enrique, Ramayo González, Morales-Ortega Roberto César, Ovallos-Gazabon David Alfredo, Carlos Andrésa, Collazos-Morales, 'Monitoring system of environmental variables for a strawberry crop using IoT tools', Elsevier, vol. 170, 1083–1089, 2020.

[15] Manasa Reddy M,M.K Saiteja, Gurupriyanka J, Sridhar N, Naveen Kumar G N, 'IOT based Crop Monitoring system for Smart Farming', Proc. In sixth Int. Conf. on Communication and Electronics Systems, Coimbatore, India, pp. 562–568, 2021.

[16] Vaishali Puranik, Sharmila, Ankit Ranjan, Anamika Kumari, 'Automation in Agriculture and IoT', Proc. In the fourth Int. Conf. on Internet of Things: Smart Innovation and Usages (IoT-SIU) Ghaziabad, India, pp. 1–6, 2019.

[17] O. Pandithurai, S. Aishwarya, B. Aparna, K. Kavitha, 'Agro-tech: A digital model for monitoring soil and crops using internet of things (IOT)', Proc. In the third Int. Conf. on Science Technology Engineering & Management, Chennai, India, pp. 342–346, 2017.

[18] Swagatam Bose Choudhury, Prachin Jain, Sujal Kallamkuth, Saranya Ramanath, Prakruti V. Bhat, 'Precision Crop Monitoring with Affordable IoT', IEEE Global IoT Summit (GIoTS), pp. 1–6, 2019.

[19] Ramya Venkatesan, Anandhi Tamilvanan, 'A sustainable agricultural system using IoT', Proc. In Int. Conf. on Communication and Signal Processing, Chennai, pp. 763–767, 2017.

[20] S.R. Prathibha, A.Hongal, and M.P.Jyothi, 'IOT Based Monitoring System in Smart Agriculture', Proc. In Int. Conf. on Recent Advances in Electronics and Communication Technology, Bangalore, India, pp. 81–84, 2017.

[21] J. Jose, B. K. Samhitha, M. Maheswari, M. Selvi and S. C. Mana, 'IoT based Smart Warehouse and Crop Monitoring System', Proc. In

fifth Int. Conf. on Trends in Electronics and Informatics, pp. 473–476, 2021.

[22] Sanwta Ram Dogiwal, Pankaj Dadheech, Ankit Kumar, Dinesh Raja, Abhishek Kumar and Mahender Kumar Beniwal, 'An Automated Optimize Utilization of Water and Crop Monitoring in Agriculture Using IoT', IOP Conference Series, vol. 1131(1), 2021.

[23] Ali, A.A Terteil, Choksi, Viraj, M.B Potdar, 'Precision Agriculture Monitoring System Using Green Internet of Things (G-IoT)', Proc. In the second Int. Conf. on Trends in Electronics and Informatics, Tirunelveli, India, pp. 481–487.

[24] B.P. Sreeja, S. Manoj Kumar, P. Sherubha, S.P. Sasirekha, 'Crop monitoring using wireless sensor networks', Materials Today: Proceedings, Elsevier, pp. 2214–7853, 2020.

[25] S. Geetha, P. Deepalakshmi, Pande, Shilpa, 'Managing Crop for Indian Farming Using IOT', Proc. In int. Conf. on Clean Energy and Energy Efficient Electronics Circuit for Sustainable Development, India, pp.1–5.

[26] C. Suresh, M. Ravikanth, G. Nikhil Reddy, K. Balaji Sri Ranga, A.A Rao, K. Maheshwari, 'Paddy Crop Monitoring System Using IoT and Deep Learning', Springer, pp. 791–807, 2022.

[27] I. Mohanraj, Ashokumar, Kirthika, J. Naren, 'Field Monitoring and Automation Using IOT in Agriculture Domain', Procedia Computer Science, Elsevier, vol. 93, pp. 931–939.

[28] S. Rajalakshmi, J. Dharani, 'Effective Crop Productivity and Nutrient Level Monitoring in Agriculture Soil Using IOT', Proc. In Int. Conf. on Soft-computing and Network Security, Coimbatore, India, pp. 1–10, 2018.

[29] A. MSD, J. Kuppili.J, N.A. Manga, 'Smart Farming System using IoT for Efficient Crop Growth', Proc. In Int. Conf. on Electrical, Electronics and Computer Science, 2020.

[30] Rao, R. Nageswara, B. Sridhar. B, 'IoT based smart crop-field monitoring and automation irrigation system', Proc. In the second Int. Conf. on Inventive Systems and Control (ICISC) - Coimbatore, India, pp. 478–483, 2018.

5

Predictive Analysis in Smart Agriculture

S. Balaji[1], J. Ashok[2], P. Selvaraju[3], and Veer P. Gangwar[4]

[1]Department of Electrical and Electronics Engineering, SRM Valliammai Engineering College, India
[2]School of Business and Management, CHRIST (Deemed to be University), India
[3]Department of Mathematics, Rajalakshmi Institute of Technology, India
[4]Department of General Marketing, Mittal School of Business, Lovely Professional University, India
Email: sribalaji06@gmail.com; ashok.j@christuniversity.in; pselvar@yahoo.com; gangwarveer@gmail.com

Abstract

Analyzing large databases for hidden connections, correlations and insights is known as big data analytics. Although many countries still use outdated farming methods, technological advancements have allowed for specific improvements (especially in developing countries). Big data analytics has the potential to expand the agricultural sector in this regard significantly. The farmers rely heavily on old methods for deciding what to plant and how to cultivate it. Walking through fields, selecting soil samples for moisture analysis, and visually inspecting plant leaves are typical examples of these time-honored practices. Understanding the significance of technology for acquiring crop information in considerable amounts and turning that data into usable knowledge is crucial for agriculturists (mainly farmers). Integration of big data could help agriculture make changes to its current practices. If used correctly, big data analytics can shed light on the most efficient crop cultivation methods. Extensive developments in three areas—crop prediction, precision farming and seed production—are reshaping the agricultural industry. There are four parts to this chapter. The first part of this paper provides an introduction to analytics on big data in agriculture. The second part will then focus on the various big data methods used in the agricultural sector. The third section provides two examples of how big data analysis methods were put to use in the field of agriculture. In the fourth section, the authors examine the several agricultural research avenues open to scholars and scientists. This chapter concludes with a brief overview.

5.1 Introduction

Agriculture is the industry that meets the needs of all people by producing enough food for them to eat. However, farmers are treated as second-rate citizens [1]. It is time for technology to assume the reins of change after failing institutions have failed to do so. Every day, people in the agriculture industry, especially farmers, encounter new problems that necessitate the adoption of cutting-edge technologies [2].

As a result of factors such as declining groundwater supplies in rural areas, monsoons or deficiency of same, climate change, floods, migration of farmers to the cities in search of better-paying workers, unfair price-fixing policies of produce, and more, agriculture in several nations lacks organized attention, banks support in for loans and suffer from a wide range of disasters [3].

Big data analytics is used to analyze large databases for hidden connections, correlations and insights. Traditional agricultural methods are still used in some nations despite significant technological advancements (especially in developing countries). Big data analytics can potentially aid the agricultural sector in this respect considerably. The farmers rely heavily on old methods for deciding what to plant and how to cultivate it. Walking through fields, selecting soil samples for moisture analysis and visually inspecting plant leaves are typical conventional practices [4].

However, agriculturists (particularly farmers) must realize the significance of technology in acquiring crop information in enormous amounts and in turning that data into valuable knowledge [5]. Integration of big data could help agriculture make changes to its current practices. Understanding effective agricultural production practices requires the use of big data analytics. We highlight three broad topics that are dramatically influencing agriculture today:

- **Accurate crop forecasts:** The success of their crops is the main source of worry for farmers. There has been a history of wrong crop predictions. With the help of cutting-edge algorithms, today's farmers can more correctly forecast crop yields. To make accurate forecasts, these algorithms process vast amounts of data on crops and climate [6]. Even before a seed is planted, farmers may now estimate a harvest. Harvesting at optimal times for maximum yields is made possible because of the insights provided by big data analytics. Accurate crop forecasting can alleviate a lot of the farmer's anxiety about the harvest.

- **Precision farming:** This is the state-of-the-art approach to cultivating plants. Several tasks in agriculture could be made more efficient with the use of the IoT [7]. Various farmers, especially in wealthy nations, are making more money from precision farming than they would be making without it. Precision farming has also enhanced the quality of harvested goods [8].
- **Producing high-quality seed and reducing food insecurity:** The number of people at risk of becoming hungry is growing at an exponential rate, as the world's population and average temperature, according to recent studies[9]. This is especially true in Africa, where the lives of 20 million people are currently in jeopardy. Many people worldwide are working to ensure enough food for everyone, but their efforts may be better directed toward ensuring that the agriculture sector makes efficient use of big data analytics [10]. Some researchers and agricultural scientists have been looking at plant data over the last few decades to develop more productive and healthy plant life. Recent advancements in big data analytics have allowed for more rapid, substantial and high-quality crop development. The elimination of world hunger may be possible with the use of genetically modified seeds and big data analytics [11].

5.2 Big Data in Agriculture

The following section elaborates on why big data is so important in agriculture. There is also a discussion of several standard methods for dealing with massive datasets. Before delving into the strategies for dealing with big data, the authors briefly examine what big data is.

5.2.1 Big data's impact on agriculture

The following five characteristics of big data are widely referred to in the literature.

- The data volume is represented by V1. In this analysis, the total amount of data was represented by the value of V1.

- Specifically, V2 is the range of times when the data is most relevant and effective. For example, some data loses its value with time if not analyzed immediately. For instance, it is possible to spot several common pests and diseases during an animal's lifetime.
- The letter V3 can stand for a wide variety of data. Data can come from multiple sources (e.g., videos, remote-sensing data and pictures), be acquired at different times and have different resolutions, all of which can contribute to the observed discrepancies (e.g., spatially varied data on image resolution). Data obtained for varied uses also has distinctive qualities.
- A V4 signifies that the information is reliable. Values 4 encompass the potential, quality, dependability, correctness and overarching confidence in the data.
- Data's "valorization," or its value in fostering appreciation, innovation and knowledge, is denoted by the notation "V5."

The five "Vs" can be used to define big data, although they are not required for a successful big data study [12]. Large data weaken fourth V because of its general tendency toward reduced accuracy and instability. According to [13–15], the authors need to add another "V" for visualization to demonstrate or illustrate the data by using complex data structures and information-dense scenarios.

According to authors [16], big data's main focus is improving the speed with which massive datasets may be referenced, searched, visualized and aggregated. It strongly emphasizes data mining, especially in areas where this was previously too costly or impossible to implement [17].

The DIKW hierarchy illustrates the connections between data, insight, expertise and knowledge by using the letters D, I, K and W to represent data, insight, expertise and wisdom, respectively. Essentially, it communicates the idea that data describes the information, knowledge describes the information and wisdom describes knowledge. Figure 5.1 depicts the DIKW hierarchy.

Data is transformed into an entity (information), as depicted in Figure 5.1. The procedures responsible for the change are also specified. Essentially, "wisdom" refers to the ability to weigh competing options and judge based on what one believes to be the most plausible. With fewer intangible resources like values, ethics, interests and preferences, wisdom is paired with the knowledge already at hand. The knowledge base is constructed from collected data. The raw material from which information and meaning can

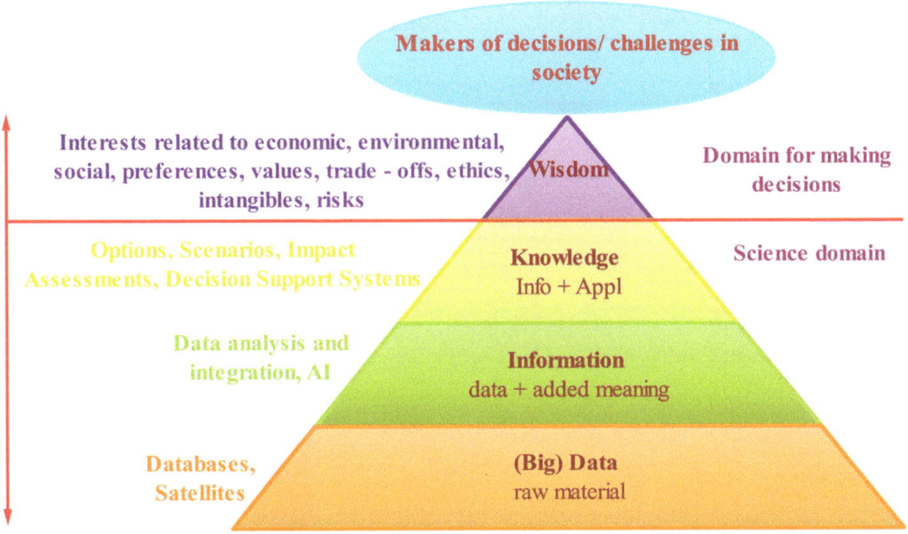

Figure 5.1 DIKW hierarchy.

be derived is data. Policy and decision applications like integrated models, impact assessments and decision-making assistance systems can all be used together to generate information that can be applied to enrich data.

Perishable species and goods (including humans, plants, agricultural products and animals) all have intricate communications with their environment, making agricultural and environmental research systems dynamic in nature. Extensive details about the status and behavior of entities and their interactions with their surroundings are required for a complete description of such setups. These structures can store both current and historical or hypothetical conditions. The spatial and temporal entity behavior data is essential because biological systems exhibit spatial and temporal variability. In addition, comprehending these interconnections calls for in-depth familiarity with the various subsystems involved. Climate, water, life and the ground itself are all potential components.

To better understand how less valuable raw data can be turned into knowledge that can be of greater use to the end users, the DIKW hierarchy has been adapted for the big data environment. In this context, "knowledge" stands for the bits of information that can be used to make sound judgments. Users rely on this accurate and timely data assuming that it was compiled using the best available sources drawn from the available big data, inferred, and processed in light of the users' specific decision context. The community of ICT specialists, domain experts and data scientists working on this problem

Table 5.1 Applications of big data in broader sections of agriculture [1].

Agriculture area	V1	V2	V3	Number of papers
Climate change	Medium	Medium	High	4
Land	High	Low	Medium	5
Animal research	Medium	High	Low	4
Soil	Medium	Low	Low	2
Crops	Medium	Medium	Low	3
Food availability and security	Medium	Low	Medium	4
Weeds	Low	High	Low	1
Farmers decision	High	M	High	2
Study of different plant and animal species	Medium	Low	High	1
Remote sensing	High	Medium	Medium	3
Farmers' finance and insurance	High	Medium	Medium	5

face a significant barrier in making such assumptions. The next steps in big data analytics can only be determined through teamwork and the capacity to combine bottom-up and top-down approaches in tandem [18].

The relationship between the three values of big data and the need for labor in agriculture is shown in Table 5.1. The third column lists the total number of articles that answer problems in specific agricultural fields. There has been a lot of research on food availability and security, insurance and financial worries for farmers, the consequences of climate and weather change and the management of land and livestock. Table 5.1, columns 4 through 6, displays the average volume, velocity and variety ratings provided by the authors [19] for the various agricultural sectors.

The researchers evaluated the first three big data [20] using low, medium and high indices based on a meta-analysis of 34 investigations. The papers were ranked from best to worst using an L, M and H scale. There was a moderate to large range in data volume and a moderate to small range in the total number of publications included in the analysis.

Exceptions to the rule where high velocity was required include projects dealing with weeds and animals, as a swift response is essential when evidence of diseases or weeds is discovered [21]. Since accurate climate, weather, etc. forecasting aids farmers in their normal operations, projects connected to these themes and applications related to decision-making by farmers require a wide range of information. Since photos are so huge, remote-sensing applications ate up the most data. Projects including land management, food availability and security, soil analysis and biodiversity had the slowest average speed.

Table 5.2 Big data is being utilized in many various ways in the agricultural industry.

Dimensions of big data	Description of "V"	Applications of agriculture
V1	Data volume	Sharing data for earth observations, weather forecasting, daily herd culling, increased farmer productivity, small farmer protection and insurance, farmer financing, remote sensing-based food security estimations, land use and land cover classification, etc.
V2	Applications of high velocity for time	Improved weather forecasting, animal food safety, weed discrimination, animal disease diagnosis, higher production, monetary transactions for farmers in remote areas and sharing of earth observation data.
V3	High variation in data from different sources	Tolerance of crop drought, the science of climate, identifying management zones, estimating food in developing countries, assessing wildlife populations, insuring and protecting farmers, increasing farmers' productivity, knowing how farmers can be sustainable and efficient in their operations, etc.
V4	High dependability for uses where accurate data is necessary	Topics covered include dairy herd culling, weed discrimination, animal food quality and safety, food availability estimation in developing countries, animal disease recognition, insurance for small farmers, increasing farmer productivity, sharing data from earth observation and assessing the viability of wildlife populations.

The areas of soil, animal studies, weed studies, crop studies and remote sensing are the least diverse. The "V" properties of big data and its relevance to agricultural applications are illustrated in Table 5.2, summarizing descriptions from the above literature review.

Access to large amounts of data is essential for weather forecasting, food security, agricultural output estimation and land mapping applications. Applications involving the detection of sickness in animals, the diagnosis of malnutrition in plants and the formulation of policy about agricultural output, product quality and consumer health all necessitate high speeds since they must be implemented in real-time. In these cases, long-term strategic planning must give way to more immediate, tactical decision-making (lower velocity). The diversity of the data constitutes the fourth dimension of big data and is crucial for uses such as insurance premiums and agricultural yield estimation. The fifth and last V, data reliability, is essential for applications such as detecting animal and plant diseases, culling herds, improving agricultural output, and forecasting harvest yields. Big data analytics is still in its infancy in agriculture, as seen by the paucity of relevant academic literature and industry-led efforts.

5.2.2 Sources and methods for big data

- When analyzing large data, it is important to remember where the data came from. The following are examples of popular places to find huge data:
- Agriculture, including ground-based chemical detection devices, biological sensors, weather stations, etc.
- Data is compiled from a variety of sources, both public and private, including company and government annual reports, data compilations, public rules and regulations, official announcements, etc.
- Internet-based libraries and other decentralized web services
- Information gathered by sensors in the sky, such as satellites, aircraft and unmanned aerial vehicles
- Online information gathered by a company is accessible through its websites, social media feeds and other online channels.

Various velocities and amounts of data are shown in the aforementioned sources. There are several potential locations where may locate such data, including static repositories, web services, live feeds, archives, files and so on. The information is structured and presented differently, but its format varies greatly.

Table 5.3 shows the many types of agriculture, big data sources and related technologies.

To handle the issues that arise with each agricultural application, authors need access to various huge datasets. Static repositories, geographical data and satellite-based remote-sensing comprise the bulk of the big data used in most agricultural fields. For their studies, scientists in soil, animals and crops use ground sensors. The data gathered by weather stations are also used in fields such as agriculture, banking and insurance, as well as for the study of climate change, the mapping of landscapes, and the formation of farmer

Table 5.3 Sources of big data and methods for interpreting big data in agriculture.

Domain	Sources of big data	Methods for big data analysis
Weather and climate change	Data from weather stations, static information studies (climate and weather records), satellite remote sensing and geospatial information	Scalable vector machine (SVM) analysis, statistics-based analysis, cloud computing, GIS analysis and MapReduce analytics are all examples of machine-learning techniques
Land	Airborne remote sensing, radar data, geographical zone data, historical datasets associated with land characterization, crop phenology, temperature and rainfall, global tree cover maps, elevation and sensors in cameras	NDVI-related vegetation indices, SVM, k-means clustering, randomized trees and other machine-learning techniques
Research in animals	Camera sensors, ground sensors (for tracking grazing activity, food intake, temperature, milk production and so on, and historical data on animals and soils in terms of their physiological parameters)	Techniques of machine learning related to decision trees, Neural Networks, SVM
Crops	Metabolite sensors on the ground, satellite imagery and archived data	Vegetation indices (NDVI, Wavelet and Fourier transforms) and machine-learning techniques (SVM, K-means clustering)
Soil	Ground-based sensors (electrical conductivity, salinity and humidity) and photosensitive cameras	Machine learning (K-means clustering)
Weeds	Information based on historical (library of digitized plant and weed photos, plant-specific data), Remote sensing (plane and drones),	Image processing methods, ML methods and the NDVI
Food safety and availability	Data gathered through surveys, statistical data, GIS geographical data, remote sensing (synthetic aperture radar) and historical information and databases (such as ENAR, CIALCA and rice crop growth figures)	Analytics based on statistics, simulation and modeling, network analysis, GIS spatial data analysis, image processing and machine learning (neural networks)
Investigations on the genetics of plants and animals	Geospatial information systems (GIS) and databases	Statistics
Farmer's decisions	Datasets containing historical information, weather stations, data based on the web, human-based sensors and remote sensing related to satellites and drones	Cloud-based systems, web-based services, mobile apps, statistical analysis, modeling, benchmarking, simulation, big data storage, message-oriented middleware
Farmer's insurance and finance	Web-based data, humans as sensors, historical information, weather stations	Cloud platforms, mobile applications, web services
Remote sensing	Data collected via planes, drones and satellites; web-based data, including maps and imaging; datasets containing historical information, including datasets of surface land containing images; MODIS surface reflectance datasets	Connected cloud-based platforms, statistical data analysis, GIS map-based spatial data analysis, image processing, NDVI-related vegetation indices, decision-making systems, big data storage, community- and web-based portals, mobile, computer vision, MapReduce-based analytics and AI-based applications

decisions. Most of these works have integrated information from multiple sources to solve various issues.

5.2.3 Methods and software for agricultural big data analysis

Methods used in specific agricultural fields are listed in Table 5.2, column 4. Thirty-four studies were reviewed by the author [22, 23], and ML was employed in 13 of them, cloud-based platforms in 9, image processing in 8, simulation and modeling in 7, statistical analysis in 6 and vegetation indices (NDVI) in 6. In addition, four publications dealt with geographic information systems, while five used Internet services (GIS). Most land-based remote-sensing applications use hybrid approaches except for a few soil-specific uses.

According to the research, the following resources are employed by the respective programs:

- Authors [24] both use machine learning to produce forecasts in the agricultural sector.
- Cluster analysis was used in farming by the authors [25].
- Scholars including researchers [26] have used classification strategies in the field of agriculture.
- The use of image processing methods including harmonic and Fourier analysis, wavelet decomposition and curve fitting has been applied by authors [27] in agricultural settings.
- Remote-sensing methods were coupled with image processing by researchers [28], and the generated image was fed into a machine-learning model.
- Literature [29] shows that cloud platforms coupled with MapReduce can be used for massive storage in agriculture. The authors [30] applied GIS to address spatial issues.

To archive huge data that is inherently heterogeneous, DBMSs that support the array data model are ideal. To manage and store massive amounts of unstructured data, NoSQL platforms might be used. Array DBMS systems developed for this very purpose handle storage for large raster datasets.

Moreover, remote-sensing applications have extensively used vegetation indices (VI) for mapping soils and crops. The VI is presented as a set of individual graphics, each representing a distinct surface reflectance

Table 5.4 Software packages widely used for agricultural big data processing.

Categories	Application software
Image processing tools	OpenCV library, VTK and IM Toolkits
Machine-learning tools	Weka tool, TensorFlow of Google, mlPy, Flavia, Mlpack, Apache Mahout, R, Oryx, SHOGUN and Mlib
Cloud-based systems that can both process and store large amounts of data for later analysis	Cloudera, IBM Pure Data system for analytics, IBM InfoSphere Big Insights, EMC Corporation, Apache Pig and Aster SQL MapReduce
Databases used in big data and GIS	ArcGIS, Autodesk, MiraMon and MapInfo
Connected middleware for message-based systems	Apache HAWQ, Post-GIS, MongoDB, Cassandra, HadoopDB, Oracle Geo Raster, MonetDB/SciQL, Google Big Table, Oracle Geo Raster, Hive and SciDB
Modeling and simulation	MQTT and RabbitMQ
Statistics tools	NorsysNetica, R and Weka
Time series data analysis	MATLAB, RATS, BFAST and Stata

pattern across two or more wavelengths. As the most popular graphical indicator for determining whether or not a target is covered in vegetation using remote-sensing analysis, the normalized difference vegetation index (NDVI) is invaluable. The rapid dissemination of alerts from event-based systems is facilitated by message-oriented middleware, which has applications in fields such as natural catastrophe prediction.

Table 5.4 displays the several software packages used for big data analysis. These packages vary depending on the type of analysis being performed.

Image analysis to approximate agricultural problems using remote-sensing photos is a relatively new method. Since 2008, there has been a meteoric rise in demand for Landsat satellite data because of its accessibility and inexpensive cost. Remote sensing is extensively utilized in agriculture because of its nondestructive nature and its ability to collect data systematically across broad areas. Several recent articles [27, 31] explore the dispersal of insurance products in the case of crop loss and the evaluation of risks connected to fire, flood, and extreme rain or drought.

The AVHRR, MODIS, MERIS and SPOT VEGETATION satellites have a coarser resolution and have emerged as the primary data sources for monitoring large regions. Combining remote sensing with other data sources (such as GIS information, field sensors and historical records) can greatly enhance analysis, especially for predictive applications like crop identification and grassland classification. The authors [32] found that because there have been so few studies in this area, big data analytics are still in their infancy in the agricultural sector. However, this area has a great deal of untapped potential because of the many different players and the high rate of change in the system. Many new businesses are emerging in recent years to take advantage of the prospects in the agricultural market afforded by big data analytics. The advent of cloud computing's scalability in storage and processing power has the potential to inspire the development of novel agricultural business models.

5.3 Case Studies of Big Data Analytical Methods in Agriculture

In this section, the authors examine examples of how big data analytics has been used in the agricultural sector. There is an emphasis on recent and relevant literature.

5.3.1 Case study 1

Datasets with varying amounts of content, storage formats and organizational structures are essential to studies in the agricultural and environmental sciences. As a result of the cloud architecture supporting linked open data (LOD), several datasets have recently been made publicly available, resulting in a plethora of disparate data sources. Unfortunately, the LOD is difficult to access and cannot keep up with the rapid developments in data science. It necessitates the availability of sophisticated big data querying infrastructure. In this paper, the authors analyze the research project Sema Grow, which was established by ERFP7 to provide effective methods of accessing remote data. Building efficient systems and techniques for searching dispersed data sources like LOD nodes was the project's primary focus. To facilitate data reuse and sharing amongst scientific programs, it implemented Semantic Web concepts including the SPARQL query language and RDF. It made possible the sharing of LOD-related cloud-based data.

The SemaGrow system overcame the unique problems associated with agricultural data by drawing on various use cases through a tailor-made set of data pilots. SemaGrow overcame obstacles associated with detecting and consolidating big data from disparate spatiotemporal datasets. Adaptation to climatic changes was an integral part of the use cases that SemaGrow managed, such as the modeling of climatic variables relevant to agriculture.

Extensive files of varying dimensions held the climate parameters needed for regional modeling. The authors used long-term temporal coverage and spatial coverage connected to various regions. Many scientific investigations aim to generate redundant data for the sake of experimenting. This causes wasteful use of resources and is especially problematic in areas with scarce data storage, processing power and communication infrastructure. The SemaGrow efficiently constructed compact datasets by combining temporal, thematic and spatial requirements via queries. The SemaGrow system allowed an agricultural scientist to access a centralized dataset of temperature, soil and precipitation parameters, even in a remote area of Ghana with few storage and networking choices.

SemaGrow's primary goal was to give agricultural experts unhindered access to large, disparate datasets so that they could easily aggregate and download them. Due to the reasonable solutions implemented, manual large data integration required less time and fewer resources. The timely delivery of relevant data through the system facilitated stakeholders' decision-making. The DIKW approach helps close the information gap by coordinating and processing information at a deeper level. As a result, less time and energy were needed to handle integration difficulties across several data sources and to extract valuable insights from raw data.

5.3.1.1 Methodology

SemaGrow's technology was integrated into the Trees4Future Clearinghouse to enhance the system's big data querying capabilities for usage in agricultural settings. SemaGrow, a stack of tools for building and deploying distributed systems and RDF databases for storing data and metadata have replaced Trees4Future's previous back-end. Consequently, semantic queries could be run on both metadata and actual data within the application. Authors have an efficient query, a cluster of mixed-type data nodes with the help of the SemaGrow Stack, which you can get on GitHub at https://github.com/semagrow/semagrow. To put it simply, The Stack is a federated SPARQL query processor. Optimizing query execution is facilitated by an in-built query planner that uses metadata from individual nodes. Statistical data, such as database histograms, were left out of the original VoID vocabulary, so the Sevod vocabulary was created to fill that gap. For further details on this project, please visit http://www.w3.org/2015/03/sevod. Because of its use of the reactive software paradigm, Stack can deal with slow data nodes. The Stack accomplishes this by allowing users to perform queries on the underlying triple stores without knowing anything about the stores' various schemas.

ISIMIP and AgMIP collaborated to establish the triple stores by syncing their separate datasets. ISIMIP has made its input/output data available in the NetCDF file format, which is frequently utilized in the domains of climate and forecasting. By bringing together economics, meteorology and agriculture experts, AgMIP can develop more accurate economic and crop models to predict how climate change will affect future generations. This effort takes advantage of existing means of communication and information storage. AgMIP employs the JSON data format for describing data, with the ICASA Variable List used to describe the data. Data from several sources are amplified into triple stores, expanding the usability of the original datasets. Since this is the case, SPARQL can now query the information.

Following the consolidation of the nodes, a spatiotemporal triple store called Strabon is incorporated to provide solutions to spatial queries such as point-in-polygon. The demo's front end connects to the SemaGrow instance on Stack using middleware to exchange data. It is a pre-processor for query results, transforming raw data into a format that the demonstration can work with. In addition, it acts as a translator, changing incoming URL requests into the correct SPARQL queries for the Stack. Strabon generates acceptable NetCDF files, allowing it to meet the needs of its consumers effectively.

5.3.1.2 Results

Users testing out the demonstrator app were few and far between. Users praised its usefulness. However, users expected faster results in the case of metadata searches, which took about 5–30 s. Also, the system's download was painfully slow (taking days) despite users' expectations that it would take no more than 30 minutes. Because of this, substantial infrastructure scalability is required. Despite taking several days, the method is faster than the manual technique, which necessitates continual consultation and coordination between remote and local experts and is thus far slower, according to specialists in the field. Therefore, fusion queries save time compared to the manual procedure by taking a few hours or days rather than many days.

5.3.2 Case study 2

To learn more about the ongoing European research effort known as "Trees4Future," visit the website www.trees4future.eurepresents. Critical forest genetics and forestry research infrastructures will be consolidated, strengthened and improved as part of this project. This initiative expands access to forestry data for use by farmers and policymakers. Researchers in forestry rely on information gleaned from their professional and academic networks. However, work has been done recently to improve data distribution in forestry, especially in genetics. Nonetheless, important data is stored in isolated locations, such as private or local databases. Metadata documentation is incomplete as well.

In many cases, researchers fail to properly document data or secure data, either because the underlying research results are not released or because they are concerned about the prospect of data exploitation. Not much good may come from sharing information, even within reliable networks. For this reason, it is difficult to access relevant data for use in agricultural studies. Researchers in forestry modeling come together as part of the Trees4Future initiative. To deal with current issues in agriculture, such as climate change, integrated modeling is necessary. Models that explain diverse subdomains and offer coverage at temporal and spatial levels must be combined to assess climate change and choose an appropriate adaptation strategy accurately. This combined model set was made possible by collaborating with Trees4Future and the modeling environments ForGEM, EFISCEN and Tosia. The Adaptation Response in Trees (ForGEM) model is a simulation tool for estimating the impact of selective breeding on forest ecosystems. The EFISCEN model predicts the growth of forest resources on a European and regional scale. The Tosia model examines the effects on forest-wood supply chains in terms of social effects, environmental modifications and economic shifts.

Heterogeneous data from fields as diverse as genetics, soil, statistics, the market and the environment are at the root of many of the difficulties associated with integrated modeling. For forest management purposes, this type of information is difficult to access. The Trees4Future initiative, by creating an infrastructure for forestry data, offers a technical solution to this issue by facilitating the release, discoverability and documentation of forestry data. In addition, this infrastructure can help with things like data sharing and documentation.

There is also the necessity to address trust and quality concerns in this endeavor. If the data's original owner is concerned about how it will be utilized after being shared, such concern is warranted. In the consumer's interest, giving true and comprehensive information is essential. Since data can come from various heterogeneous sources and domains, it is important to deal with the diversity challenges that arise when processing this data. The metadata and the underlying data must have a solid semantic connection.

5.3.2.1 Methodology and implementation

Technology for finding and navigating data has greatly improved users' access to it. The primary objective was to provide widespread access to high-quality

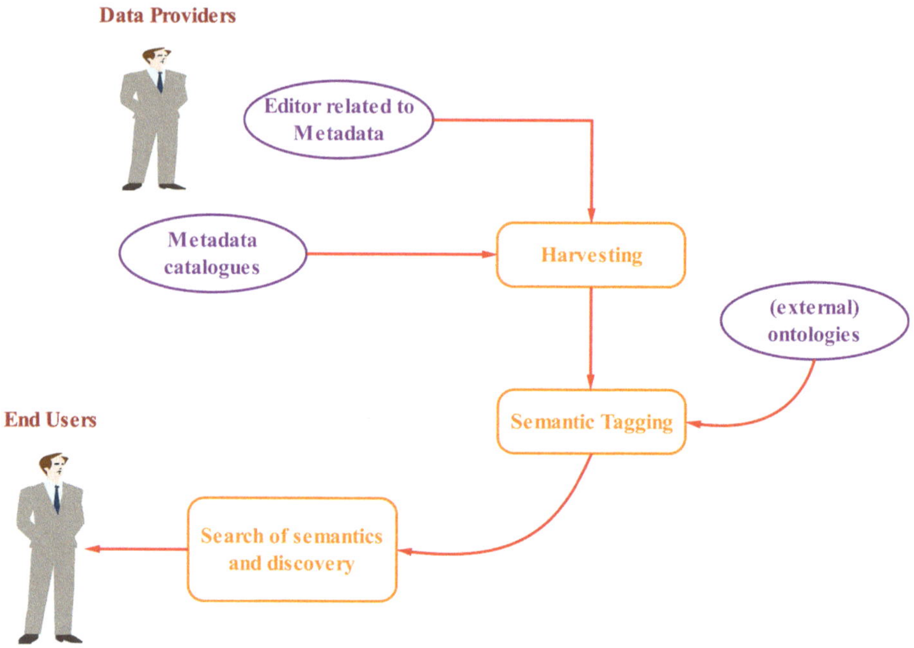

Figure 5.2 Trees4Future method for managing the publication process.

datasets that were either poorly documented or unavailable. It is now easier for users to navigate through datasets and locate the forestry data that has been made public. This search function was deemed to be both user-friendly and straightforward by both end users and data publishers/owners.

Figure 5.2 depicts the Trees4Future project's established procedure.

- Metadata that is unique to the forest industry will benefit greatly from a concise schema based on the Dublin Core standard.
- A metadata editor (online) and embedded repository are provided by the Open Archives Initiative Protocol for Metadata Harvesting (OAI-PMH) to make the metadata records uniform and harvestable.
- The forestry ontology provides a conceptual framework for the datasets and establishes links to external ontologies, such as AGROVOC, which describe genetic features.
- Metadata forestry schema and metadata (standardized) are used in conjunction with an annotation, harvesting and simplification technique to facilitate metadata collection.

- Metadata can be broken down into ontology ideas with the help of the supplied INSPIRE and ISO standards and the NLP techniques used therein. These are then saved to an RDF database, and associations are made between ideas in the generated dataset and external ontologies.
- Quickly access registered datasets via queries based on their semantics as included in the RDF store with this system and user interface.

5.3.2.2 Results

With the newly established infrastructure, finding forestry-related study material is now easier and accessible to more people than ever before. Integrated modeling in forestry, including the use case mentioned above, incorporates climate adaptation, and this infrastructure is excellent because of its flexibility to support multiple heterogeneous data sources. The system's primary benefit is that it can gather all of the forest-related information that is currently spread out in many places and organize it in one central location. The platform has made available over 300 datasets from European data archives. Small businesses and individuals are also provided with a publishing portal for their metadata. In addition, new modelers are allowed to submit reference datasets for use in integrated modeling.

Due to a lack of quantity and quality, linking concepts in metadata with ideas in an external ontology is difficult. The system may also automatically choose datasets that academics will find useful, such as those from simulation and modeling-related technical disciplines. Insufficient depth and structure in metadata fail to convey the intricate nature of scientific datasets. Commonly used but crucial domains like lineage lack the necessary structure to handle the intricate data generation operations. The shallowness of the metadata structure prevents the systematic addressing of data. The described system fixed this problem by establishing associations between the separate AGROVOC vocabulary pieces and a comprehensive subdomain with domain-specific semantics. According to appearances, this is not a one-size-fits-all answer but rather shows a complex and individualized system.

5.4 Agricultural Big Data Analytics: Related Research Fields

In this part, the authors list some problems that still need fixing when analyzing massive amounts of agricultural data.

5.4.1 Open problems

Problems facing the agricultural sector may become much more severe if big data analysis techniques are more widely used. Here are only a few of the major issues:

- Due to the high cost of storing and disseminating agricultural data, only the largest agriculture corporations can afford to maintain a monopoly on this information. The author notes that this results in a significant amount of reliance on the part of farmers on huge agricultural groups. Large companies' data monopolies could limit the spread of new technologies to smaller competitors.
- The ownership difficulties, stemming from the uncertainty about who would control the data and monetization issues relating to that data may occur. The seed companies and farms may mishandle agricultural data. Hedge funds may cause unnecessary speculation in commodity markets due to access to real-time agricultural data.
- Some potential problems with the privacy, availability and veracity of big data can develop during analysis and collection.
- The digital divide between developing and developed countries is exacerbated by unequal access to resources including software, processing power, Internet speed and qualified professionals. However, important big data benefits may only be realized if farmers have access to high-quality education and digital technology, neither of which are typically present in developing countries.
- Insufficient real-world data is available to product designers. However, real-world data is crucial for testing and evaluating products and services in various environments, including those that experience extremes of temperature, humidity and other environmental factors. Large data visualization is similarly intricate.

5.4.2 Restrictions on big data analysis methods in agriculture

Incorporating big data analysis techniques presents the following difficulties:
Due to a dearth of knowledge and tools,

- Unreliable systems for acquiring and interpreting large amounts of data. High-dimensional datasets and big data demand infrastructures for efficient storage and management.

- Those tasked with analyzing big data often lack the specialized knowledge and training necessary to do so accurately.
- There is a lack of effective governance and organization in big data.
- Appealing business models are necessary to ensure that all parties involved in agriculture receive a fair share of the benefits.
- Due to the lack of definitive semantics, organizations and scholars find it challenging to reuse and understand large data.

Table 5.1 of the authors' research [3] shows that they zero attention on extremely big datasets, such as those generated by applications in climate and weather change research, land identification, farmer decision-making, financial and insurance markets and remote sensing. Big data is valuable not just because of its sheer amount but also because of its velocity, authenticity, valuation and diversity. The use of big data in agriculture has been viewed through a biased and limited lens because of the neglect of these considerations.

Sample errors, a lack of data with acceptable spatial resolution, and the collecting and distributing pictures within the allowed time limit are only some of the real-world issues being faced by remote-sensing systems for farm management. Also persistent are difficulties associated with data extraction, image interpretation and the influence of the weather. Finally, many researchers in the field of agriculture have noted that a lack of data (or an insufficient amount of data) and the unreliability of that data are two of the most common challenges they face. Big data analytics in agriculture has significant hurdles due to agricultural data's time sensitivity and heterogeneity.

5.4.3 Resolving complex issues and overcoming obstacles

Many farmers and agriculturalists from different parts of the world have recently come together to form online forums and cooperative societies where they can share their knowledge and experiences with using big data in agriculture while also discussing the social and political issues that affect their profession. Data copyrights must be protected, and user access to data must be governed by clear policies on who owns what. Protecting big data and maximizing its usefulness by employing the right technologies calls for policies to regulate the security and administration of the data. Spending is necessary to build farm cloud infrastructures to save, display and analyze agricultural data. In addition, it is necessary to execute analytics more rapidly and on a bigger scale. Price agricultural data reasonably, or better yet, make it open source, so it can be used to further important studies. By combining and simplifying data and applying the right analysis, it is possible to build intuitive platforms for users.

Data semantics technologies like RDF, linked data and ontologies could be utilized to facilitate information exchange. Open-source tools, such as crop-recognition maps, crop calendars, vegetation index, yield prediction models, crop acreage estimations and seasonal weather reports, could be useful in agriculture. Combining these resources should help with both large-scale activities and the analysis of different types of data. In agriculture, large datasets must be made publicly available to conduct in-depth studies.

5.4.4 Potential use of big data analysis in agriculture

This section discusses the potential of big data analytics to address a wide range of problems in the agriculture sector. Here are a few examples of feasible contexts for this technology:

1. For agriculture to meet international requirements, it is crucial to access platforms that guarantee all participants in the supply chain have the highest-quality commodities and procedures.
2. A more accurate forecast of future demand and improved methodologies for predicting future yields are both necessary moving forward.
3. For improved crop growth, farmers need accurate knowledge of how to use herbicides, fertilizers and pesticides.
4. Tools for plant-level scanning, shipment tracking and customer-level purchase tracking are all areas that might use improvement in food supply chain tracing.
5. Big data analysis solutions in agriculture are needed to reduce the prevalence of food-borne illnesses.
6. Since agricultural goods typically deteriorate after harvesting, effective optimization approaches are required. This is crucial to improve food quality and decreasing waste. Food processors increasingly turn to popular optimization tools including neural networks, genetic algorithms and meta-heuristics.
7. Using advanced remote-sensing techniques is essential for accurately mapping large agricultural fields. More precise monitoring of farmed land worldwide is required for increased yields.

8. To accurately simulate environmental processes, state-of-the-art scientific and simulation models must be built. Policymakers can use these models to their advantage when making agricultural decisions contributing to ecological sustainability.
9. High-throughput screening methods are needed to accurately capture the interplay between the environment and plants. These methods need to be more precise and accurate, and they must be able to conduct quantitative analyses of the data.
10. Agricultural robots could be used for automatic weed detection and removal, pest identification and prevention, and harvesting. These robots can help agriculture gradually shift from its current uncertainty towards a more predictable future.
11. "Genome editing," a subset of genetic engineering, can be used to alter the genetic code of a plant or animal down to the level of individual genes. It can simulate the mutation process essential to plant breeding. It also does not result in the offspring being of a different species. This technology can be useful in enhancing epigenetics studies.

The above applications can generate copious amounts of big data that policymakers may use to strike a better equilibrium between supply and demand for food crops. Making available agricultural data would also foster important smart farming research and development opportunities.

5.5 Conclusions

This chapter delves deeper into the approaches of big data analysis and their applications in farming. Several inquiry strategies and their respective outcomes across all applications are discussed at length. The reader will learn from this chapter about current agriculture projects that use big data analytics approaches. Extensive discussion is given to the features of large data, the methods of analysis employed, and the situations in which they prove most useful. The authors also discuss the difficulties and unanswered questions associated with applying big data methods to agriculture. As has been shown, significant shifts in farming practice are possible due to the widespread availability of open-source software and hardware, big data analytic tools and techniques and agricultural datasets. There is a proliferation of new businesses dedicated to using big data for agricultural purposes. Adopting open standards for datasets can improve agricultural development and research possibilities. The authors can efficiently balance food demand and supply using big data analytical methods and end food scarcity.

References

[1] M. Manideep, R. Thukaram, and M. Supriya, "Smart Agriculture Farming with Image Capturing Module," in *2019 Global Conference for Advancement in Technology, GCAT 2019*, 2019.

[2] B. Kalantar, S. B. Mansor, H. Z. M. Shafri, and A. A. Halin, "Integration of template matching and object-based image analysis for semi-Automatic oil palm tree counting in UAV images," in *37th Asian Conference on Remote Sensing, ACRS 2016*, 2016, vol. 3, pp. 2333–2340.

[3] Dhanalakshmi R., Jose Anand, Arun Kumar Sivaraman, and Sita Rani, "IoT-based Water Quality Monitoring System using Cloud for Agriculture Use", *Cloud and Fog Computing Platforms for Internet of Things*, Edited by Pankaj Bhambri, Sita Rani, Gaurav Gupta, Alex Khang, Routledge Taylor & Francis Group, May 2022.

[4] C. Bruno *et al.*, "Embedded Artificial Intelligence Approach for Gas Recognition in Smart Agriculture Applications Using Low-Cost MOX Gas Sensors," in *2021 Smart Systems Integration, SSI 2021*, 2021.

[5] C. Huang and Y. Chen, "Agricultural business and product marketing affected by using big data analysis in smart agriculture," *Acta Agric. Scand. Sect. B Soil Plant Sci.*, vol. 71, no. 9, pp. 980–991, 2021.

[6] N. Jain and Y. Awasthi, "WSN-AI based Cloud computing architectures for energy-efficient climate-smart agriculture with big data analysis," *Int. J. Adv. Trends Comput. Sci. Eng.*, vol. 8, no. 1, pp. 91–97, 2019.

[7] R. Popli, D. Singh, R. Kumar, G. S. Saini, and K. Garg, "Role of Contemporary Techniques in Agriculture Development: A Systematic Review," in *2022 2nd International Conference on Advance Computing and Innovative Technologies in Engineering, ICACITE 2022*, 2022, pp. 1677–1681.

[8] W. Liu, "Smart sensors, sensing mechanisms and platforms of sustainable smart agriculture realized through the big data analysis," *Cluster Comput.*, 2021.

[9] M. Kumarasamy, B. Pattanaik, J. N. Dwivedi, B. R. Ramji, M. Ponnusamy, and V. Nagaraj, "Predictive analysis of smart agriculture using IoT-based UAV and propagation models of machine learning," *Int. J. Eng. Syst. Model. Simul.*, vol. 14, no. 1, pp. 16–23, 2022.

[10] I. Marcu, A.-M. Drăgulinescu, C. Oprea, G. Suciu, and C. Bălăceanu, "Predictive Analysis and Wine-Grapes Disease Risk Assessment Based on Atmospheric Parameters and Precision Agriculture Platform," *Sustain.*, vol. 14, no. 18, 2022.

[11] S. Rajkumar, M. Arun, J. Hirwani, and S. S. Sanjeev, "Predictive analysis of crops cultivation for a smart green environment using azure services," *Int. J. Recent Technol. Eng.*, vol. 7, no. 5, pp. 295–298, 2019.

[12] M. Salah Uddin, M. Asaduzzaman, R. Farzana, M. Samaun Hasan, M. Rahman, and S. M. Allayear, "Implementation of smart indoor agriculture system and predictive analysis," in *Communications in Computer and Information Science*, 2019, vol. 1045, pp. 424–435.

[13] K. Sharma, C. Sharma, S. Sharma, and E. Asenso, "Broadening the Research Pathways in Smart Agriculture: Predictive Analysis Using Semiautomatic Information Modeling," *J. Sensors*, vol. 2022, 2022.

[14] S. Bouarourou, A. Zannou, A. Boulaalam, and E. H. Nfaoui, "IoT Based Smart Agriculture Monitoring System with Predictive Analysis," in *2022 2nd International Conference on Innovative Research in Applied Science, Engineering and Technology, IRASET 2022*, 2022.

[15] K. Priyadharshini, R. Prabavathi, V. B. Devi, P. Subha, S. M. Saranya, and K. Kiruthika, "An Enhanced Approach for Crop Yield Prediction System Using Linear Support Vector Machine Model," in *2022 International Conference on Communication, Computing and Internet of Things, IC3IoT 2022 - Proceedings*, 2022.

[16] Y. Tang, S. Dananjayan, C. Hou, Q. Guo, S. Luo, and Y. He, "A survey on the 5G network and its impact on agriculture: Challenges and opportunities," *Comput. Electron. Agric.*, vol. 180, 2021.

[17] S. M. Patil and R. Sakkaravarthi, "Internet of things based smart agriculture system using predictive analytics," *Asian J. Pharm. Clin. Res.*, vol. 10, pp. 148–152, 2017, doi: 10.22159/ajpcr.2017.v10s1.19601.

[18] C. Kulatunga, L. Shalloo, W. Donnelly, E. Robson, and S. Ivanov, "Opportunistic Wireless Networking for Smart Dairy Farming," *IT Prof.*, vol. 19, no. 2, pp. 16–23, 2017.

[19] M. Pyingkodi *et al.*, "Sensor-Based Smart Agriculture with IoT Technologies: A Review," in *2022 International Conference on Computer Communication and Informatics, ICCCI 2022*, 2022.

[20] V. R. R. Kolipaka, "Predictive analytics using cross-media features in precision farming," *Int. J. Speech Technol.*, vol. 23, no. 1, pp. 57–69, 2020.

[21] J. Bastos, P. M. Shepherd, P. Castillejo, M. S. Emeterio, V. H. Diaz, and J. Rodriguez, "Location-Based Data Auditing for Precision Farming IoT Networks," in *IEEE International Workshop on Computer Aided Modeling and Design of Communication Links and Networks, CAMAD*, 2021, vol. 2021-October.

[22] A. I. Rokade, A. D. Kadu, and K. S. Belsare, "An Autonomous Smart Farming System for Computational Data Analytics using IoT," in *Journal of Physics: Conference Series*, 2022, vol. 2327, no. 1.

6

Machine Learning in Smart Agriculture

A. Thirumurthi Raja[1], A.S. Arunachalam[2], and R. Gobinath[3]

[1]Directorate of Distance Education (DDE), SRM Institute of Science and Technology (SRMIST), India
[2]Department of Computer Science, Vels Institute of Science Technology & Advanced Studies (VISTAS), India
[3]Department of Computer Science, Christ University, India
Email: thirumurthiraja@gmail.com; arunachalam1976@gmail.com; iamgobinathmca@gmail.com

Abstract

Agriculture is the cultivation of the soil, the growth of crops and the raising of livestock. Agriculture is critical to the economic development of a country. Farming generates nearly 58% of a country's primary income. Previously, cultivators had accepted conventional farming practices. Because these methods were imprecise, they produced less and took longer time. Precise farming boosts productivity by precisely determining which steps must be completed at what time. Precision farming entails forecasting the weather, analyzing soil, recommending crops for cultivation and calculating the amount of fertilizer and pesticides that must be used. Precise farming uses advanced technologies such as IoT, data mining, data analytics, and machine learning (ML) to collect data, train systems and predict outcomes. Precision farming employs technology to reduce manual labor and boost productivity. Farmers have recently faced several difficulties, such as crop failure due to insufficient rainfall, soil infertility and so on. The proposed work in determining the soil, managing crops and harvesting efficiently can solve the problems caused by environmental changes. It guides a person's farming strategy to produce better results through a proper prediction process. The goal of this research is to assist an individual in efficiently cultivating crops, resulting in high productivity at a low cost. It also assists in estimating the total cost of cultivation and forecasting the likely economic barriers. This would help a person plan activities prior to cultivation, resulting in an integrated farming solution.

6.1 Introduction

Farmers and agriculture are regarded as the backbone of a country's economic growth. Agriculture not only contributes to the improvement of a country's wealth; it also contributes to various subsequence businesses that rely on agricultural products. Proper crop growth raises livelihood in the human era and creates the best yielding conditions for farmers. Skilled farmers cultivate crops more productively, increasing the likelihood of producing new breeds in agriculture. Many ideas in modern agriculture and science are taught to new technologies based on scientific principles to produce better crops.

Food products produced by agricultural industries are extremely beneficial in resolving many economic crises and crop waste. The tomato crop is an example of such an agricultural crop that quickly reaches the wastage stage after yielding from agricultural farms. Many laborers, merchants and food product industries rely on farms for a living. Farmers spend the majority of their time on farmland to produce higher yields and place little emphasis on water conservation. Agriculture fields are the most important resources for producing healthy lifestyles in humans and cattle. The waste products collected from agricultural fields are mostly used for fertilizer and cattle feed. The agricultural field officers and researchers are very interested in learning about the significance of crop development growth that is not affected by various crop diseases.

There is a greater likelihood that disease-affected crops will spread the disease to neighboring plants and fields, potentially reducing crop yield. Crop disease effects are becoming more prevalent, resulting in a variety of changes in agricultural field studies. The disease identification procedure used during disease identification involves several hypothetical steps and creates several barriers for farmers to take the necessary precautions on time. The clinical approach to overcoming such barriers is critical in evacuating many complex disease identification procedures available in agriculture.

The researchers work hard not only to identify the type of disease infections on the plant but also to solve the problems from the ground up. Many agricultural research studies are significant in explaining the importance of the time-consuming disease identification procedure. Clinical scientists and researchers from various fields are collaborating to achieve better results in identifying disease infections on agricultural crops. One of the fields that plays a significant role in identifying the type of disease infection on the plant and providing appropriate solutions to many unidentified problems in agriculture farmland is computer vision and image processing.

In the field of agriculture, a computer vision-based approach is most useful in identifying the type of infection that occurs in agricultural crops. There should be a complete solution providing an area that serves as a disease identification as well as a proper solution to the type of infection. Image processing and sensing are two fields that can help solve problems that arise after collecting image samples from agricultural fields. Many techniques and methodologies are used in the process of identifying disease types and predicting disease types. The algorithmic representation and incorporation of many computerized techniques can solve issues in the image processing identification procedure.

Machine learning (ML) and image processing are frequently used in tandem to solve many agricultural problems. Smart agriculture devices and applications are being developed for the benefit of farmers and researchers who have spent more time determining the type of crop disease. The machine learning procedure can solve many problems in farmers' fields, but it can be a time-consuming strategy for agricultural officers to identify the type of infection and provide a proper timely remedy.

Machine learning plays a significant role in achieving magnificent self goals and improving identification processes in the agricultural field. The algorithms' major effects are critical in providing a proper solution in disease prediction procedures. The activation function in machine learning and deep learning can provide us with the best solution for transferring data into the best form suitable for the final prediction stage. The smart devices created for the agricultural revolution require proper training for the end users. Many subsequent areas, such as big data, cloud computing, and the Internet of Things, are also very useful in creating an adequate advantage in implementing smart devices in agriculture. Adequate communication and training provided to end users can raise awareness among farmers and agriculture officers about the importance of protecting crops from harmful and deadly diseases.

The developed agricultural applications provide the greatest benefit of an accurate tracking system, a timely monitoring system and a perfect prediction mechanism. Many research works using machine learning provide a prediction procedure as well as a calculation of the greenhouse effect occurring in agriculture. The applications are designed to withstand a variety of climatic issues as well as new disease attacks. Some research works are also used to calculate gas emissions and gas emission possibilities, which is also a trend for farmers. Implementing machine learning procedures in applications will always be a better way to produce product results and a better economy for farmers.

The use of various sensor devices in the computer vision-based approach predicts a wide range of issues, including soil temperature, harmful radiation emission, air humidity, soil water potential, air temperature, solar radiation and luminosity. The computer vision-based approach also provides accurate results in a timely and user-friendly manner. The applications created using a computer vision-based approach are highly efficient and cost-effective. The ground control station, which is located near the field, collects accurate results from the fixed sensors and perfectly predicts the emission rate. This system also aids in the generation of warning messages for field workers regarding radiation levels and effects.

Deep learning is very similar to machine learning in that it produces manual activation functions while dividing pixels. Deep learning and machine learning follow very similar procedures in the initial stage of the prediction process. In converting images to 512 × 512 size, the image resizing and feature extraction procedures used in deep learning and machine learning are very similar. Both learning strategies use image comparison and resizing technologies that are similar. The feature extraction process used in the learning procedure may differ from one another. Convolution methods are commonly used during the classification process to divide images into various levels for better identification.

Algorithms like ReLu and convolutional neural networks are extremely useful in the classification process and provide the best triggering function during the activation process. During the identification stage, artificial intelligence in the image object prediction process aids in determining variations in the pigments of the objects present on the surface of the leaves. In the artificial intelligence process, many heterogeneous filtering techniques are used to find the best solution with the fewest effects. In some cases, automatic trigger functions are used to activate functions automatically during level divisions. The training and testing data are examined from the images collected from the various sensors used to monitor the fields.

6.2 Start of Agriculture

People used to spend the majority of their time in the forest looking for food. They do this by hunting and eating various wild animals found in the forest. A few centuries ago, people decided to get involved in agriculture and learn how to grow cereal and root crops on their land. After that, they found it convenient to settle down in the same place for the rest of their lives through farming. It is extremely difficult for living beings to survive in this world without agriculture and farmlands [15].

When people became involved in agriculture, they began to grow natural crops and practiced herding and breeding wild animals for their agricultural work. Domestication is the process of adapting wild plants and animals for human consumption. Rice or corn was most likely the first domesticated plant cultivated by Chinese farmers after domestication.

People were used to eating mostly plants and crops before the discovery of domestication. As a result, wild animals are being domesticated, which will benefit people. Dogs were the first domesticated animals, and they are still used by humans to hunt other animals. Likewise, sheep and goats have been domesticated, following cattle and pigs. The majority of these domesticated animals used to hunt for food. However, milk, cheese and butter are now produced by all domesticated animals. People eventually domesticated various animals such as oxen for ploughing, pulling and transportation. Agriculture enables people to produce more food than they require. As a result, people could use the extra food to trade for other goods if crops failed. This scenario will require people to perform tasks unrelated to farming and cultivation [15].

Agriculture caused a revolution, causing people to live near their fields and cultivate in the same spot regularly. As a result, agriculture led to the growth of permanent villages. They conduct trade with the goods of other villages from these villages. Based solely on agriculture, new economies have been developed and have proven to be very successful in some areas, resulting in the growth of cities and civilizations. According to history, the first civilizations based on extensive agriculture arose near the Tigris and Euphrates Rivers in Mesopotamia (now known as Iraq and Iran) and along the Nile River in Egypt [15].

6.2.1 Technological advancement

The agricultural sector's progress was glacial. Agriculture had a tool called the fire in ancient times. The Americans were the first to use fire as an agricultural tool to manage the growth of berry-producing plants, as the plants grew quickly whenever a wildfire occurred. So they gather the berry plants and set fire to them. The farmers will then cultivate small plots of land on their own using axes, which are useful for cutting trees as well as digging sticks into the soil. After a few centuries, farmers began to use various farming tools such as bone, stone, bronze and iron to perform cultivation. They are developing new farming methods. Farmers then consider the future and decide to store the crops safely. So they are attempting to create jars and pits out of clay soil

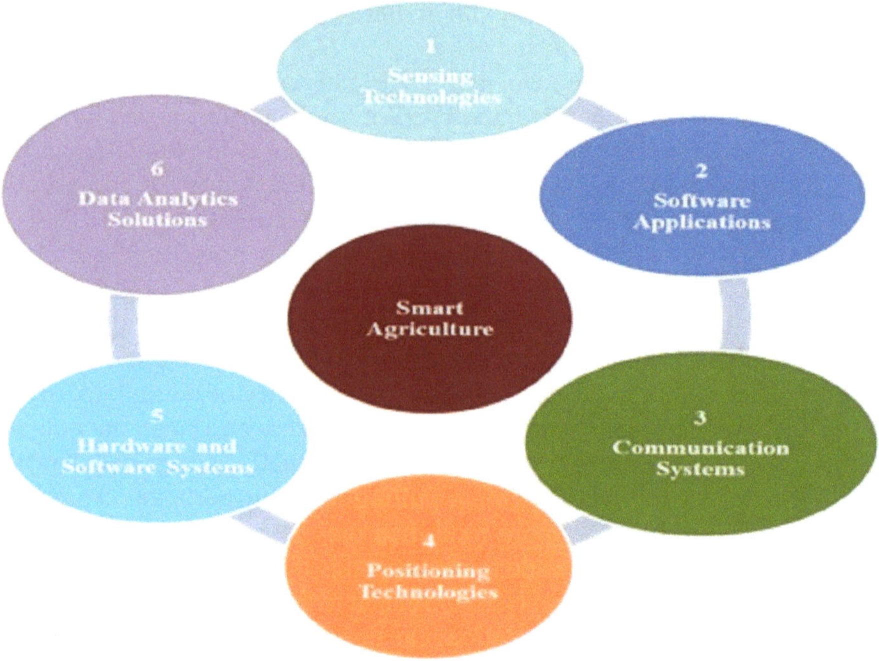

Figure 6.1 Technologies used in smart agriculture.

to store food in. That food is extremely useful in times of scarcity or natural disaster. They were also interested in making clay pots and other containers to transport food from one location to another. They are also used to cook food with the help of fire.

Farmers in Mesopotamia developed simple irrigation systems for cultivation around 5000 BCE. That irrigation system involved channeling water from nearby streams onto their fields. Farmers decide to settle in one area and practice agriculture after irrigation. The people then decided to join forces and work together to perform agriculture over a large area while also efficiently maintaining better irrigation systems [15].

Smart applications in agriculture are useful in managing the practical difficulties of collecting information from agriculture fields and sending it to ground control in time to avoid the complexity of predicting the type of disease attacks in agriculture crops. The communication between the farmers and the control room is very simple to achieve the desired result. The process of verifying the solution to the problem is very quick, and many practical issues in missed communications are resolved (Figure 6.1).

The technologies available for today's smart farmers are (i) sensing technologies, (ii) software applications, (iii) communication systems,

(iv) positioning technologies, (v) hardware and software systems, and (vi) data analytics solutions.

(i) Sensing technologies

In precision agriculture, several sensing technologies are used to provide data that helps farmers monitor and optimize crops, as well as adapt to changing environmental factors, including:

- Location sensors use GPS satellite signals to determine latitude, longitude and altitude within a few feet. To triangulate a location, at least three satellites are necessary. Precision placement is the foundation of precision agriculture. The NJR NJG1157PCD-TE1 is an example of a GPS-integrated circuit used in location sensors.
- Optical sensors utilize light to assess soil qualities. The sensors detect near-infrared, mid-infrared and polarized light reflectance frequencies. Sensors may be installed on automobiles, aircraft and even satellites. Soil reflectance and plant color are only two examples of optical sensor data variables that may be pooled and analyzed. Optical sensors have been created to measure the clay, organic matter and moisture content of the soil. For instance, Vishay offers hundreds of photodetectors and photodiodes, which serve as the basis for optical sensors.
- Electrochemical sensors offer crucial data for precision agriculture, including pH and nutrient levels in the soil. In the soil, sensor electrodes detect particular ions. Currently, sensors placed on specifically constructed "sleds" are utilized to gather, analyze and map chemical data from the soil.
- Mechanical sensors monitor soil compaction, which is often referred to as "mechanical resistance." The sensors use a probe that enters the soil and load cells or strain gauges to assess soil resistance. On big tractors, a comparable version of this technique is used to forecast the pulling needs for ground-engaging equipment. Tensiometers, such as the Honeywell FSG15N1A, detect the force produced by the roots during water absorption and are extraordinarily beneficial for irrigation interventions.
- Dielectric soil moisture sensors detect the dielectric constant (an electrical characteristic that varies with the amount of existing moisture) in the soil to assess moisture levels.
- Airflow sensors detect the soil's air permeability. It is possible to collect measurements at particular sites or while in motion. The

targeted output is the pressure necessary to push a predefined quantity of air into the earth at a particular depth. Soil characteristics, such as compaction, structure, soil type and soil moisture level, result in the production of unique identifying units that are strategically scattered across agricultural areas. These stations are equipped with a range of sensors that are suited to the local climate and crops. Air temperature, soil temperature at different depths, rainfall, leaf wetness, chlorophyll, wind speed, dew point temperature, wind direction, relative humidity, solar radiation and atmospheric pressure are monitored and recorded at predefined intervals. This data is collated and transferred wirelessly to a central location.

(ii) Software applications

Specialized software solutions for certain farm kinds or IoT platforms that are application independent. Based on the junction of cybernetics and agronomy, agriculture software is a recent, advantageous digital world invention. The lives of farmers are facilitated by digital apps. Their company is more comfortable, predictable, confident and prosperous due to their innovative ideas.

FarmLogs, Climate FieldView, Farmers Edge, Agrian, Trimble, Agrivi, Granular, FarmShots, Strider, Proagrica, AgriEdge and EOS Crop Monitoring are among the most effective agricultural applications now available.

The intelligent agricultural system facilitates the execution of several complicated tasks. Diverse tools are developed to assist individuals in making crucial choices quickly. In addition to accelerating the process, farming applications also give background information. This improves the probability that the answer is right. Agronomy software facilitates intelligent agriculture management by encompassing a vast array of agricultural activities, including: loss prevention, track farming, pest control, risk management, weather precipitation, crop rotation, resources optimization, fertilizer/water saturation, yield prediction, seeding/harvesting time and record-keeping/maintenance.

(iii) Communication system

As information technology has improved, several wireless communication systems have been suggested for a range of purposes. In general, these technologies are categorized as short-distance wireless communication (10 m), medium-distance wireless communication (10–100 m), and long-distance wireless communication (more than 100 m) (distance 100 m). RFID, Bluetooth, and ultra-wideband (UWB) are examples of typical technologies for short-range wireless communication. Wi-Fi and ZigBee are the two most popular medium-distance wireless communication technologies. In addition to the well-known cellular networks (2G/3G/4G), Low Power Wide Area network (LPWA) is a new kind of technology with several applications for long-distance wireless communication technologies. In recent years, many LPWA technologies, including LoRa, NB-IoT and Sigfox, have developed.

In contrast to conventional long-distance wireless communication technologies (2G/3G/4G), LPWA technologies feature low power consumption, low data rates, deep/wide coverage, and many connections. Consequently, LPWA technologies cannot be utilized for streaming music or video data, but they are suitable for connecting devices that only need to transfer modest quantities of data over vast distances while preserving battery life. The enhanced performance of LPWA technologies, such as extended transmission distance and low power consumption, is at the price of a low data rate and significant latency. Consequently, common LPWA application scenarios are delay-tolerant, do not need large data speeds, use little power and are inexpensive.

(iv) Positioning technologies

The global positioning system (GPS) and geographic information systems (GIS) enable the simultaneous gathering of real-time data and precise location information, leading to the effective mapping and analysis of vast quantities of geospatial data. For farm planning, field mapping, soil sampling, tractor navigation, crop scouting, variable rate applications, and yield mapping, precision agriculture employs GPS-based applications. GPS receivers gather position data to map field borders, roadways, irrigation systems and agricultural trouble areas such as weeds and disease. GPS precision enables farmers to produce farm maps with specific acreage for field regions, road locations and distances to objects of interest.

GPS helps farmers to travel accurately to particular spots in the field to gather soil samples year after year. In agriculture, satellites are utilized in several ways, first to estimate crop yields. Optical and radar sensors can offer an accurate depiction of agricultural areas, as well as differentiate between crop varieties and assess crop health and maturity. This information helps to educate the market and provides early warning of crop failure or famine. In precision agriculture, satellite photographs are used to precisely define a farmer's fields, sometimes in combination with geographical information systems (GIS), in order to provide more intense and effective farming operations.

NDVI (normalized difference vegetation index) is a typical agricultural remote sensing metric that quantifies how much more near-infrared light is reflected than visible red. The rationale behind NDVI is that leaves reflect a great deal of near-infrared light (NIR). When a plant is dehydrated or under stress, its leaves reflect less NIR light but the same quantity of visible light. This facilitates distant plant monitoring.

(v) Hardware and software systems

Recent technology advancements have led to the creation of a number of the robots mentioned below. Several advances in agricultural robotics depend on machine vision technology for avoiding risks, identifying crops and determining when they are ready to be harvested.

Typically, a camera or numerous cameras are utilized to provide information to the robot, enabling it to find and access nearby crops. Using machine vision, robots can now conduct jobs such as weed picking, growth monitoring, harvesting, sorting and packaging.

(vi) Data analytics solutions

Farmers assess harvest yields, fertilizer demand, cost savings and even future crop optimization plans using data. Smart farming incorporates data analytics technologies to gather information from diverse farming practices to generate algorithms that can be utilized by various farms to provide productive and sustainable crop yields. With this information, farmers will be able to more accurately estimate agricultural activities and use strategies that are not only healthier for their crops but also more ecologically beneficial.

The capacity to extract insights from data, develop algorithms and build new technologies continues to expand at a dizzying rate. The mix of shared knowledge, intelligent technology and audacious innovation has the potential to enable the agriculture sector to accomplish extraordinary achievements. To obtain a competitive edge via data analytics, it is essential to evaluate the project's strategy and long-term objectives. Before establishing a data analytics strategy, an agribusiness must have in place the following tools:

- **Collect data**. This enables you to collect data from reliable, carefully chosen sources and store it in a single, safe place to streamline processes.
- **Standardize data**. The capacity to merge various datasets into a single data structure enables real-time comparisons, trend monitoring and the finding of patterns in the data, which assists in the identification of new possibilities.
- **Clean data**. Ensuring that your data is clean, accurate and thorough will still trust in your decision-making abilities.
- **Enrich data**. Connectivity to external information, such as meteorological data, local soil analytics and insect monitoring.
- **Analyze data**. The capacity to analyze data is crucial for deriving value from the collected information. Determine the available tools for building analytics and verify that they support the desired results. Once the required technology and communication tools have been implemented, it is time to consider business strategy. These are the most important measures to take before.
    - Comprehend how the data analytics platform will help the entire business plan of the corporation.
    - Establish a vision for analytics and maturity targets for essential processes.
    - Prioritize and develop a strategic road plan with both short- and long-term goals.
    - Create a plan for the ultimate architectural objective.

6.2.2 Benefits of smart agriculture

Increasing production control improves cost management and reduces waste. The capacity to identify abnormalities in crop growth or animal health, for instance, decreases the risk of product loss. Moreover, automation increases productivity [2].

6.3 Machine Learning

Machine learning (ML), a subfield of artificial intelligence (AI), is the theory that a computer program can pick up new information and adjust to it without the help of a person. A computer's complicated algorithm or source code enables the device to recognize data and generate predictions based on that data. To help with decision-making, ML is effective for sorting the vast amount of information that is consistently and easily accessible in the world. ML can be applied to a wide range of industries, including lending, news reporting, investment, advertising and more [7].

In other terms, machine learning (ML) is an application of AI that enables systems to learn from their own experiences and develop without being explicitly programmed. Making computer programs that can access data and utilize it to learn on their own is the goal of machine learning [14].

Starting with observations or data, such as real-world examples, personal experience or instructions, machine learning can be initiated. It looks for patterns in the data to conclude from the given examples. The main objective of machine learning is to make it possible for computers to learn on their own, without aid from humans, and to modify their behavior accordingly [14].

6.3.1 Approaches of machine learning

Based on the type of input or feedback that the learning system has access to, traditional machine learning algorithms can be divided into three major types [4]. They receive supervised instruction.

6.3.1.1 Supervised learning

Algorithms for supervised machine learning make predictions about the future by utilizing tagged examples to apply lessons from the past to fresh data. By examining a known training dataset, the learning algorithm creates an inferred function to forecast output values. The system may provide targets for any new input once it has received enough training. Additionally, to identify faults and correct the model, it can compare its output to that which is intended [14].

6.3.1.2 Unsupervised learning

Unsupervised machine learning techniques are employed when the data used to train is neither categorized nor tagged. Unsupervised learning studies how systems can infer a function to describe a hidden structure from unlabeled data. The system never has a certain understanding of the proper output. Instead, it uses datasets to infer what the outcome should be [14].

6.3.1.3 Reinforcement learning

Algorithms used in reinforcement machine learning interact with their environment by taking actions and identifying rewards or errors. The most crucial components of reinforcement learning are delayed reward and trial-and-error search. With the help of this technique, robots and software agents may autonomously decide how to act in a situation to maximize their performance. The reinforcement signal, which is a straightforward reward signal, is necessary for the agent to learn which behavior is optimal [14].

6.3.1.4 Other approaches in machine learning

ML has the ability to apply additional approaches that were not frequently used in the existing ways, and occasionally the same machine learning system will use more than one approach [4].

6.3.1.5 Self-learning

Another strategy that applies the principles of machine learning is self-learning. With the aid of a neural network, it was originally presented in 1982. The crossbar adaptive array neural network has self-learning capabilities (CAA). It is a process of learning, and there are no outside incentives or counsel from others. The CAA self-learning algorithm was implemented as a crossbar, which limits the consideration of actions and emotions (feelings) decisions to scenarios having consequences. The system is determined by the way intellect and emotion interact [5].

6.3.1.6 Feature learning

A machine learning-based strategy is known as feature learning algorithms, often referred to as representation learning algorithms. As a pre-processing stage, this algorithm both safeguards and transforms the information in their input. In other words, pre-processing comes before categorization or prediction. Using feature learning methods, it is possible to recreate inputs from unknown data-generating distributions, demonstrating that the input is not real data being utilized for improbable configurations. This approach thus eliminates manual feature engineering to prevent erroneous input data, enables a machine to learn the features, and uses them to carry out a given task [6].

Feature learning is applicable to both supervised and unsupervised learning. In supervised feature learning, features are trained using labeled input data. Examples include supervised dictionary learning, multilayer perception and artificial neural networks. Features are learned using unsupervised feature learning from unlabeled input data. Examples include lexical learning, independent component analysis, autoencoders, matrix factorization and other kinds of clustering [6].

6.3.2 Machine learning in agriculture

Agriculture is utilizing machine learning principles to raise crop output and quality. The agriculture sectors where ML ideas are used are as follows [8]:

- Farmers are increasingly implementing ML concepts and improving them because they have faith in it. By implementing AI and ML concepts in the food industry to increase efficiency and production predictability.
- The farmers' business network is a social network created primarily for farmers to use for internal communication. For better pricing of the data, this network will make use of ML principles and analytical tools.
- Machine learning (ML) creates robots that are used to monitor the growth of crops.
- By utilizing ML ideas, sensors are used to collect data on crops.
- A recent study found that in order to meet human requirements, AI and machine learning are essential in the agricultural sector.

Smart farming is a newly emerging idea that uses highly precise algorithms to improve the resourcefulness and productivity of agriculture. Machine learning (ML) is a branch of science that gives computers the capacity to learn without being explicitly programmed. ML has partnered with big data technologies and high-performance computing to create new chances to unravel, compute and comprehend data-intensive processes in agricultural operating environments [12].

Every step of the growing and harvesting process involves ML. ML helps with soil preparation, seed breeding, water feed metering and seed planting. Robots will be used by ML to gather the harvest and use computer vision to assess the ripeness of the crops [12].

6.4 Applications of ML in Smart Agriculture

Artificial intelligence (AI) and machine learning (ML) have vast applications. Farmers may now access sophisticated data in big volumes and numerous analytics tools, which will help us achieve better outcomes, and superior efficiency, decrease food waste and ultimately minimize detrimental environmental repercussions.

The following are some of the innovative methods specialists in the agricultural business are employing ML.

6.4.1 Plant breeding

Plants, like people, contain genes that influence their appearance. Some plants have genes that enable them to absorb more soil water and nutrients than others. While certain plant genes are more successful at resisting illness than others. Some plant genes even influence how a plant tastes, smells, softens and so on. The whole agricultural economy focuses upon the production of successful seed products that combine the greatest qualities of numerous plant strains. Scientists would have to wait 10 years without the help of ML technology for a single hybrid development cycle. (However, it is still quicker than nature's pace of completion of the process.)

A vast quantity of data was utilized to analyze plant performance over time under diverse settings. Now, ML algorithms are effective for analyzing performance, giving improved optimization and identifying biotech behaviors necessary to maximize yields profit, given the probability of detrimental environmental elements in a certain season, such as poor weather conditions and insect populations. ML algorithm improvement may increase the stability of these immensely advantageous and expensive-to-create biotech behaviors by lowering resistance accumulation. ML may be used by scientists to predict which gene combinations would result in desired behaviors in new plants, giving an ideal starting point for plant development [9].

6.4.2 Species management

We learn about species identification, breeding and recognition in species management.

6.4.2.1 Species breeding

Because most plants have similar leaf compositions, color and forms, identifying them with the naked eye may be challenging. Farmers may now quickly differentiate between plants by analyzing complicated patterns and precisely recognizing related plant and weed species using machine learning. Farmers would considerably benefit from computerized plant species identification since it will save time and boost production in other crucial areas [9].

6.4.2.2 Species breeding

The process of selecting species takes time. Because we need to find particular genes that influence the efficacy of water and nutrients in that species, adaptability to climate change, disease resistance, nutritional content or a better favor of that specific species. ML algorithms are effective in species selection because they analyze crop performance in various climates and identify novel traits using decades of field data. Based on the ML data, they can build a probability model to forecast which genes are most likely to give a favorable attribute to a plant [12].

6.4.2.3 Species recognition

The human approach to species identification was to categorize plants exclusively based on their leaf color and form. ML can deliver more exact and quicker answers by analyzing the species leaf utilizing vein morphology, which includes more accurate information about the unique species leaf and its properties [12].

6.4.3 Field conditions management

We are adamant about the nature of soil and water in field management. In field management, we are adamant about the nature of soil and water.

6.4.3.1 Soil management

Soil is a valuable agricultural resource that comprises complicated processes and unclear mechanisms. The temperature of the soil alone will provide information about the impacts of climate change on local production. In agriculture, machine learning algorithms are utilized to determine evaporation processes, soil moisture, and temperature to better understand ecosystem dynamics [12].

ML is experienced in analyzing data on soil conditions, moisture levels in soil, soil temperature, soil compounds, and the influence of soil on crop development and animal well-being [9].

6.4.3.2 Water management

Agriculture relies heavily on water management. Hydrological, climatological and agronomical balances are determined by water management. The most modern ML applications are used to determine water level evapotranspiration on a daily, weekly or monthly basis, allowing for more disciplined irrigation system usage. The ML applications then forecast the daily dew point temperature, which assists in predicting weather and measuring evapotranspiration and evaporation [12]. ML may be used to analyze precipitation and evapotranspiration. It is the process by which water travels from the soil and via plant transpiration to the atmosphere, following which technologists develop ML applications for better resource management and irrigation systems [9].

6.4.4 Crop management

Crop management teaches us about yield production, crop quality and crop assessment.

6.4.4.1 Prediction of yield

Agriculture's final outcome is yield. Only by precisely estimating yield can supply and demand be balanced. Precision agriculture is a novel machine learning paradigm that will define yield mapping and estimating, crop supply and demand matching, and crop management. State-of-the-art ML techniques for prediction have been deployed, using computer vision technologies to give a data-like multidimensional analysis of crops, weather, and economic situations to maximize production for farmers [12].

6.4.4.2 Crop quality

In crop management, we must determine crop quality as an important factor. We must determine the different crop quality levels by precise detection of all crops and categorization. The uniqueness of crop quality will only boost product pricing and minimize waste. In contrast to human specialists, ML approaches will conduct more accurate identification and classification of crops from apparently meaningless data, as well as interconnect with other crops to expose and discover new features playing a part in the overall quality of the crops [12].

6.4.4.3 Quality assessment

Farmers may utilize ML technology to determine which crops are harvestable and which are not on a particular day. By enrolling into a customized dashboard on a computer or tablet, the farmer may instantly receive an accurate evaluation. Harvestable crop weight and maturity level may also be measured and anticipated. Image analysis methods are used to assess crops both before and after harvest for the presence of desirable characteristics, the level of damage, nutritional content, and other aspects that may impact the final viable yield and product pricing [9].

6.4.5 Disease management

We learn about crop disease identification and weed detection in disease management.

6.4.5.1 Detection of crop diseases

Crops are required to guard against pests and illnesses. So we will spray insecticides consistently across the cropping area, either in the open or in the

Figure 6.2 Identify crop disease detection.

greenhouse. Furthermore, for success, this strategy requires a large volume of pesticides to be sprayed, which is costly for farmers and has an impact on the environment. ML implements precision agricultural management, in which agrochemical inputs are focused specifically on afflicted plants, and the time necessary to spray pesticides is reduced. As a result, it will benefit farmers [12] (Figure 6.2).

ML applications may be used to analyze crops for disease detection, offer an accurate disease diagnosis, and propose the best treatment method. ML approaches assist farmers in recognizing agricultural diseases in precise plants and locations, allowing the farmer to avoid treating the whole farm. If the treatment is implemented on all farms, it may have unforeseen implications such as polluting the environment or reducing bee populations. ML helps businesses produce crop disease treatment solutions more efficiently and provide better service to their clients [9].

6.4.5.2 Detection of weeds

Weed is a plant similar to a crop, but its development is unappreciated and harms the original plant. Weeds in crops are difficult to identify and eliminate. Weeds are so dangerous risk to agricultural output. Weeds are detected and differentiated using computer vision and machine learning techniques at a reasonable cost and with no environmental hazards or negative effects. ML algorithms are being used to create robots that will eliminate weeds, decreasing the future demand for pesticides [12]. Farmers may employ machine learning-based image processing algorithms to detect weed species and assess which crops are disease-free and which are afflicted with fungus, bacteria or viruses. The capacity to categorize weeds using digital technologies enables farmers to command mechanical machines to clear weeds from fields, sparing the environment from the harmful effects of pesticide usage while also saving farmers time, labor, and money [9].

6.4.6 Livestock management

The production of livestock, the upkeep of livestock health, the herding of animals and their overall well-being are all topics that are covered in livestock management.

6.4.6.1 Livestock production

The ML allows exact forecasting and evaluation of many agricultural factors in order to improve the economic efficiency of livestock production systems such as cattle and egg production. Weight prediction systems utilized by ML can forecast future weights roughly 100 days before slaughter, allowing farmers to adjust diets and circumstances accordingly [12].

6.4.6.2 Animal welfare

Because of the inefficiencies of their living conditions, the animals in livestock today do not have liberty or happiness. They are just used as food containers. Classifiers in machine learning are used to recognize animal behaviors such as chewing signals to anticipate diet changes and movement patterns such as standing, moving, eating and drinking. The ML will evaluate and inform on the animal's stress, illnesses, weight increase, and productivity [12].

6.4.6.3 Upkeep of livestock health

ML assists farmers in maintaining a happy and healthy herd of farm animals. We can learn about it from the examples below. In one case, an organization employed sensors to monitor cow behavior. Using machine learning, the gathered data will be used to forecast reproductive trends, identify eating habits and diseases, and warn farmers of indicators of heat stress. In another scenario, a cow's health may be determined by applying deep learning algorithms to photos of white blood cells collected from the cow's blood or milk. Before a basic checkup, these methods reveal signs of major health conditions. Farmers may then undertake proper health measures prior to the need for antibiotics [9].

6.4.6.4 Animal herding

Autonomous robots are utilized at several phases, such as monitoring agricultural fields, detecting weeds between crops and identifying damaged plants before being taught to group farm animals. It also assists in physically herding animals on their route to a desired goal; these ML-powered gadgets may pull large goods from one spot to another and work with drones to convey important information to farmers [9].

6.4.6.5 Selective breeding

Selective breeding is a method that uses genetic data optimization to support the preservation of desirable attributes such as milk quality, disease resistance, fertility and others in cattle. Ranchers have practiced selective breeding since the beginning of time. Ranchers have anticipated certain monitoring criteria to generate cattle lines with desired qualities for generations. With the use of ML technologies like sensors and apps, we can investigate genetic molecular markers, environment, feed mix, and birth trends utilizing enormous quantities of data, and ranchers may formulate livestock-related choices with considerably more accurate findings [9].

6.4.7 Ranching

Farmers that deal with animals may utilize machine learning to save ranching time and expenses. ML technology helps agricultural and ranching activities like livestock health care, dairy and egg production, animal herding, and selective breeding [9].

6.4.7.1 Production of dairy products and eggs

Machine learning data analysis will aid ranchers by providing more accurate and efficient outcomes. They are in charge of producing milk, eggs and other foods. Data on dairy cow activity is collected and analyzed using machine learning technologies. It also gives farmers the ability to make operational choices that boost milk output by up to 40% [9].

Ranchers rely on machine learning-enhanced technologies to:

- Examine cattle behavior trends and give suggestions for better care and feeding.
- Predict illness and administer suitable remedies.
- Improve milk and egg production.
- Manage farm animals such as cattle and sheep.
- Analyze and disseminate farm-condition data.
- Predict reproductive trends to improve breeding choices and generate hardier cattle.

- Boost productivity, save time, and operate more smoothly [9].
- Carry out physical work.

6.4.8 Agrochemical production and application

Predictive analysis is the main strength of ML technology. ML technology is largely used to make pesticides, crop disease remedies, antibiotics, microbials and other chemical and biochemical goods for farmers and ranchers. This technology will assist them in manufacturing the product as efficiently as possible while also having no negative impact on the environment. Furthermore, machine learning simplifies and rationalizes activities like crop disease detection, diagnosis, and treatment [9].

Farmers will be able to use a smartphone application to learn about different plant diseases and how to diagnose them. Again, the ML mobile application is utilized for digital activities such as determining when to spray pesticides on crops and identifying afflicted plants on the farm without disturbing the nearby, healthy or unaffected plants. Spray drones are trained using ML technology to spray pesticides selectively on the plants that need treatment. ML can detect not just plant diseases but also other dangers to crops and animals.

To detect additional risks, ML applies an image analysis algorithm, and pest control firms equip their personnel with a trustworthy, real-time tool for recognizing bugs, enabling them to deliver focused extermination services [9].

6.4.9 Remote weather monitoring

Conditions of the weather and the natural environment are the most challenging aspects for farmers. Farmers will be able to familiarize themselves with different meteorological variables such as temperature, rainfall, humidity, moisture and specific chemical composition to maximize yields throughout the year. This is done to increase crop production. Using ML approaches, researchers have found several customizable sensors that may aid farmers in monitoring the state of their crops as well as the weather from a distant place. Because of these sensors, the amount of time that farmers need to spend physically present on their farms will be cut down. The use of sensors may, in some respects, double productivity by enabling farmers to increase agricultural production while expending less labor [11].

6.4.10 Farmer's little helper

This is a mobile device application that may be considered a bonus. For example, a farmer may remain at home and consider what steps to do next in the management of his harvested crops. He is unsure whether to sell the crops to a local producer or a regional enterprise. He wants some advice on the numerous possibilities available before making a final selection. Industries are currently focusing on building customized chatbots to support him with this objective. Chatbots will be able to converse with farmers and offer them useful data and insights. Farmers' chatbots are predicted to be wiser than consumer-oriented Alexa and similar assistants since they will be able to not only provide numbers but also evaluate them and advise farmers on difficult issues [12].

6.5 Conclusion

Agriculture is an essential skill in this world. Agriculture is the source of nourishment for all living beings. People used to spend much of their time looking for food. Following that, people found crops and were able to produce them, resulting in excess food that was exported from one region to another. Population expansion and the devastation of agricultural areas for human existence have raised food demand. Land degradation occurs as a consequence of climate change and the growth of numerous businesses. To prevent the issue of hunger, we must conserve and enhance food supplies using cutting-edge technology. One such technology that may assist us in overcoming all of these food shortage challenges is machine learning. Smart agriculture is now taking place with the use of machine learning, which will assist farmers in doing their chores more effectively and producing more food crops and other products. This chapter explains how machine learning may assist farmers enhance production while also providing precise and predetermined responses.

References

[1] J. Angel Ida Chellam, M. Siva Sangari and R. Rajendra Kumar, "Drought Prediction Using AI Based Multiple Linear Regression Technique", *Journal of Cardiovascular Disease Research*, vol. 12, pp. 305–310, 2021.

[2] C. Encinas, E. Ruiz, J. Cortez and A. Espinoza, "Design and implementation of a distributed IoT system for the monitoring of water quality

in aquaculture", *2017 Wireless Telecommunications Symposium (WTS)*, pp. 1–7, 2017.

[3] G.K. Ganjegunte, Z. Sheng and J.A. Clark, "Evaluating the accuracy of soil water sensors for irrigation scheduling to conserve freshwater", *Appl Water Sci*, vol. 2, pp. 119–125, 2012.

[4] A.Gupta, S. Mishra, N. Bokde and K. Kulat, "Need of Smart Water Systems In India", *International Journal of Applied Engineering Research*, vol. 11, no. 4, pp. 2216–2223, 2016.

[5] S. Das and S. Kar, "Group decision making in the medical system: an intuitionistic fuzzy soft set approach", *Appl. SoftComput*, vol. 24, pp. 196–211, 2014.

[6] D. Stanujkic, B. Djordjevic and M. Djordjevic, "Comparative analysis of some prominent MCDM methods: a case of ranking", *Serbian banks. Serbian J Manag*, vol. 8, no. 2, pp. 213–241, 2013.

[7] A. De Gelder, J.A. Dieleman, G.P.A. Bot and L.F.M. Marcelis, "An Overview of Climate and Crop Yield in Closed Greenhouses", *J. Hortic. Sci. Biotechnol*, vol. 87, pp. 193–202, 2012.

[8] M.K. Gupta, P. Chandra, D.V.K. Samuel, B. Singh, A. Singh and M.K. Garg, "Modeling of Tomato Seedling Growth in Greenhouse", *Agric. Res*, vol. 1, pp. 362–369, 2012.

[9] https://objectcomputing.com/expertise/machine-learning/machine-learning-in-agriculture

[10] M. Azaza, K. Echaieb, F. Tadeo, E. Fabrizio, A. Iqbal and A. Mami, "Fuzzy Decoupling Control of Greenhouse Climate", *Arab. J. Sci. Eng*, vol. 40, pp. 2805–2812, 2015.

[11] https://readwrite.com/smart-farming-solutions-how-iot-transforms-agriculture-sector/

[12] https://www.kdnuggets.com/2019/05/machine-learning-agriculture-applications techniques.html

[13] R. Ben Ali, S. Bouadila and A. Mami, "Experimental Validation of the Dynamic Thermal Behavior of Two Types of Agricultural Greenhouses in the Mediterranean Context", *Renew. Energy*, vol. 147, pp. 118–129, 2020.

[14] M.H. Liang, L.J. Chen, Y.F. He and S.F. Du, "Greenhouse Temperature Predictive Control for Energy Saving Using Switch Actuators", *IFAC PapersOnLine*, vol. 51, pp. 747–751, 2018.

7

Deep Learning in Smart Agriculture

K. Nisha[1], C. Augustine[2], A. Jose Anand[3], and M. Kamesh[4]

[1]Department of CSE, Chennai Institute of Technology, India
[2]Department of ECE, GRT Institute of Engineering & Technology, India
[3]Department of ECE, KCG College of Technology, India
[4]Department of ECE, Velammal Institute of Technology, India
Email: nishak.mecse@gmail.com; austynineeleven@gmail.com; joseanandme@yahoo.co.in; kameshm90@gmail.com

Abstract

Deep learning (DL) is an innovative technology of machine learning (ML); it is a subset of artificial intelligence (AI). Since neural networks (NN) mimic the activities of the human brain, DL will perform the same task. In DL, nothing is explicitly programmed. It is an ML class that uses many nonlinear processing units to perform feature extraction as well as transformation. The production of every preceding layer is engaged as the input of every succeeding layer. DL models are capable enough to focus on the exact features themselves that require less guidance from the programmer and are very helpful in solving dimensional problems. Smart farming in agriculture is a new idea that makes agriculture a productive and structured method that inherits the new technologies. Nowadays, DL involves various research activities that are used to help the farmers to lessen the fatalities in farming. This technology is used to predict which crop is suitable for which weather conditions. Right now, DL, PC vision, picture handling, advanced mechanics, and Internet of Things (IoT) innovations are extremely useful to farmers. Computer-based intelligence controlled drone innovation is extremely valuable for horticulture as it gives top notch pictures to make it more straightforward to screen, examine, and break down crops. This method is helpful for deciding the advancement of a harvest. Additionally, the farmer can conclude regardless of whether the harvest is ready for yield. There is no restriction to the portrayal of DL involvement in agribusiness. The study examines typical agricultural uses of deep learning. DL algorithms of convolutional neural networks (CNNs), recurrent neural networks (RNNs), long short-term memory (LSTM) networks, and generative adversarial networks (GANs), radial basis function networks (RBFNs), multi-layer perceptrons (MLPs), self-organizing map (SOM), deep belief networks (DBNs), and restricted Boltzmann machines (RBMs) have been extensively premeditated and are useful in numerous areas of agriculture. Specifically, the applications of CNN-based supervised learning (CNN-SL) and transfer learning advance the development of modern agriculture.

7.1 Introduction

Agriculture guarantees food security and is therefore the backbone of the country. It plays an important role in most foreign trade in the country. In most parts of the world, about 75% of the population depends on agriculture for their livelihoods. Population growth requires higher yields in the agricultural sector and better agricultural conditions. Farmers are looking for effective ways to increase crop production at lower cost and more efficiently using available resources. This is contributing to the new use of digital technology in the agricultural sector to help farmers make better decisions and increase yields. Today, we use in-depth learning methods to overcome various problems and challenges in the agricultural sector. With the growing need to increase food production efficiency, technology can provide and perform farm operations to increase yields [1]. Modern agriculture can collect and process data to automate the entire farming process, from land preparation and planting to post-harvest harvesting. It means that large amounts of agricultural/farm data can be analyzed using new information and communication technologies. This chapter employs precise agricultural technology in the form of an unmanned aerial vehicle equipped with optic and radiometric sensors to capture the most accurate images of the field environment during the normal production/growth cycle. Then, an in-depth study of the convolution neural network is used for image training to develop a multi-class organization scheme to regulate the complete health prominence of the plant. Using a pre-trained new images has shown that the model can accurately forecast any plant disorder with an accuracy rate of 99% [2].

7.1.1 Applications of DL algorithms

Incorporated with other approaches such as support vector machine (SVM) and random forest, in-depth learning models anticipate the yield of soybean crops. This work uses an in-depth study method to predict crop yields. First, they created a soybean yield prediction model where the available data is sufficient for the purpose. They use transmission erudition to forecast the yield of soybean crops when data is restricted. They have demonstrated the cost-effectiveness of remote sensing data and the usage of knowledge transmissions to effectively forecast crop yields to extents where few data are obtainable. How to forecast climate indicators to identify the effects of drought on agriculture? This learning uses a synthetic NN to predict the normal indication of plant differences using the normalized difference vegetation index (NDVI). The foretold NDVI based on satellite indications was used to predict agricultural drought and its upcoming impacts. This prototype works best in desert environments. The results helped make informed decisions about plants and livestock [3].

Their forecasts are built on the local construction of the farm and the quantity of nitrogen manure used during the mounting period. The convolutional neural network has been developed to capture the local properties of the arena characteristics and show the crop yield reaction to nutrient control and seed proportion. Prototypes are qualified from a database shaped from nine cornfields. Subsequently, over-treatment of nitrogen fertilizers and nutrients can diminish crop yields and crop excellence, and consequences are recycled to comprehend and regulate nutrient utilization. Automatic analysis platform measuring yield-related phenotypes for the production of lettuce from high-resolution aerial imagery. They recycled ultra-scale NDVI airborne photographs to extent yield-related phenotypes. Leading, a copy construct on the assembly of the CNN was developed to calculate lettuce. At that time, an unlocked ML process is recycled to divide the acquired lettuce into three dimensions [4]. Lastly, a global positioning system (GPS) map on behalf of harvest areas created on the size of the lettuce size was produced. They have shown this by performing real-time crop observation throughout the various phases of growth that allow for improved crop management under changing agricultural conditions to increase productivity. Importantly, mask R-CNN (mask region-based convolutional neural network) is the newest expansion in R-CNN household of article finders and includes a segmentation system and a discovery system. The R-CNN mask can be end-to-end to keep it unswervingly from distant hearing aids, but improved presentation and quicker conversion can be attained through transfer knowledge. In transmitting knowledge, the NN begins training in a different database such as COCO and ImageNet to study the simple topographies of an image. While preparing a remote data sensor network, techniques such as weight loss or opt-out controls can be recycled to ensure that the system has the lowest quantity of requirements for remote sensing images [5].

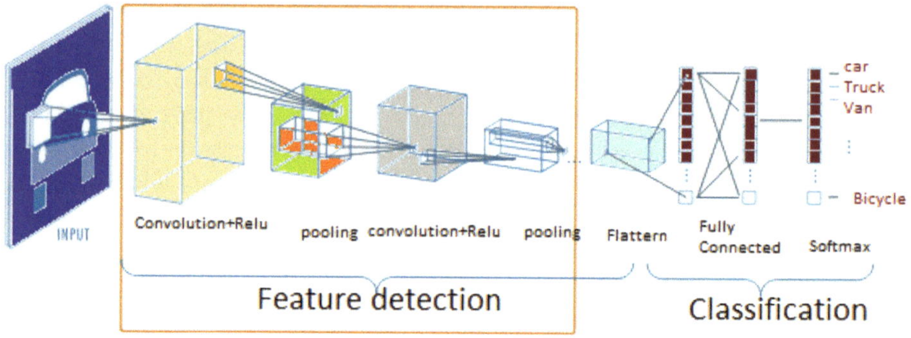

Figure 7.1 CNN structure.

7.2 Deep Learning Algorithm

The following are the various DL algorithms.

7.2.1 Convolutional neural networks

Deep NN operates in parallel, combining multiple layers of nonlinear processing with simple elements derived from the biological nervous system. It has an input layer, numerous hidden layers, and an output layer. Layers are linked together by nodes or neurons, and every hidden layer receives the previous layer's output as input. CNN and ConvNet are two of the most common DL algorithms. A CNN, like other NNs, has an input layer, an output layer, and many hidden layers in between [6]. Figure 7.1 shows the process flow of the CNN architecture.

Feature detection layers: These layers operate on the data in one of three ways: convolution, pooling, or a rectified linear unit (ReLU). Convolution applies a series of convolution filters to the input image. Each filter activates a different aspect of the image. Pooling diminishes the number of limits the network must learn by simplifying the output with nonlinear downsampling. For faster and more effective training, the ReLU converts negative standards to zeros and keeps positive standards. These three processes are recurring hundreds or thousands of times, and each layer learns to recognize different functions [7].

Classification layer: The CNN architecture then moves on to classification after detecting the feature. The completely associated layer, which yields a K-layered vector, is the penultimate layer. K addresses the quantity of classes that the organization can anticipate. This vector contains the order

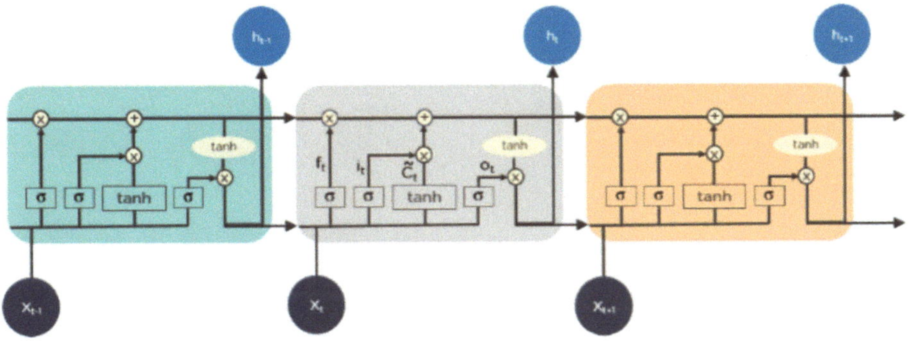

Figure 7.2 LSTM architecture.

probabilities for each picture class. The softmax task is utilized in the closing layer of the CNN design to give characterization yield [8].

7.2.2 Long short-term memory networks

LSTMs can be characterized as RNNs that are modified to learn after some time and adjust to conditions. Past data can be stored and retrieved for an extensive period, which is the only behavior by default. LSTMs are designed for long-term storage and can limit memory or previous inputs; so they will be used primarily in time series forecasting in the future. This analogy derives from a chain-like structure consisting of four interacting layers that communicate in different ways. In addition to time series prediction applications, it can also be used to build speech recognition, pharmacy development, and music loop creation. LSTMs operate on a series of events. First, they tend not to remember the irrelevant details they achieved in their previous state. It then selectively updates a particular cell state value and finally produces a particular part of the cell state as output. Following is a diagram of their behavior [9]. The structure of LSTM is shown in Figure 7.2.

7.2.3 Recurrent neural networks

RNN consists of several directed connections that form a cycle that allows the input provided by the LSTM to be used as an input in the current phase of the RNN. Since these inputs are deeply embedded as inputs, we remember the LSTMs so that they are absorbed into the internal memory for some time. Therefore, the RNN depends on the input held by the LSTM and operates under the LSTM synchronization phenomenon. RNNs are primarily

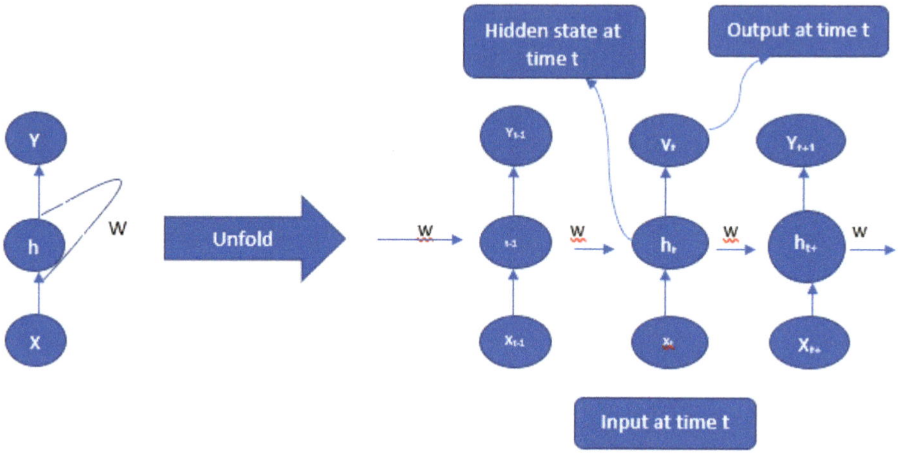

Figure 7.3 RNN architecture.

used for image captioning, time series analysis, handwritten data recognition, and machine translation of data. The RNN follows a working approach by adjusting the timing of the output feed ($t1$) when the time is defined as t. The output determined by t is then fed at input time $t + 1$. Similarly, these processes are repeated for all inputs of any length. There is also the fact that the RNN stores historical information and the input size does not increase even if the model size is increased [10]. Figure 7.3 shows the architecture of RNN.

7.2.4 Generative adversarial networks

GANs are described as deep learning algorithms that are used to gener¬ate new information for the process. GAN commonly includes additives specifically a generator that learns to generate fake information and a discriminator that adapts itself with the aid of using studying from this fake information. Over a few times, GANs have received massive utilization considering that they are regularly getting used to making clear astronomical pictures and simulated lensing of the gravitational darkish matter. It is likewise utilized in video games to boom snapshots for 2D textures with the aid of recreating them in a better decision like 4K. They also are utilized in developing practical cartoons individual and additionally rendering human faces and 3D item rendering. GAN appears in simulation with the aid of producing and knowledge of the faux and actual information. During the process, to apprehend that information,

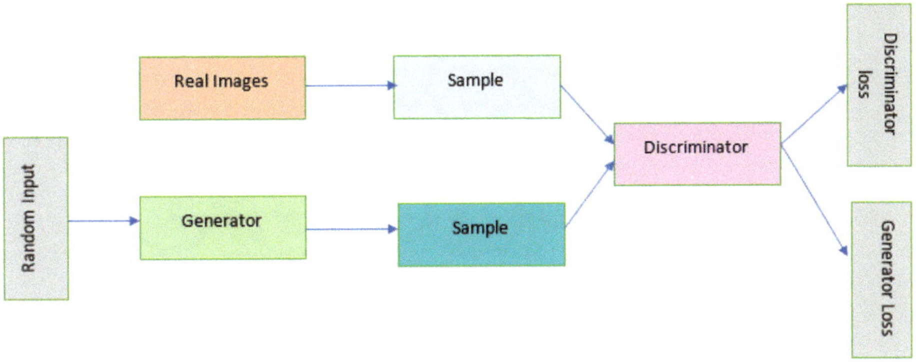

Figure 7.4 GAN architecture.

the generator produces one-of-a-kind types of faux information wherein the discriminator fast learns to conform and reply to it as fake information. GANs then ship those identified outcomes for updating. Consider the following picture to visualize the functioning [11]. Figure 7.4 shows the architecture of GAN.

7.2.5 Radial basis function networks

RBFNs are particular sorts of NN that observe a feed-ahead technique and employ radial features as activation features. They encompass three layers, specifically the enter layer, hidden layer, and output layer, which might be on the whole used for time-collection prediction, regression testing, and classification. RBFNs do those obligations with the aid of measuring the similarity gifts within the education information set. They normally have an enter vector that feeds that information into the enter layer, thereby confirming the identity and rolling out consequences with the aid of using evaluating preceding information sets. Precisely, the enter layer has neurons that can be touchy with that information, and the nodes within the layer are green in classifying the magnificence of information. Neurons are initially gifted within the hidden layer even though their paintings are in near integration with the enter layer. The hidden layer includes Gaussian switch features that can be inversely proportional to the space of the output from the neuron's center. The output layer has linear information wherein the Gaussian features are exceeded within the neuron as parameters and output are generated. Consider the following figure to recognize the manner thoroughly [12]. Figure 7.5 shows the architecture of RBFN.

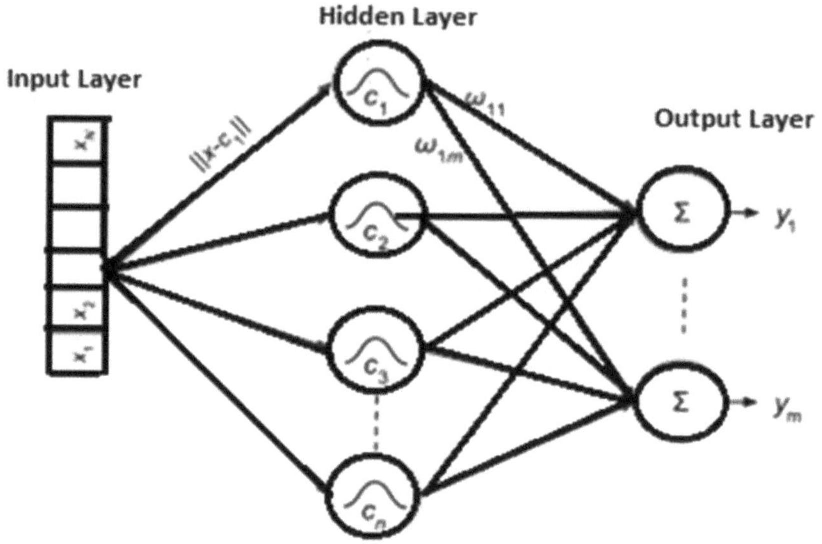

Figure 7.5 RBFN architecture.

7.2.6 Multi-layer perceptrons

MLPs are the foundation of DL innovation. It has a place with a meeting of feedforward NN having different stores of perceptrons. These discernments have different impelling errands in them. MLPs likewise have associated info and result layers and their number is something very similar. Likewise, there is a layer that stays concealed in the midst of these two layers. MLPs are for the most part used to fabricate picture and discourse acknowledgment frameworks or a few different sorts of interpretation programming. The working of MLPs begins by taking care of the information to the info level. The neurons present in the sheet structure a chart to lay out an association that passes in a single bearing. The heaviness of this info information is found to exist between the secret layer and the information layer. MLPs use actuation capabilities to figure out which hubs are prepared to fire. These initiation capabilities incorporate the tanh capability, sigmoid, and ReLUs. MLPs are for the most part used to prepare the models to comprehend what sort of connection the layers are accomplishing the ideal result from the given informational index. See the following figure to see better [13]. Figure 7.6 shows the design of MLP.

7.2.7 Self-organizing maps (SOMs)

SOMs were imagined by Teuvo Kohonen for accomplishing information perception to figure out the elements of information through counterfeit and

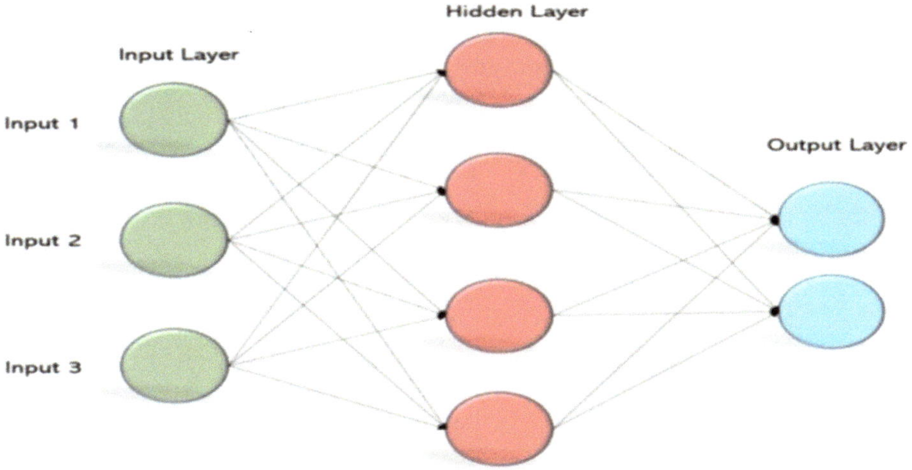

Figure 7.6 MLP architecture.

self-sorting out NN. The endeavors to accomplish information representation to take care of issues are principally finished by what people cannot envision. These information are for the most part high layered; so there are lesser possibilities of human contribution and cause less mistake. SOMs help in imagining the information by introducing the loads of various hubs and afterward picking irregular vectors from the given preparation information. They inspect every hub to find the relative loads; so conditions can be perceived.

The winning node is chosen and that is called best matching unit (BMU). Afterward, SOMs find these winning nodes; however, the nodes lessen over the long haul from the example vector. Subsequently, the nearer node is to the BMU, the almost certain it is to recognize the weight and take greater action. Different emphases are additionally performed to guarantee that hubs near the BMU are not lost. This example is a combination of RGB colors used in everyday work. Consider the following images to understand how they work. [14]. Figure 7.7 shows the architecture of SOM.

7.2.8 Deep belief networks

DBNs are called generative models since they have different layers of dormant as well as stochastic factors. The idle variable is known as a secret unit since they have twofold qualities. DBNs are likewise called Boltzmann machines in light of the fact that the RGM layers are stacked over one another to lay out correspondence with past and successive layers. DBNs are utilized in applications like video and picture acknowledgment as well as in catching profound articles. DBN is upheld by a voracious calculation. The

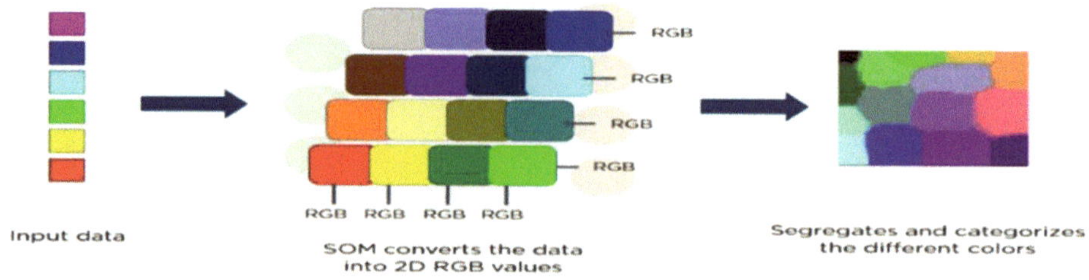

Figure 7.7 SOM architecture.

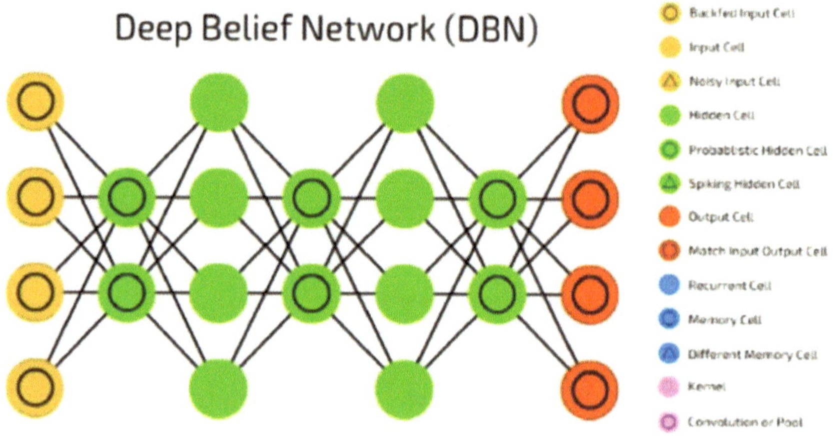

Figure 7.8 DBN architecture.

layer-to-layer approach, which utilizes a hierarchical way to deal with produce loads, is the most well-known way DBN works. DBN utilizes a bit-by-bit way to deal with Gibbs inspecting on top of two secret layers.

These stages then use a model that follows the sampling method of their ancestors to extract samples from the units displayed. DBN learns from the values present in the potential values of each layer by following a bottom-up path approach [15]. Figure 7.8 shows the architecture of DBN.

7.2.9 Restricted Boltzmann machines

DBMs gain from the qualities present in the idle worth from each layer following the base-up pass approach as shown in Figure 7.9. Restricted Boltzmann Machines (RBMs): RBMs were created by Geoffrey Hinton and look like stochastic brain networks that gain from the likelihood dissemination in the given info set. This calculation is basically utilized in the field of

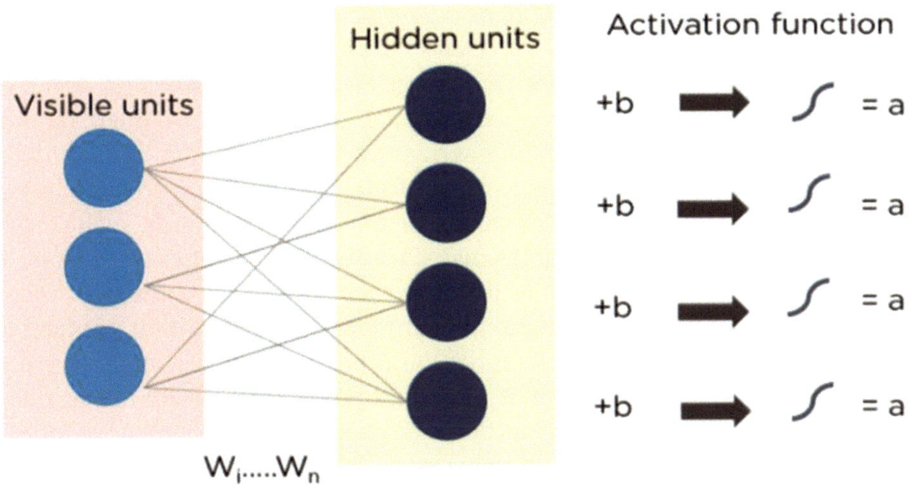

Figure 7.9 RBM architecture.

aspect decrease, relapse and order, and subject demonstrating and is viewed as the structure blocks of DBNs. RBIs comprise two layers, specifically the noticeable layer and the secret layer. Both of these layers are associated with secret units and have predisposition units associated with hubs that produce the result. Normally, RBMs have two stages, in particular forward pass and in reverse pass. The working of RBMs is completed by tolerating inputs and making an interpretation of them to numbers so that data sources are encoded in the forward pass. RBMs consider the heaviness of each and every info, and the regressive pass takes these information loads and makes an interpretation of them further into reproduced inputs. Afterward, both of these interpreted contributions, alongside individual loads, are joined. These information sources are then pushed to the noticeable layer where the initiation is done, and the yield is produced that can be handily recreated. To figure out this cycle, consider the underneath picture. Autoencoders are a unique kind of brain network where information sources are yields that are seen as generally indistinguishable. It was intended to basically take care of the issues connected with unsupervised learning.

7.3 DL Applications in Agriculture

The list identifies relevant activities, showing agricultural research area, specific problem, DL models and facilities used, data sources used, classes and data labels, pre-processing data and/or additions, and overall performance, achieved in terms of accepted metrics, as well as in comparison with other strategies, wherever available. Sixteen total sites, identified for weed identification and weed identification, crop identification, fruit counting, and crop classification, have been identified [17]. DL applications in smart farming have many possibilities, including agricultural info handling. Observing the condition of flora and fauna is essential for agricultural construction. Some plant and animal state variables cannot be unrushed straight. In such scenarios, different plant state variables may have some dependencies; so you can use DP to regulate uncountable info from the leisurely information. Plants and animals interact with environmental factors in agricultural production. This makes it easier to process agricultural information. Optimal control of agricultural production system. Agricultural production system management strategies are often based on the farmer's experience and the knowledge of experts who do not consider the physiological status of plants (animals) or real-time demands. This inevitably means that the strategy is not optimum. The benefit of DP is that it allows you to perfect complex schemes deprived of much reliance on information of the mechanism. Therefore, modeling an agricultural production system using DP enables the development of optimal control strategies using real-time measurements of the plant (animal) biological states and ancient data. Agricultural manufacturing contains many types of errands. These errands are frequently labor-intensive and the employed situation is very difficult. The use of DP to impersonate human performance and powered agricultural apparatus has found widespread use in numerous areas of agriculture, includ¬ing sowing, reaping, and post-harvest dispensation. A robot that produces apples, for example, is extremely valuable. Background light, branches, and leaves can all significantly interfere with computer vision signals in natural conditions. The path of the robot arm to reach the target also needs to be optimized. The DP method is very useful for all these problems.

Agricultural monetary gadget management. Agriculture yield itself is not always sufficient for agriculture. Along with the cost and quality of agricultural products, there are numerous other factors to consider. Expect agriculture product fees to be significant. However, fees are affected by a variety of factors. In this case, DP could be used to make charge adjustments based on specific variables. Most DL applications in agriculture are classified into certain categories. It is required for pest control, harvesting robots, yield forecast, disaster observing, and other applications. Manual detection of plant diseases can be time-consuming. Fortunately, advances in artificial intelligence (AI) can detect plant diseases through image processing. Models for detecting plant diseases are primarily conducted on leaf image classification and pattern recognition. Using a new DL framework, we created a crop disease detection model. This model can detect 13 diverse kinds of plant ailments in healthy leaves and discriminate plant leaves from their surroundings. Another study on detecting plant diseases using DL found that overall accuracy that was possible could reach 95.8th or 100th training iteration and could be further fine-tuned to 96.3th improvement. The results are superior to manual detection. All of this demonstrated that DL performed admirably in detecting plant diseases. Crop classification and weed detection are especially important as many countries develop initiatives to build national agricultural monitoring network systems, which have an impact on agricultural automation. Because image recognition can be used to recognize many features of plants, CNNs have been widely used for weed identification and plant classification. A novel approach for weed identification and control that combines CNN and K-mean trait learning has been proposed. Weed detection's manual design capabilities can result in unstable detection results and poor feature extraction generalization. As a result of using DL and K-means pre-training, the recognition accuracy was 92.89%. AlexNet is one of the most widely used pre-trained CNN architectures for plant taxonomy. Experiment results at Istanbul Technical University show that ML processes are based on handcrafted structures in distinguishing phenomenological stages. Another study uses a Self-Organizing Map (SOM) to segment optical images and then restore misplaced content from images received from the satellite. The particular method includes a postprocessing way that includes multiple rectification methods and geospatial analysis based on the available information. The most important crops were classified with 85% accuracy. However, obstacles have hampered the application of CNNs to plant classification. Each pixel in a space-based synthetic aperture radar (SAR) image.

7.4 DL Information Retrieval Methods

This monograph gives an impression of generally speaking DL mythology and the implementations on different signaling and data regulation tasks. The demand parts are chosen in different gestures of the resulting three principles: comprehension of data; the product districts that have previously been changed over utilizing the productive utilization of DL innovation, like discourse affirmation and PC vision; and the application regions that can possibly be affected essentially by DL. We encountered DL and research and development including NLP and text dispensation, data recovery, and DL-approved multimodal data processing to perform multiple tasks. Later, in 2006, DL, or graded learning, has arisen as a novel extent of ML exploration. For numerous years, the methods established through DL exploration are impacting an extensive variety of gesture and info dispensation efforts inside the conventional and the novel broadened choices counting vital features of ML and AI; see overview articles added to the media coverage of such progress. In recent years, many workshops, tutorials, and conference sessions have been and are being conducted on DL and its requests to numerous signal and information dispensation areas [18].

The creators are effectively associated with DL research and arrange or give a few of the above events, instructional features, and publications. Specifically, we conducted teaching training in various situations and invited them to lectures on that point. Much of this monograph is based primarily on her information exercises and instructor materials. Commonly, the various high-level metaphors of DL are having the following two crucial features:

1. models that include numerous layers or phases of nonlinear evidence records;
2. techniques for learning feature depiction at sequentially advanced or extra nonconcrete layers, either supervised or unsupervised.

DL is at the crossroads of NN, AI, graphical modeling, optimization, pattern recognition, and signal processing research. The radically augmented chip processing capabilities (e.g., general-purpose graphical processing units or GPGPUs), the knowingly augmented size of data used for training, and modern improvements in ML and signal/evidence dispensation research are three critical causes for the acceptance of DL currently. These advancements have allowed DL procedures to successfully develop multifaceted, compositional nonlinear purposes, learn dispersed and graded characteristic representations, and effectively use labeled and unlabeled data [19].

7.5 Classes of DL Networks

As previously stated, DL denotes a fairly broad class of ML practices and structural design that use numerous layers of nonlinear facts treating that are graded everywhere. Based on the structure and concepts proposed for usage, e.g., synthesis/generation or recognition/classification, the majority of the work in this area can be roughly classified into three major classes. Information about the final class label is not accessible, but a deep unsupervised or generative knowledge system is used to intern high-level associations of practical or observable data for pattern research or synthetic resolution. Such deep systems are referred to in the literature as uncertain traits or expression learning. In generation mode, it is suggested to describe the general statistical distribution of detectable fact and its corresponding classes, which is accessible and stored as visible data. In the second case, applying Bayes' theorem to this type of generational network can turn it into a discriminatory network for learning [20].

1. Deep learning networks proposed for supervised learning unswervingly offer discriminative control for outline arrangement resolutions, frequently by distinguishing the subsequent deliveries of classes hardened on the visible statistics. For such supervised learning, end tag data is continually obtainable in direct or indirect procedures, known as discriminative deep networks.
2. Hybrid deep networks, in which the objective is discernment, and is regularly significantly aided by the results of unsupervised or procreative networks.

This is made possible by improving the optimization of DL networks within the group (2). The objective is achieved when supervised learning discriminative standards are useful to guess the limitations in some of the deep procreative or unsupervised deep networks in category (1) above. The term "hybrid" is used differently in the above than it is in the nonfiction, where it denotes mixture schemes for speech detection that feed the output likelihoods of an NN into an HMM. Deep discriminative prototypes (e.g., DNNs, RNNs, CNNs, etc.) and generative/unsupervised models (e.g., constrained Boltzmann machines or RBMs, deep belief networks or DBNs, deep Boltzmann machines (DBMs), regularized autoencoders, etc.) are commonly used ML techniques. This two-way characterization technique keeps on neglecting a vital tracking down in profound deep learning investigation: how regenerative or unsupervised learning reproductions can fundamentally work on the preparation of DNNs and other profound supervised knowledge copies through superior enhancement. Additionally, profound organizations for unaided learning may not be guaranteed to, should be probabilistic, or have the option to genuinely test from the model (e.g., customary autoencoders, inadequate coding organizations, and so on). We note here that later investigations have summed up the customary vilifying autoencoders so that they can be proficiently inspected from and consequently have become generative models. However, the conventional two-way organization has some important changes between DNN for both unsupervised and supervised learning. Comparing the two highly monitored models, such as DNN, is often useful for training and testing. Some important changes between deep networks for unsupervised and supervised learning. Comparing two deeply supervised models, DNN, they are often effective for training and testing, have elastic concepts, and are suitable for end-to-end training in complex schemas (for example, crazy propagation of beliefs, etc.). Deep unsupervised learning models, especially stochastic generative models, on the other hand, are easy to understand, integrate, configure, and deal with uncertainties but are naturally stubborn in the adoption and education of complex systems. These differences are used in this monograph because they are reserved for the projected three-way arrangement [21].

7.6 Deep Autoencoders

Deep autoencoders are a rare variety of DNN and its output vector has similar dimensions as the input vector. It is widely useful to study, explain, or efficiently encode innovative data in the input vector way on the secret layer. Note that the autoencoder is a nonlinear quality removal method that does not use class labels. Therefore, the extracted features are intended to store information and better represent it, rather than performing a classification task, but these two objectives may be correlated. By definition, an autoencoder has an input layer that embodies the input parameter vector of the invention, with multiple invisible stages representing the modified features, and the output session that competes with the input layer for modernization. If the amount of secreted stages exceeds one, the autoencoder is said to be deep. The measurement of the secreted stages is small or large compared with input parameters [22].

A back propagation variant has typically a stochastic slope descent technique, which is used to train an autoencoder. With back-propagation the systems can be trained with multiple hidden layers, to solve basic problems to be addressed. Pre-training every layer as a modest autoencoder solves the problem. This method was used to create a deep autoencoder that maps imageries to fast tiny binary code, also image-based retrieving systems, to encode the reports (called semantic hashing), added to predict exposure-like speech topographies [23].

At this point, consider a series of tasks to develop an autocoder that extracts duplicate voice code from raw voice spectrogram data in an informal way (that is, without voice class labels). The various binary cipher representations mined by this prototypical is used to recover speech info or as a bottleneck feature for speech appreciation. Figure 7.10 depicts a deep autoencoder system architecture that covers enclosing 256 frequency containers and 1, 3,

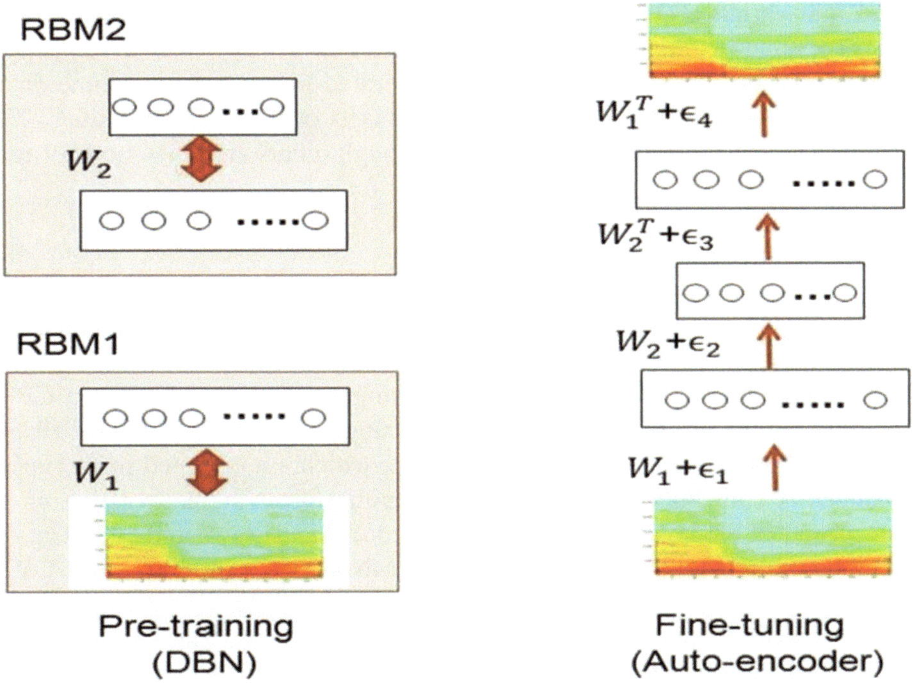

Figure 7.10 Deep autoencoder architecture to extract binary speech codes.

9, or 13 edges. A Gaussian–Bernoulli RBM is an undirected graphical representation with one noticeable A layer of linear variables with Gaussian noise and a single discrete layer of 500–3000 binary static variables. The creation likelihoods of its unseen elements are preserved as data for training additional Bernoulli–Bernoulli RBM behind learning the Gaussian–Bernoulli RBM. The two RBMs are self-possessed to form a deep belief net (DBN) and the situations of the second layer of binary secreted elements can be easily inferred from the contribution in a solitary pass. The DBN is shown on the left side of the figure, with the two RBMs in different boxes. The deep autoencoder is created among three secreted layers by "unrolling" the DBN with its own weight matrix. The matrices are used to encode the input in the inferior layers of deep autoencoder, and the conditions are used to decode the input in the upper stage. The deep autoencoder is tweaked to minimize reconstruction mistakes using error back-propagation, as shown on the right side of Figure 7.10. Following completion of knowledge, any variable-length spectrogram is prearranged and reconstructed. Each feature's first N successive overlap edges of 256-point log power ranges are regularized to zero-mean and unit-variance slantwise examples [24]. Perceiving the foundations of data recycled to train the DL model, great datasets of images are used primarily, with a huge amount of images or generated by the authors. Some datasets are derived from famous and wide datasets of Plant Village, LifeCLEF, MalayaKew, UC Merced, and Flavia, although others are collections of real images gathered by researchers.

7.7 Information based on Images

TensorFlow represents operations that change calculations, shared states, and states using data flow diagrams. This function represents a node in a data flow diagram that spans multiple computing devices. B. Multi-core CPU, general-purpose GPU, and tensor processing unit custom ASIC (OPUS). This model has many mathematical layers, which are operated in a priority manner. A convolutional layer typically applies an indigenous gathering to all subsets of the layer's inputs. The Caffe model and associated optimization method are provided in the text form rather than code. Before training, Caffe provides model definition, optimal settings, and weights. Caffe rapidly processes large amounts of data. Caffe is also modular and can be effortlessly protracted to novel responsibilities. Customers create their representations by combining the different categories of neuron layers delivered by Caffe, a clean and modifiable framework for a collection of cutting-edge deep learning algorithms and reference models for multimedia scientists and practitioners [25].

7.8 DL Frameworks

Satellite data is invaluable and critical for sustainable land use planning to reduce carbon emissions, maximize economic benefits, and minimize land dilapidation. The difficulty in exhausting the data is interpreting the collected images. Satellite image translation utilizing CNN, genetic algorithms (GA) [26], and other optimization algorithms [27, 28] has proven to be a particularly valuable approach for precision agriculture and agribusiness decision making. You can populate the grid form with country types and other information. We used the grid model to evaluate the target and a genetic algorithm to find the best solution. A similar concept was used for flower grading. CNN can also be used to provide weather forecasts, which are critical in agriculture. Farmers, consumers, and governments rely on pre-harvest yield forecasts to develop policies for vending, acquiring, marketplace interference, and mitigating nutrition shortages. CNN uses it to forecast crop yields. It can be used for animal research as well as plant research. CNN, for example, is widely used to categorize animal behavior. The use of RNN in smart farming is extremely helpful for processing time series data and is widely used in agriculture, animal testing, and estimating event dates.

7.9 Land Cover Classification (LCC) Methods

LCC is recognized as an important and difficult task in agriculture, and determining which class a typical land belongs to is critical. In the past, many applications relied on single-time observations, ignoring the effects of time series on some problems. Vegetation, for example, frequently changes its spatial appearance, which can throw the mono-temporal approach off. The single-time approach, on the other hand, is susceptible to biases such as weather. As a result, a deep sequence model is used, and LSTM is a popular RNN variant. They conclude that SVMs work best with LSTM units. Furthermore, the RNN model not only detects the land cover class but also includes two sets of image inputs (same area, different timestamps), multiple LSTM units, and a graphic output depicting the area being modified. It is also used to spot changes. REFEREE was able to learn stable and rational change rules for both binary and multiclass changes using the RNN model. Plant phenotypes have become a hot but difficult topic due to the growing need for precision agriculture. In a nutshell, a plant's phenotype is used to identify the species of the plant based on its appearance and characteristics. During normal periods, most ML approaches rely on a single static observation, resulting in similar plant false positives [29]. CNN created a new DL structure for detecting plant phenotypes and combined it with the LSTM unit. CNN extracted DL about smart farming concepts, tools, applications, and opportunities and fed the output to the LSTM unit to build a sequence model, according to its structure. The sequence model significantly improved accuracy from 76.8% to 93% when compared to the previous CNN-only model, according to experimental results. RNNs, in addition to CNNs, are used to estimate crop yields with reduced bias using time series data. The RNN received the input, and the prediction Y became part of the input for the next time step. A new LSTM model has also been developed to forecast the city's 24- and 72-hour meteorological attributes (temperature, humidity, and wind speed).

Because NN can infer complex functions and time series inputs, they are naturally used for this task. When an image is compressed, some functionality is lost or becomes inaccurate, and photorealistic textures must be restored. To accomplish this, they devised a perceptual loss function composed of opponent loss and content loss. When compared to a wide range of pixel-by-pixel MSE losses, the contented loss purpose is approached by perceptual similarities. The model was able to restore highly compressed images after training with 350,000 images, outperforming some cutting-edge replicas at the time. This task is so used widely including image analysis, which is especially useful in farming, or remote sensing images. According to the results, the composite image was successfully converted to local features such as color, light scatter, texture, and so on, resulting in a global feature conversion [30].

7.10 Conclusions and Future Works

The results from this study can be used to solve many agricultural problems, including classification and prediction, and those associated with computer vision, image analysis, and data analysis. In this brief overview, DL concepts, tools, limitations, and algorithms are summarized. This article reviews the applications of DL in agriculture. DL has been used extensively in various areas of agriculture, such as detecting plant diseases, classifying plants and weeds, counting fruits, classifying land, predicting yields, and tracking animal behavior. While DL research is gaining in popularity, the research conducted on DL for agriculture is highly varied depending on the data analyzed. There is a great demand for agricultural products in the world due to the population growth. That is why we need to increase agricultural production to meet global food needs. State-of-the-art technologies such as machine learning, deep learning, IoT, and robotics are being used to increase agricultural production, reduce production costs, and increase revenue. Rain, cyclone-like stormy weather, floods, and climate change all have a negative impact on agricultural land crops. In order to mitigate the effects of these events, farmers often turn to advanced technologies like deep learning. In the future, deep learning will be very successful in the agricultural field. The overall benefits of deep learning in agriculture are encouraging for its further use toward ensuring food security and smarter farming.

Future efforts could also focus on creating mechanisms to accelerate and strengthen farmers and plant pathologists to participate in the image collection and labeling process. Another possibility that deserves further investigation is using plant parts other than leaves. It is also worth noting that, even with the current CNN architectures working well for plant disease recognition, it may be possible to develop new, leaner configurations that are better suited for this application. Deep learning is the need for large datasets, which would serve as input during the training procedure. Despite data augmentation techniques that can preserve the labels of images, in reality, it is often necessary to use hundreds of images to capture the data accurately. Images from different types of sensors could make deep learning algorithms suitable for a much broader range of agricultural applications. Another limitation of DL models is that they can learn some problems well, but they cannot generalize beyond the boundaries of the dataset's expressiveness.

References

[1] Firdaus, Y. Arkeman, A. Buono, and Hermida, "Satellite Image Processing for Precision Agriculture and Agroindustry using Convolutional Neural Network and Genetic Algorithm", IOP Conference Series: Earth and Environmental Science, Vol. 54, 25–26 Oct. 2016.

[2] P. Samuel S., K. Malarvizhi, S. Karthik and M. Gowri S.G., "Machine Learning and Internet of Things based Smart Agriculture," 6th International Conference on Advanced Computing and Communication Systems, pp. 1101–1106, 2020.

[3] R. Nikhil, B. S. Anisha, and R. Kumar P., "Real-Time Monitoring of Agricultural Land with Crop Prediction and Animal Intrusion Prevention using Internet of Things and Machine Learning at Edge," IEEE International Conference on Electronics, Computing, and Communication Technologies, pp. 1–6, 2020.

[4] J. Anand, and J. R. P. Perinbam, "Automatic Irrigation System using Fuzzy Logic", AE International Journal of Multidisciplinary Research, Vol. 2, Issue 8, pp. 1–9, August 2014.

[5] P. S. Nishant, P. S. Venkat, B. L. Avinash and B. Jabber, "Crop Yield Prediction based on Indian Agriculture using Machine Learning," International Conference for Emerging Technology, pp. 1–4, 2020.

[6] S. Albawi, T. A. Mohammed and S. Al-Zawi, "Understanding of a Convolutional Neural Network," International Conference on Engineering and Technology, pp. 1–6, 2017.

[7] Z. Lin and L. Lei, "Feature Extraction of Face Image Based on Convolutional Neural Network," International Conference on Virtual Reality and Intelligent Systems, pp. 427–429, 2020.

[8] S. I. Shahid and M. Shahjahan, "A New Approach to Image Classification by Convolutional Neural Network," 3rd International Conference on Electrical Information and Communication Technology, pp. 1–5, 2017.

[9] N. S. Malinović, B. B. Predić and M. Roganović, "Multilayer Long Short-Term Memory Neural Networks in Time Series Analysis," 55th International Scientific Conference on Information, Communication and Energy Systems and Technologies, pp. 11–14, 2020.

[10] A. N. Michel, "Recurrent Neural Networks: Overview and Perspectives," International Symposium on Circuits and Systems, pp. III–III, 2003.

[11] A. Creswell, T. White, V. Dumoulin, K. Arulkumaran, B. Sengupta and A. A. Bharath, "Generative Adversarial Networks: An Overview," IEEE Signal Processing Magazine, Vol. 35, No. 1, pp. 53–65, Jan. 2018.

[12] M. George, "Radial Basis Function Neural Networks and Principal Component Analysis for Pattern Classification," International Conference on Computational Intelligence and Multimedia Applications, pp. 200–206, 2007.

[13] R. Rezvani, M. Katiraee, A. H. Jamalian, S. Mehrabi, and A. Vezvaei, "A New Method for Hardware Design of Multi-Layer Perceptron Neural Networks with Online Training," 11th International Conference on Cognitive Informatics and Cognitive Computing, pp. 527–534, 2012.

[14] P. Klement and V. Snášel, "SOM Neural Network - A Piece of Intelligence in Disaster Management," World Congress on Nature & Biologically Inspired Computing, pp. 872–877, 2009.

[15] Yuming H., Junhai G., and Hua Z., "Deep Belief Networks and Deep Learning," International Conference on Intelligent Computing and Internet of Things, pp. 1–4, 2015.

[16] M. Tanaka and M. Okutomi, "A Novel Inference of a Restricted Boltzmann Machine," 2nd International Conference on Pattern Recognition, pp. 1526–1531, 2014.

[17] Dhanalakshmi R., J. Anand, A. K. Sivaraman, and S. Rani, "IoT-based Water Quality Monitoring System using Cloud for Agriculture Use", Cloud and Fog Computing Platforms for Internet of Things, Edited by P. Bhambri, S. Rani, G. Gupta, A. Khang, Routledge Taylor & Francis Group, May 2022.

[18] A. M. Elbir, "DeepMUSIC: Multiple Signal Classification via Deep Learning," IEEE Sensors Letters, Vol. 4, No. 4, pp. 1–4, April 2020.

[19] D. Xie, L. Zhang, and L. Bai, "Deep Learning in Visual Computing and Signal Processing", Applied Computational Intelligence and Soft Computing, Hindawi, 19 Feb 2017.

[20] M. Sun, T. Zhang, Y. Wang, G. Strbac and C. Kang, "Using Bayesian Deep Learning to Capture Uncertainty for Residential Net Load Forecasting," IEEE Transactions on Power Systems, Vol. 35, No. 1, pp. 188–201, Jan. 2020.

[21] L. D. C. S. Subhashini, Y. Li, J. Zhang, and A. S. Atukorale, "Integration of Fuzzy and Deep Learning in Three-Way Decisions," International Conference on Data Mining Workshops, pp. 71–78, 2020.

[22] J. Zhai, S. Zhang, J. Chen and Q. He, "Autoencoder and Its Various Variants," IEEE International Conference on Systems, Man, and Cybernetics, pp. 415–419, 2018.

[23] W. Wang, Y. Huang, Y. Wang, and L. Wang, "Generalized Autoencoder: A Neural Network Framework for Dimensionality Reduction," IEEE Conference on Computer Vision and Pattern Recognition Workshops, pp. 496–503, 2014.

[24] B. Hou and R. Yan, "Convolutional Auto-Encoder Based Deep Feature Learning for Finger-Vein Verification," IEEE International Symposium on Medical Measurements and Applications, pp. 1–5, 2018.

[25] F. Ertam and G. Aydın, "Data Classification with Deep Learning using Tensorflow," International Conference on Computer Science and Engineering, pp. 755–758, 2017.

[26] J. Anand, J. R. P. Perinbam, and D. Meganathan, "Design of GA-Based Routing in Biomedical Wireless Sensor Networks", International Journal of Applied Engineering Research, Vol. 10, No. 4, pp. 9281–9292, 2015.

[27] J. Anand, J. R. P. Perinbam, and D. Meganathan, "Performance of Optimized Routing in Biomedical Wireless Sensor Networks using Evolutionary Algorithms", Comptes rendus de l'Academie bulgare des Sciences, Tome 68, No. 8, pp. 1049–1054, 2015.

[28] J. Anand, J. R. P. Perinbam, and D. Meganathan, "Q-Learning-based Optimized Routing in Biomedical Wireless Sensor Networks", IETE Journal of Research, Vol. 63, Issue 1, pp. 89–97, 2017.

[29] E. Balouji, I. Y. H. Gu, M. H. J. Bollen, A. Bagheri, and M. Nazari, "An LSTM-based Deep Learning Method with Application to Voltage Dip Classification," 18th International Conference on Harmonics and Quality of Power, pp. 1–5, 2018.

[30] Liu, T. Wu, M. Wang, M. Fu, J. Kang, and H. Zhang, "Recurrent Neural Networks based on LSTM for Predicting Geomagnetic Field," IEEE International Conference on Aerospace Electronics and Remote Sensing Technology, pp. 1–5, 2018.

8

Image Analysis for Better Yield in Farming

G. Anbarasi[1] and B. Vishnupriya[2]

[1]Department of Biotechnology, PSGR Krishnammal College for Women, India
[2]Department of Biotechnology, Kongunadu Arts and Science College, India
Email: anbarasi.gk08@gmail.com; bvishnupriya_bt@kongunaducollege.ac.in; vinubiotech28@gmail.com

Abstract

In India, 70% of farmers have less than 0.5 hectares of land, especially small and marginal farmers. In high altitude, people are unable to utilize their land for agriculture. In this context, people are forced to approach multi-story vegetable cropping, which could be an effective and exceptional strategy in today's context. Crops of varying heights, rooting pattern, and duration are cultivated at the same time in the same space. This innovative practice helps the small farm holders earn much more than the existing one. In addition, it is also practiced as kitchen garden in urban regions and the produce are used for household purposes. People in rural areas and farmers with big plantations are also adopting this system to expand the availability of arable land and cultivate varied crops. Like the traditional farming system, it also requires water, manure, and land to obtain production per unit area. The gardens are appealing because they generate farm food at a minimal cost of management and maintenance. This innovative technique fetches higher price and performance meets the needs of the customers. The foundational components for conducting image processing are elucidated in Figure 8.1.

8.1 Introduction

The current scenario in low- and middle-income countries are that population growth requires higher intake of meat, fruits, and vegetables over grains, and creates necessity in improving food availability. On the other hand, food

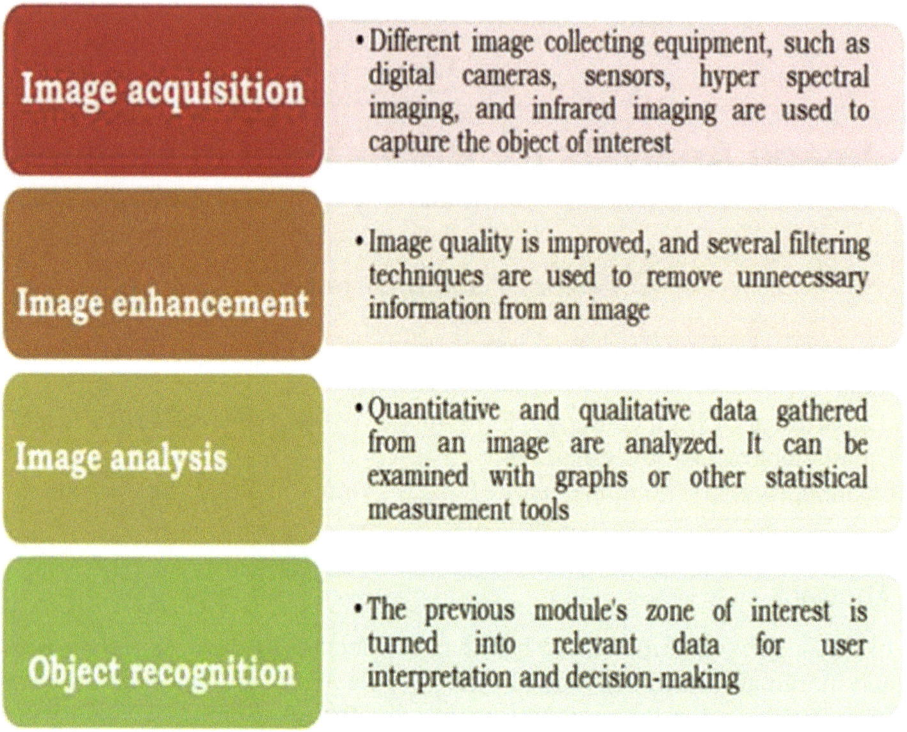

Figure 8.1 Basic modules of image processing.

scarcity and malnutrition continue to be a major hurdle in many countries According to the recent reports from the Food and Agriculture Organization (FAO), around 13% of the people in developing countries are malnourished [1]. As per the fact, the global population was around 7.6 billion in 2018 [2] and it was expected to reach 9.2 billion by 2050, which results in 59%–102% of the rise in food consumption [3, 4]. In order to boost economic growth and productivity, optimization and the introduction of novel agricultural practices for increased crop yield are considered as important practices for a country like India to meet the demand for the exponentially growing population.

India's core concern of overpopulation is rapidly developing in today's world. As the population rises, farmland becomes increasingly scarce. India had 93 million small farms, and it accounts for 23% of the world's total. While 92 million agricultural households are classified as marginal farmers, with an operating holding size of less than 1 hectare, 70% of farmers have less than 0.5 hectares. Farmers in small and marginal altitudes, in particular, who maximize the use of their property by the agricultural community, have become an important part of the community [5]. Rapid urbanization, hydropower projects, dams and rivers, highway highways, soil erosion (120.72 million hectares), soil salinity, and other factors will reduce land availability for agriculture in the future. In this situation, a multi-story cropping system would be a better option to provide people with food, nutrition, and income security [6]. The available small cultivated land can be effectively used for the production of upland and horticultural crops to achieve food security for the growing population and sustainable income for the agricultural community. Since land is a renewable resource and needs to be maximized to grow crops, diverse cropping systems maintain soil fertility by recycling crop by-products while maintaining sustainable productivity. A multi-storied cropping system is a modern method for horticultural crops and most plantation crops in particular, including coconut, areca nut, and coffee. Tall growing perennials with appropriate semi-perennial and annual crops make the system more adaptive. This farming strategy allows farmers to earn income from the same plot of land while growing different height plants in the same field at the same time. Land management with this system can maintain a sustainable ecological balance with efficient use of all natural resources [7].

A multi-story farming system would be a preferable solution in this situation to supply people with food, nutrition, and financial stability [6]. Small cultivated land can be efficiently used to produce upland and horticultural crops, ensuring food security for the rising population and long-term revenue for the agricultural community. Diverse cropping systems preserve soil fertility by recycling agricultural by-products while maintaining sustainable productivity, which is important because land is a renewable resource that must be exploited to grow crops. These technologies are critical for marginal areas due to space constraints on most farms. A multi-story cropping system is a modern method of growing horticultural crops in general and plantation crops in particular, such as coconut, areca nut, and coffee. The method becomes more adaptable when tall growing perennials are combined with appropriate semi-perennial and annual crops. Farmers can earn money from the same plot of land by growing different height plants in the same field at the same time with this farming approach. This strategy of land management can maintain a healthy ecological balance while maximizing the utilization of all natural resources [7, 8].

In addition, farmers can evaluate agricultural data predictions and produce crop yield estimates using an image analysis of improved yield in farming on a multi-storied cropping system. As a result, the farmer will be able to keep track of the yield. Image processing technology produces an intriguing discovery that could aid farmers in increasing crop productivity. Imaging is used to correctly identify each fruit in a plant using a normal digital camera and machine learning skills. As soon as the input is inserted, the acquisition of tree fruit image, image processing (including image enhancement, image extraction, and image element separation), image fragmentation, polygon equation, fruit detection, and calculation is completed in a fraction of a second, and it will be extremely fast, and farm laborers will benefit from it [9].

Advantages of multi-story cropping system with IoT:

- It maximizes production from small plots. This can help farmers manage with land scarcity while also dramatically improving income per unit area.
- Farmers can respond quickly to any significant change in weather, humidity, air quality, and crop or soil health in the field using real-time monitoring and forecast systems. As a result, there is less risk of agricultural output loss.
- Improves the physical properties and health of the soil.
- Cropping patterns with sensors help maintain soil fertility by allowing nitrogen to be fixed in the soil.
- A variety of crops can be cultivated in one site, ensuring that the family's nutrition is well-balanced.
- Dense planting reduces the number of weeds.
- Prevents the crop from being damaged by bad weather, such as drought.
- Protects the crop against climatic extremes including severe rainfall, soil erosion, landslides, and other natural disasters.
- Data collection and wireless monitoring and control are aided by robots and drones.
- Promotes winter crops by creating a suitable environment.
- High-quality crops are produced by smart gadgets.
- Contributes to the maintenance of ecological balance.

8.2 Image Processing of Crop Yield

Calculating the yield for fruits or vegetable requires more time and effort. Farmers can benefit from forecasting technologies to increase output. Crop yields can be forecasted using an image processing model. The backdrop of each image is removed using a different method in order to distinguish the fruit area in the input image. The centroid of the fruit is used to compute its value. The goal is to change the staffing system by using a machine-based technique to filter fruit by evaluating maturity level. The program uses novel computational methodologies such as machine learning, artificial intelligence, and deep learning to perform pre-image processing, feature extraction, and fruit classification. It serves as a source of information. It offers a computerized concept and approaches for machine learning in tree fruit acquisition, counting, and sorting. With the advent of the Internet of Things, picture analysis can assist farmers in predicting production. The results can be kept in a database, allowing the farmer to migrate as needed while maintaining a record of the production [9].

8.2.1 Crop yield detection

Farming solutions are a sensor-based system for monitoring crop fields. Farmers can benefit from IoT-enabled smart agriculture applications by assuring optimum yields, profitability, and environmental protection. Precision agriculture is the application of Internet of Things (IoT) technology to ensure that all resources are efficiently utilized and agricultural yields are maximized while operational and administrative costs are kept to a minimum. Agricultural IoT includes specialized hardware, wireless communication devices, software, and IT-enabled services. Sensors, devices, machines, and information technology have all played a role in the evolution of modern farm and agricultural practices.

The block system's design is split into two parts: front end and back end. In the front end, a graphical user interface (GUI) is built with buttons like refresh and detect. The software will run and the crops will be detected when you push the detect button on the backend, with an estimate result presented at the end. The illustration of the Block Diagram for Crop Yield Detection can be seen in Figure 8.2.

8.2.2 Special features

The frames of an input video are loaded into this GUI. The refresh and detect buttons are the accessible feature buttons. The video starts over when you press the refresh button [9].

8.2.2.1 Image processing

Image processing techniques are routinely utilized for crop growth monitoring. A digital color charged coupled device (CCD) camera was used to

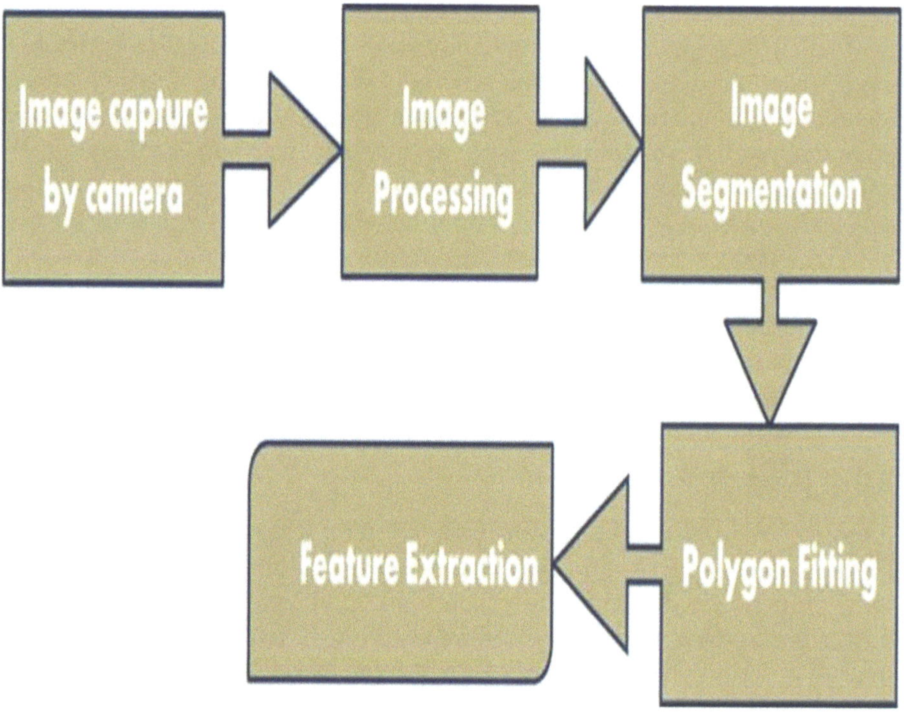

Figure 8.2 Block diagram for crop yield detection.

capture the image. The weather and natural lighting settings reported include full sun with front and back lighting, full sun with fruits in the shade, and foggy circumstances. A single tree is taken from several angles. All of the photos were transferred to the computer using a USB cable. The image sensor in a charge coupled device (CCD) used to snap images of tree fruits detect the number of light rays striking the region of the CCD sensor and the amount of light changed into an electric signal. Small cameras (GoPro) are sometimes preferred over larger cameras (Canon) with optical lenses that are more flexible [10].

Hyperspectral cameras, such as those manufactured by Resonon (USA), are now commercially accessible. Light in both the visible and non-visible spectrums is captured by hyperspectral cameras. For crop manufacturing, a few research initiatives in recent years have used depth or 3D cameras.

8.2.2.2 Image segmentation

Image-based precision agriculture relies heavily on crop segmentation and identification. This method entails dividing image portions; ideally, the plant should be separated from other elements like dirt, foreground objects such as machinery, and weeds. The Otsu technique [11] is a popular segmentation method. Crop localization in images is sometimes achieved by combining object recognition with picture segmentation to compel identification of a crop's physically viable structure [12].

8.2.2.3 Machine learning

In recent years, machine learning algorithms that automatically determine the optimal parameter values from manually annotated training data have gained popularity. Typically, the training data is often a subset of the overall image dataset. The support vector machine (SVM) is the most widely utilized mathematically based algorithm for parameter selection [13]. The introduction of high-performance graphics processing units (GPU) has lately made deep neural networks conceivable [14]. As a result, researchers are beginning to investigate the use of deep neural networks in the agriculture industry [15].

8.2.3 Smart farming with autonomous movers

The autonomous ground vehicles were taught, redesigned, and reassigned roles for tree trimming and flowering stage processes such as fruit thinning, mowing, pesticide spraying, sensing, fruit harvesting, and post-harvesting. The agricultural environment is detected, regulated, and navigated using a combination of sensors and robotics. Using self-guided vehicles and autonomous execution of farm activities including spraying, pruning, mowing, thinning, and harvesting, the automation system improves farm proficiency. Weed detection and management are typically carried out by autonomous robots. A path-planning robot's dynamic and kinematic restrictions were effectively used to predict and harvest yields.

8.2.3.1 Agricultural drone

Agricultural drones give agricultural information by providing a bird's-eye view perspective with utilizing multispectral images and regular surveys of the area. It gives you a bird's eye view of the agricultural world. Aerial pictures are taken with drones outfitted with a 4K resolution high-definition camera and GPS. In the agricultural industry, ripeness assessment is crucial for identifying the crop's quality and maturity level.

8.2.3.2 Unmanned aerial vehicle

The method of applying the procedure to individual plots in order to suit the needs of breeding programs is known as UAV. Algorithms for imaging employing machines for locating and counting agricultural items were applied to photographs of grapes, tomato, apple, mango, and citrus fruits using harvesting robots and ground surveillance vehicles. From aerial photos, certain UAVs have been developed to distinguish and count sorghum heads. It was utilized to extract 10 morphological sorghum features [16].

- **Area:** The number of pixels in each candidate's head region: the actual number of pixels.
- **Eccentricity:** The ellipse's eccentricity is the same as the candidate head region's second moments.
- **Extent:** The percentage of pixels in the prospective head area that make up the overall bounding box pixels.
- **Perimeter:** The total amount of pixels surrounding the prospective head region's perimeter.
- **Minor axis length:** The minor axis length of an ellipse with the same normalized second central moments as the candidate head region.
- **Convex area:** The smallest convex polygon with enough pixels to fit the candidate head region.
- **Filled area:** When all holes have been filled, the total number of pixels in each potential head area.
- **Equiv diameter:** The circumference of a circle with the same area as the candidate's head region.
- **Solidity:** The proportion of convex hull pixels that are also in the candidate head area.
- **Circularity:** The candidate's head area's circularity.

8.2.4 Applications of image processing

Image processing's importance and impact in society can be measured by its uses in a variety of sectors, such as medical imaging and aerial photography and industrial inspection, as well as satellite imaging applications for law enforcement, defense, and agriculture sectors. Every field, whether in the corporate or public sector, has a direct or indirect relationship with image processing techniques. In one field, an image processing technology that produces superior performance accuracy may provide inferior performance in another. As a result, choosing the right modules for the application is crucial. This section discusses many image processing applications in the agricultural sector (Figure 8.3).

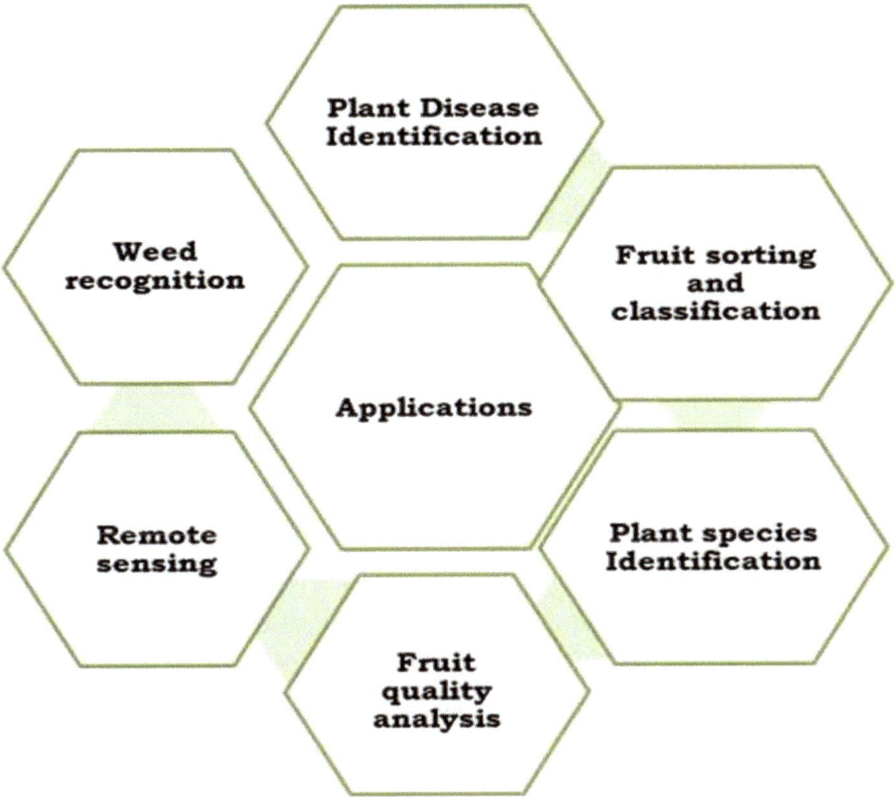

Figure 8.3 Various applications of image processing in agricultural sector.

8.2.5 Multi-storied cropping system

Two or more crops of varied heights are planted on the same plot of land at the same time in a multi-storied cropping system. To maximize sunshine, fertilizer consumption, sustainable land use, and ecological balance, a multi-story cropping system supports crops of varied heights, canopy patterns, and root systems. In this cropping system, the potential of more efficient use of resources like sunlight, water, soil, and fertilizers leads to increased biological variety, more crops per unit area, and production sustainability [17].

8.2.5.1 Principles of multi-cropping system

Coriander, radish, beetroot, cluster bean, cowpea, and other crops with varying growth habits, root depth, and crop length can be used to create multi-tier cropping systems. Farmers reduce soil erosion by cultivating multiple crops in the same area at the same time, which is a key drawback of monoculture cropping.

The basic principles of multi-storied cropping system are as follows:

(i) Crop diversification options based on scientific, ecological, and economic grounds.

(ii) Increases system productivity.

(iii) Improves system productivity.

(iv) Makes use of high-efficiency resources.

(v) Intense input utilization.

(vi) Long-term sustainability of farm resources and the environment.

8.2.5.2 Basic types of multi-storied cropping system

1. **Monoculture:** Refers to the year-round cultivation of the same crop in the same field. Rice after rice, for example, or jute after jute.
2. **Duoculture:** On a parcel of land, two types of crops are cultivated alternately every year. Jute, for example, comes after the vegetable, and rice comes after the veggie.
3. **Polyculture:** When more than two types of crops are cultivated in succession on a plot of land over the course of a year, it is called polyculture.

In addition, there are several types of multiple cropping techniques, which are exciting:

- **Pure stand multiple cropping** is a type of multiple cropping in which crops are produced in succession over a certain length of time in a unit of land owned by farmers. Each crop is sown and harvested separately in this practice, with each crop having its own land preparation; Aus paddy *T. Amon* Lentil, for example.
- **Multiple cropping of a mixed type:** It is a type of multiple cropping in which two or more crops are produced in the same season and on the same piece of land at the same time. Depending on the maturation period of each crop, it is harvested one after the other in mixed cropping; for example, Aus+B, Amon, Mustard+lentil, and Rai+Barley.
- **Multiple cropping of the intercrop type:** It is a type of multiple cropping when the same minor crop is grown between the major crops.

Figure 8.4 Steps to develop multi-storeyed system in agroforestry.

8.2.5.3 Key points to design multi-story cropping

- Choose plants that offer a variety of advantages.
- Plants that give both short- and long-term products should be used.
- Plants that compete for the same resources should be avoided.
- Make use of the various canopy layers to expand your options.

8.2.5.4 Steps to develop multi-storied system in agroforestry

Hirall Jana of Burdwan, West Bengal, suggested the following procedures for developing a multi-storied agroforestry system [18].

The above design illustrated in Figure 8.4 was followed at Mt. Kilimanjaro, The Chagga people. Home gardens are multi-story collections of over story forest trees prized for lumber, a middle layer of tiny trees such as coffee and banana, and a rich array of understory herbs and vines utilized for food and medicinal.

8.2.6 Tools to evaluate the performance of multi-story cropping system

To provide an overview of a cropping system framework for assessing cropping system changes. The framework consists of a rule-based rotation generator and a set of algorithms for computing impact indicators. It is organized as three steps:

(i) Plan crop rotations.

(ii) Evaluate crop production activities using environmental, economic, and phytosanitary indicators.

(iii) Designing a cropping system and assessing its impact. Cropping system planning and evaluation are aided by the framework. Some examples of multi-tier crop farming have been effective.

8.2.6.1 Land equivalent ratio (LER)

The relative and solitary crop that would be required to provide the equal yield produced by intercropping [19] is a key instrument for the evaluation of intercropping systems. An LER greater than one implies a yield advantage, an LER equal to one indicates no grain, no gain, or no loss, and an LER less than one indicates a yield loss. The LER is determined using the formula LER = (Ypi/Ymi), where Ypi represents the yield of each crop in a polyculture and Ymi represents the yield of each crop in a monoculture.

8.2.6.2 Relative yield total (RYT)

Replacement series yield benefits can be quantified using the relative yield total. RYT = Xmix/Xsole + Ymix/Ysole when two species are intercropped, where Xmix and Ymix are mixture yields of species X and Y, whereas Xsole and Ysole are sole crop yields of species X and Y. RYT values more than 1 imply that the species compete differently for resources or avoid competition in some way, whereas RYT values less than 1 indicate mutual antagonism. RYT values of 1 imply that the components fully share the same limiting resource, i.e., they fully compete and have no antagonistic relationships [20].

8.2.7 Reason for the need of multi-story cropping system in India

The average size of a holding in India was 2.28 hectares in 1970–1971, 1.82 hectares in 1980–1981, and 1.50 hectares in 1995–1996. With the limitless subdivision of land holdings, the size of the holdings will shrink even further.

Figure 8.5 Field layout for a multi-storey coconut cropping system.

The problem of tiny and fragmented holdings is exacerbated in highly inhabited and intensively cultivated regions such as Kerala, West Bengal, Bihar, and the eastern part of Uttar Pradesh, where the average size of land holdings is less than 1 hectare, and in some areas, less than 0.5 hectare. Our inheritance laws are the primary cause of this terrible state of affairs. The land that belongs to the father is divided equally among his sons. This distribution of land does not imply a collection or consolidation; rather, its nature is fragmented. The configuration of fields in a multi-storey coconut cropping system, attributed to the Philippines' Department of Agriculture and the Philippine Coconut Authority, has been visually presented in Figure 8.5.

In the recent years, marginal farmers of in a mid-altitude Himalayan village of the Uttarakhand state have pioneered the practice of multi-layer vegetable cultivation in their small fields based on extensive understanding of the local habitats and ecological diversity (reported by Prakash Singh and

GCS Nagi). To make better use of limited land resources by judicious use of soil and water resources, farmers in the Bastar district, in the southern portion of Chharrisgarh state, have embraced a multi-layer horticulture-based farming strategy and are reaping the rewards.

According to [21], the study evaluated the yield of elephant foot yam and companion crops within a multi-layer vegetable cropping system. The assessment factored in market prices of Rs. 10 per kg for bitter gourd, Rs. 6 per kg for ridge gourd, Rs. 5 per kg for bottle gourd, and Rs. 15 per kg for elephant foot yam.

8.2.7.1 Some of the successful multi-story cropping systems [17]

Coconut-based cropping:

- Coconut + black pepper + carrot + pineapple
- Coconut + jackfruit + coffee + papaya + pineapple
- Coconut + papaya + pineapple
- Coconut + coffee + papaya + pineapple
- Coconut + coffee + blackpepper
- Coconut + banana + ginger
- Coconut + pasture
- Coconut + papaya + pineapple + peanut

Philippines' Department of Agriculture Philippine Coconut Authority is the source of this information.

Some other cropping:

- Maize + greengram + groundnut
- Amaranth + lady's finger + colocasia
- Spinach + radish + onion
- Brinjal + lady's finger + basella + colocasia
- Eucalyptus + papaya + berseem

8.2.8 The multi-story garden farming

Kenyan farmers have recently implemented multi-story gardens, also known as "wonder multi-story gardens farming technique," for year-round food planting, especially in areas where space is restricted. These gardens are made of high-density polyethylene (HDPE) and are conical in shape. With improved access to water and manure, the multi-story gardens can hold 50 strawberry plants and up to 120 vegetable plants. They also require little technical and financial assistance and are especially suited to dry, non-fertile places where soils are unsuitable for traditional gardening. Because both rural and urban farmers may grow their own food, the innovation encourages sustainable cities and communities. This ensures family food security and, as a result, food security for cities [21].

8.2.9 Challenges of multi-story cropping

Despite the fact that multi-storied cropping is a very effective innovative technique in today's scenario of faster degrading farming lands, farmers face many challenges in adopting it due to drought conditions, a lack of funds, a lack of technical knowledge of cropping systems, timely availability of inputs, pest and disease incidents, a lack of irrigation facilities, and a lack of labor.

8.3 Conclusion

In our country, majority of farmers are in small and marginal scale and they primarily grow seasonal crops. As a result, after a specific period of interval, they expect to earn profitable yield from cultivation. Unfortunately, the variable weather conditions have a negative impact on food production quantity and quality, resulting in poverty. Farmers are also obliged to supply low-quality veggies at a high cost. In this regard, multi-story cropping system paves a novel opportunity for farmers to earn a good amount of returns throughout the year with a minimum risk of full crop failure, while also ensuring that natural resources are utilized efficiently and soil fertility is maintained. Hence, this technique is discovered to be a perspective approach for long-term production in agricultural business. This report suggests that along with the modern agriculture system of cropping, farmers can utilize recent construction of prediction tools using image processing to analyze agricultural data and are able to predict the production of crop yield forecasts.

References

[1] Our World in Data, Hunger-and-undernourishment, May. 2020.
[2] Annual Population, Food and Agriculture Organization of the United Nations Statistical Database, May. 2020.

[3] Global Demand for Food Is Rising. Can We Meet It? Harvard Business Review, April. 2016.
[4] E. Fukase, W. Martin, 'Economic Growth, Convergence, and World Food Demand and Supply'; Policy Research Working Paper, World Bank Group, Development Research Group Agriculture and Rural Development Team.,Washington, PP. 8257, 2017.
[5] O. P. Awasthi and O. P. Pareek, 'Range Management and Agroforestry', PP. 67–75. 2008.
[6] Asheesh Sharma, N. Chaudhary, author. S. Anjanawe, Multi storied cropping system in horticulture - An approach for more crop per unit area of land, Journal of Pharmacognosy and Phytochemistry, SP6; PP. 25–28. 2020.
[7] S.G.S.P. Yadav, et. al., 'Crop Yield Prediction Using Image Processing'. Nat. Volatiles & Essent. Oils, PP. 4248–4259. 2021.
[8] S. Rastia et. al., 'A survey of high resolution image processing techniques for cereal crop growth monitoring', Information Processing in Agriculture, PP. 300–315, Jun., 2022.
[9] M.H.J. Vala, A. Baxi. 'A review on Otsu image segmentation algorithm', Int. J. adv. res. Comput, Eng Technol., PP. 387. 2013.
[10] E. Hamuda, M. Glavin, E. Jones. 'A survey of image processing techniques for plant extraction and segmentation in the field'., Comput Electron Agric., PP. 184–199. 2016.
[11] Thays Falcari, Osamu Saotome, Ricardo Pires, Alexandre Brincalepe Campo. Evaluation of multi-class support-vector machines strategies and kernel adjustment levels in hand posture recognition by analyzing sEMG signals acquired from a wearable device. Biomed Eng Lett. 10(2): PP. 275–284, 2020.
[12] S. Aich et. al., 'Deepwheat: Estimating phenotypic traits from images of crops using deep learning'. In: 2018 IEEE Winter conference on applications of computer vision (WACV). 2018.
[13] W. Guo et. al., 'Field-based individual plant phenotyping of herbaceous species by unmanned aerial vehicle'. Ecology and Evolution, PP. 12318–12326. 2020.
[14] Lotus Arise. 'Cropping System and Cropping Pattern in India – for UPSC IAS. 2021.
[15] Jin Zhang, Zhao-Hua Li, Kun Li, Wei Huang, Lian-Hai Sang. Nitrogen Use Efficiency under Different Field Treatments on Maize Fields in Central China: A Lysimeter and N Stud. Journal of Water Resource and Protection, 2012

[16] A.P. Srivastava et.al., 'Livelihood Enhancement through Improved Vegetable Cultivation in Backward Districts of India', National Agricultural Innovation Project Indian Council of Agricultural Research New Delhi, PP. 1–110. 2013.
[17] P.K. Nimbolkar, C. Awachare, S. Chander, F. Husain. Multi storied cropping system in Horticulture-A Sustainable Land Use Approach. International Journal of Agriculture Science. 8(55):3016–3019. 2016.
[18] P.R. Mirjha, D.S. Rana. Yield and yield attributes, system productivity and economics of mango based intercropping system as influenced by mango cultivars and nutrient levels. Indian Journal of Agronomy. 61(3):307–314. 2016.
[19] Krishna, Hare, et al. "Fruit-based cropping systems for sustainable production." ICAR, 2013.
[20] R. Arya, O.P. Awasthi, J. Singh, C.K. Arya. Comparison of fruit based multi species cropping system under arid region of Rajasthan. Indian Journal of Agriculture Science. 80 (5):423–426. 2010.
[21] Innovative Multi-Storey Gardens enhancing food security in Kenya. https://www.icfi.nl › News. 2020.

9

Precision Farming for Crop Prediction

E. Duraiarasu[1], S. Mahalakshmi[2], A. Jose Anand[3], T. Manikandan[1], and Abdennour El Rhalibi[4]

[1]Department of Electronics and Communication Engineering, Rajalakshmi Engineering College, India
[2]Department of Computer Science and Engineering, Chennai Institute of Technology, India
[3]Department of Electronics and Communication Engineering, KCG College of Technology, India
[4]Department of Computer Science, Faculty of Engineering and Technology, Liverpool John Mooris University, UK
Email: 200801064@rajalakshmi.edu.in; mahalakshmi.ramesh07@gmail.com; joseanandme@yahoo.co.in; manikandan.t@rajalakshmi.edu.in; a.elrhalibi@ljmu.ac.uk

Abstract

Precision farming is an approach that makes use of information technology (IT) for optimum to guarantee that crops and soil have health and productivity. Precision farming developed in this work makes use of the data acquisition (DAQ) device. The DAQ consists of sensors, actuators, signal conditioning equipment, and a computer running data acquisition software that make up the whole information security architecture. Cloud observing may be a strategy for checking, watching, and overseeing the operational workflow in a cloud-based IT foundation. Manual or mechanized administration procedures affirm the accessibility and execution of websites, servers, applications, and other cloud infrastructure like monitoring cloud information over disseminated areas. Disposing of potential breaches by giving permeability to records, applications, and clients by ceaselessly observing the cloud to guarantee real-time record checks due to standard inspecting and detailing to guarantee security measures are carried out using the information security. Information security is the process of analyzing signals that indicate real-world physical conditions and translating them into advanced numeric values that can be managed by a computer. Direct attached storage (DAS), DAQ, or direct active user (DAU) information procurement systems frequently convert analog waveforms into advanced values for processing. Information procurement frameworks have several components. Sensors convert physical variables into electrical impulses. Sensor signals are converted into a format that can be translated into advanced values by signal conditioning circuitry. Analog-to-digital converters are used to convert over-conditioned sensor signals to computerized values. There are also open-source software packages that include all of the necessary tools for extracting data from various, frequently specialized equipment. These gadgets are derived from the logical community, where complicated trials are common. The digital data are acquired from the real-time analysis of the system and situation in the agricultural fields, and these data can be stored and be used to train machine learning algorithms to train the functionality of the autonomous system for different situations and different case scenarios. With AI-enabled prediction and precision farming systems, this, in turn, plays a major role in developing the technologies to turn up with no human requirement even to change the physical parameters for proper automation of the monitoring systems. Even those data are being sold as the datasets on the specific scenarios for training artificial intelligence (AI) models. Shortly, we will have a large requirement for data acquisition and the immediate evaluation of the data to train the machine learning models to automate the process to the next level. Hence, there is a need for a data acquisition system that is convenient and portable and which can react to different soil types and vegetation of any kind of place and is suitable to acquire data under any climatic conditions. This sort of acquisition system will have an individual combination of separate sensors. In this system with the data collected through the help of humidity, soil consistency, soil moisture, and water level sensor are used in the acquisition system. Thus, the acquisition system is placed where the specific data of that vegetation is required or needs to be acquired. For real-time analysis, the data can be downloaded and can be used in different formats. These systems are again very helpful not only in the farming lands but also in the hydrophobic-based farming techniques because the data only will help the controlling system to keep the parameters under range. Thus, it is a portable data acquisition system, which is suitable for all sorts of farming and agricultural lands with real-time cloud projection and the data will be shared through the mail to the respective centralized monitoring systems to the government representatives. And, thus, this will be helpful for the agricultural ministry to develop more solutions and can sort out the problem that occurs in the field of agriculture.

9.1 Introduction

Food is one of the basic essential things required for human survival. As the population increases, the demands and needs for food also increase. To increase our food supply, our agricultural sectors need to improve and increase food production as well as its quality drastically. And food production should increase to achieve the demands of day-to-day life. Therefore, a proper system is needed to improve not only the overall aspects of agricultural yield but also the standards of quality and one such system is a data acquisition system. The data acquisition system consists of a unit of software and hardware that can measure or control the physical parameters and properties of the actual object. In addition, measurement data and parameters can be easily saved and retrieved. Data acquisition hardware (DAQ), sensors and actuators, signal processing and tuning gear, and a computer running DAQ software make up the entire data acquisition system package. The process of collecting data/sampling signals that measure real-world parameters and transforming the samples/data into digital values that a computer can understand and alter is known as data acquisition. For processing and tuning, the data acquisition system converts the received analog waveform into the desired digital value.

The DAQ applications are often powered by software programs written in assembly C, BASIC, Java, C++, and Fortran, among other general-purpose programming languages. It is free software that comes with all the tools you will need to collect data from a wide range of hardware devices. The majority of these packages are made-to-order, although there are a few exceptions. Using a signal conditioning circuit the sensor signals are converted into a format with tuned data and advanced values. These conditioned electrical signals obtained as results from the sensors can be converted and computerized into digital format of data using analog-to-digital converters so that they could be displayed as an output to track and monitor the parameters. The data that is acquired from the real-time analysis of the system in the agricultural fields using various types of sensors can be stored and that data can be used to train the machine learning algorithms to train the functionality of the autonomous systems and to improve it efficiently for different situations and different case scenarios and it could be implemented with AI-enabled prediction and precision farming systems. This, in turn, plays a major role in terms of developing the farming system technology to turn up with no human requirement, without manual operation, and even change the physical parameters for proper automation of the monitoring systems. Along with the help of some well-authorized and entrusted cloud server platforms such as Amazon Drive, one could store, maintain, access, and use the data easily.

Cloud storage is presented as a storing model/system of data for computers in which the data is collected and stored digitally in the form of logical heaps. The actual physical storage contains multiple and various servers, especially those that are owned and managed by the hosting company. These cloud storage providers ensure that your data is available and accessible 24/7 in a secure, operational, and private location. Usually, we know that we can approach platforms of servers of the cloud through various consolidated cloud computing assistants. For instance, in this case, we have some authorized web services like APIs, or one could simply make use of software that uses APIs such as cloud dynamic IP generators. This type of data acquisition system could be designed in such a way as to contain an individual set of various types of sensors. Therefore, the data collected with the help of humidity, soil consistency, soil moisture, and water level sensors are used in the acquisition system so that when the acquisition system is placed where the specific data of that vegetation is required or needed to improve, the factors of agricultural fields can be easily fetched and procured. With the help of procured data, we could study the various parameters of soil and agricultural fields such as moisture, humidity, and other such parameters.

For the real-time analysis and manipulation, the data can be downloaded and it can be used in different formats. By using such cloud-based acquisition systems, we have the advantage of accessing the data from anywhere in the world with a properly working internet connection and making use of these data in many ways such as in the agricultural fields to monitor the soil quality, soil moisture, humidity, and other such factors. With the help of these acquired data, we could feed or deploy these data to the machine learning modules and systems, to train the machines not only to enhance and improve the overall yield in the agricultural fields but also to deploy these well-trained machine learning modules into other such systems for an efficient manner of usage. With such an advantageous method, one could use devices like smartphones to turn on and off the sprinklers, and we could monitor the yield's quality and measure harvest volumes, with the help of the soil moisture and density substrates, infrared crop health monitoring device substrates, and so on. Considering the versatility and flexibility, there is a large need for data acquisition systems and for the real-time evaluation of the data to train the machine learning models to automate the process to the next level. Therefore, a proper and portable system is needed, which can react to different soil types and vegetation of any kind of place and it should be suitable to acquire data under any climatic conditions.

9.2 Literature Review

The agricultural sector is constantly facing challenges due to numerous factors such as climatic conditions, soil wealth, and economic shortage. So to overcome these challenges, a proper and smart agriculture system is required. With that in mind, we have an incredible way to monitor and manipulate the various parameters involved in smart agriculture, that is, a data acquisition system. These data could be stored in that developed system by fetching over from various sensors. These smart tech-based devices, combined with data-based agriculture, would help farmers to minimize prices and thereby can increase yields and profits. The most indispensable things needed for smart and precision farming are some of the parameters and affecting factors such as monitoring environmental conditions and monitoring the availability of water resources. These measurements associated with the farming market data become the data-based driver. Many researchers have developed various devices and technologies for detecting, tracking, and collecting the essential affecting factors, and with their help, they have understood the assignment to use this model as the basic custom system for smart farming [1–3].

The use of Internet of Things (IoT) helps basically in creating attractive as well as problem-solving innovative applications and software by establishing a common platform for different domains and sectors. This system is manually designed and developed to measure some of the most needed factors such as moisture content of the soil, its sensitivity, temperature, humidity, and the overall sunlight intensity. It also helps in automating the irrigation process that works on crops in agricultural fields. And, additionally, this currently existing system gives us an enhanced as well as enlightened future for monitoring and visualizing processes of the required available data and also helps to control the irrigation process in real-time entities without any loss in data during transmission. The watering system can be controlled in two ways using a smartphone from anywhere – either by the smart automatic control or by manual control. This system is very cheap since it makes use of solar energy to boost the autonomy as well as the novelty to perform complex operations [4]. This data acquisition system is also executed in an aeration strategy for controlling the storage of grain based on simulation. The data acquisition system was used during the field trial and it was based on digital addressable devices. This strategy includes a set of sensors that monitor environmental conditions such as temperature, ambient air conditions, and other parameters that can be stored in data acquisition systems, allowing us to monitor grain bulk storage and turn the fan on or off automatically based on simulation results. This control method can also be easily adapted to other data collection system devices since it can be used to monitor and control many storage systems [5].

Precision agriculture is nothing but a comprehensive system that is sculpted (in the form of accuracy level) to uplift agricultural production by carefully modifying the management of soil and crops according to the uncommon conditions detected in every field while maintaining environmental quality and stability simultaneously. By making use of Internet of Things, the information gathered through sensors can be uploaded wirelessly to the cloud server and can be viewed and controlled by the people who use them simply from anywhere in the world with the help of a device that has a properly working internet connection. This DAQ unit comprises sensors and microcontrollers that properly procure the environmental parameters accurately. Later, the processed information converted by the data acquisition system (DAQ) along with machine learning modules could be used to automatically control the switching buttons to turn the control system actuators on or off using relays, based on user-defined thresholds. These designed DAQ units are to be used under supervision and controlled conditions to ensure and monitor environments such as greenhouses, and it lacks the hardness to be used outdoors for considerably a longer time [6].

An Arduino-based automated DAQ has been developed to easily diagnose and rectify the errors and failures existing in the present hydroponic system. This, in turn, measures relative humidity, water temperature, water level, pH level, and so on. The obtained results have been properly calibrated with the help of accurate standard instructions. In addition to that, plot curves and graphs of each parameter have been generated via the collected data in the system. The data acquisition system has been considered as an important and turnabout aspect in the place of success for the hydroponic system as it indeed helped significantly with the proof of many systems' research and analysis. The data gathered via the sensors has been stored in the SD card or any other storage devices and when it occurs as such, certain fluctuations have been found, which occur in the real-time system which sometimes proves the easy use of such a system in system monitoring. The DAQ's performance is measured in terms of necessary parameters such as moisture and light intensity. The output graphs confirm accurate results, which help in the validation process of the DAQ system by comparing it with some standard instruments. Several influencing factors have been measured with the help of sensors, and also the error percentage is found to be relatively very small at only 5% [7].

The data acquisition system is used along with an unmanned aerial vehicle (UAV) in agricultural fields to substantially improve the current method of farming. This innovative platform simplifies data acquisition with high spatial and temporal resolution and high accuracy at an affordable cost and is hence accessible to wider communities. The goal was to create a product capable of analyzing multiple crucial soil parameters via one unit and transfer the accumulated preprocessed data to an airborne mobile subsystem. This system consists of two working modules that work in harmony with each other. The soil data monitoring probe (SDMP) is a stationary unit housing various soil probes, whereas the airborne data acquisition system (ADAS) is a mobile unit that can be placed on a UAV. The SDMP captures soil metrics, preprocesses them, and stores the data on an SD card, to be delivered using an NRF24 transceiver. The unit itself is battery- and solar-powered; regulated power is fed through a custom-designed motherboard to the Arduino Mega Controller (ATMEGA-2560), and internal/external modules. It is estimated that, with its low-power design and the complimentary solar power, it can work for months without interruption. The ADAS is a lightweight device and has been mounted on a drone to commence the data collection. It initiates data collection by interrogating sleeping SDMPs based on geological stamps by waking them up from deep sleep mode with an interrupt. The ADAS is supported by a custom motherboard to support a raspberry pi zero with Wi-Fi capabilities [8].

For monitoring the important parts that help with the growth and income, a sensor-based modern technology with IoT has been developed. In addition, an optimization technique for identifying crop loss has been implemented to maximize the system boundary and convey data to farmers about the boundary of the system in various locations. To demonstrate the validity of the suggested method, certain existing methods were examined, and it shows that the proposed strategy improves the result by 58%. Despite many solutions proposed by many authors, there is still a gap in farming answers to many parametric targets such as the height of crop, weight, and thickness. Moreover, no suitable formulations or methods for implementing such real-time applications using sensors have been developed. As a result, the authors have collaborated to develop a new, unique model for monitoring all important factors for crop growth, which then gives sufficient daily information to all farmers through an online-based monitoring application. As an additional benefit of the suggested system, information can be sent to all farmers in their local languages, which can be preserved. Additionally, the proposed work was prompted by an analysis of the issues that farmers experience in their day-to-day lives, where the amount of revenue that their network/system was initially framed. Therefore, certain limits are set before applying the sensor, which is of great benefit to all farmers. At the same time, basic payments are made when farming with smart devices that monitor sustainability [9].

Precision agriculture demands a large set of data whose availability is considered to be an important factor. It is believed that precision agriculture data acquisition methods can evolve from truth collection to sensors. Remote sensing is an important and convenient way to collect precision agriculture data. This system acquires precision agriculture data by extracting indicators such as NPP, NDVI, and LAI, which builds a model of management decision support system in relation to Land Truth and informs us of the ease of use information like how, when, and where to use these precision agriculture techniques to help end-users build irrigation, fertilizers, and spray pesticide systems. In addition, programs will be created that provide functions such as data update, data editing, data output, and management decision support. This is a quick way to provide precision agriculture data to end-users [10].

For monitoring information, several types of sensor gear from different manufacturers will be employed for different crop physiological situations. However, there appears to be no uniform standard for the numerous Internet of Things (IoT) communication protocols and application layer protocols, resulting in the diversity of the Internet of Things basic devices and gate sources. With the advancement of the agricultural Internet of Things industry, it is necessary to upgrade the existing monitoring information collection system architecture and create a reliable data collection and control system that can overcome the limitations of current collection links processing heterogeneous data. This helps improvise the use of current information resources and data. The recent growth of agricultural climate parameter monitoring systems and the IoT for real-time data collection has been developed for the benefit of farmers, meteorologists, and more. The various modules are integrated to play an efficient role in interoperability, and real-time remote monitoring on the ubiquitous platform. The system also used Restful Web Services application layer protocol and Wi-Fi technology for communication between field sensors and remote users. The artificial intelligence is utilized to analyze critical soil nutrient data in future techniques to determine soil fertility [11].

Precision agriculture today relies on reliable information about production using physicochemical parameters and their environmental conditions, and a sensor connected to a wireless sensor network (WSN) can give a range of atmospheric parameters, including a few points, temperature, and humidity. WSN is utilized for tracking and monitoring in a variety of fields due to its inexpensive cost and low power consumption. Environmental monitoring, precision agriculture, process control and machinery, and system and plant automation are some of the well-known uses of these sensor networks and traceability systems. One of the applications is that the sensor node is placed only at the point of interest by the farmer and the robot automatically takes over the data collection. Since the robot can travel near the sensor node, the transmission cost and data loss rate of the node can be minimized. The analysis will be completed automatically if the autonomous system can georeference and map the transmitted sensor data. In addition to assigning values, the position of the node helps to reduce the robot's trip duration and expand the autonomous machine's motion radius [12].

The integration of aircraft-based robotic platforms with networks of ground-based intelligent sensor systems has recently emerged as a robust solution for data acquisition, analysis, and control in a variety of specialized applications. This system represents a hierarchical structure for precision agriculture crop monitoring based on collaboration between a team of unmanned aerial vehicles and the framework of a coalition radio sensor network. The key benefit is that data is collected and transferred online to a central monitoring point, while network load and delay are well-managed thanks to optimized UAV trajectories and field data processing where different crops and methods are improved and developed. The results show that the collaborative UAV–WSN approach implemented in the Romanian project MUWI improves performance in both precision and organic farming. Potential applications in modern agriculture help farmers gain better insights into crop development through targeted intervention and reduced inputs, disease prevention, and significant improvements in economic efficiency increases [13].

The Internet of Things (IoT) is a network of Internet-connected devices that communicate with web services. RFID (radio frequency identification), sensors, and cellphones are all part of the Internet of Things. The idea of this IoT is to connect almost every device in the world to make it act like a connection as a system using the internet. First, to connect objects and devices to large databases and even in networks, a simple, efficient, reliable, and cost-effective system for identifying objects is available. Only the shared data provides information about what is the status and process going on. Radiofrequency identification (RFID) provides this functionality. Second, data acquisition benefits from the ability to use sensors to detect and track changes in the body. The intelligence built into the thing itself can further enhance the functionality of the internet-connected devices by enhancing the information processing capabilities at the edge of the network. The combination of all these technological developments will enable the Internet of Things to connect different devices in the world in a sensory and intelligent way. The system collects data via wireless modules, guarantees reliable data transmission over long distances, and uses wireless transmission technology to access the internet [14].

Over the last few decades, embedded IoT devices have changed the agriculture sector. With an embedded system that can automate and dynamically adjust the gardening process, a smart garden system based on today's Internet of Things (IoT) technology is a breakthrough undertaking. Agricultural and different technologies combine to connect a large number of small devices through different communication protocols. Agricultural production is expected to double by 2025. One of the biggest challenges in agriculture is manually collecting and organizing data. This human data collection can be replaced by an automated data collection system that sends this raw data collected to cloud-based mobile applications. This task can be performed by an application available on mobile devices called Blynk. This has the benefit of quick and seamless data collection as well as reduced development and programming effort for a personalized web server. In addition, the ability to transfer raw data in real time has proven successful in developing highly efficient portable embedded systems [15].

Due to the recent developments, hardware and software tools can be modified for giving some intelligence to the data acquisition systems, which are by machine learning modules. With this feature, multisensor data fusion will combine data from many sensors, as well as relevant information from associated databases, to obtain higher accuracy and more specific interferences than a single sensor could provide. This is now possible due to the improved processing hardware and the advanced processing techniques available. The different wireless links provided and the open architecture of the network allow its easy integration with other systems giving great flexibility and functionality to high-demanding agricultural uses. Control tasks are also enhanced since they can be distributed remotely or locally executed. Also, the results of its application to the environmental control of a set of greenhouses effects are efficient. This proposed architecture has been deployed in several commercial explorations to assess its effectiveness and make any changes required by users [16].

Zigbee is one of the emerging wireless technologies that are used in the agricultural sector. This technology has been used effectively by China as a wireless system for remote sensing, dispersion, and variability detection. As we know, many data acquisition systems exist, which are old-fashioned and time-consuming. Due to these limitations of the existing systems, modern wireless technology along with a smart application is needed to improve the agricultural fields. This Zigbee module is made up of a variety of micro-sensor components that have been installed in the monitoring portion. This Zigbee is a better wireless system since it is tiny in size, consumes less power, has a slower data transfer rate, and is less expensive. Because it is a connectivity solution that combines wireless technology with Bluetooth, it is mostly used to send data over short distances. The data acquisition system was used as a remote system to improve drip irrigation and soil moisture. The basic concept behind this improvement is to deploy a soil moisture sensor to detect and measure moisture content and then process the data using a data acquisition system to control irrigation according to the crop's needs. A signal conditioning circuit is used to improve the quality of the signal, and the analog output is conditioned and processed communicating among many tiny sensors. This type of communication has high efficiency since the signals are passed through radio waves in the form of a relay. Two main things, which are network ability to store and network latency, must be considered while deploying nodes. In Zigbee, there are nearly 65,000 supported network points, where every two nodes require at least 15 ms to progress the data transmission. Under practical conditions, another main thing to be considered is its network coverage and response time. Node capacity is directly proportional to network coverage but not the system with a longer response time. Hence, this requires a suitable network design structure to accommodate various applications. Thus, these are the main factors to be considered when developing a new system by the environment [17].

A remote RF module has been used in the agricultural sector for the transmission and reception processes of control and data signals. After that, this data can be processed further to reduce the difficulties faced during irrigation and also one could automate this process using artificial intelligence with an optimal database system. This system can indeed be implemented in various operational fields and can be used efficiently to save time, money, water, electrical energy, man labor, and the required cost [18]. One of the main problems in the livestock industry is that heat stress has a major impact on the production of domestic animals. And after many developments and research, a certain hog cooling pad has been developed and the heat stress has been reduced significantly. But to study how the heat flows from the system to the livestock, a data acquisition system is needed to collect data and study the heat transfer characteristics. Temperature data was collected from two systems to track accuracy and responsiveness. These two temperature sensing systems operated at different intervals and correlation data were collected at the start of the experiment. After determining the validity, the same experiment has been run multiple times to create a dataset that combines the development of a specific coolant flow system with the performance of digital and thermocouple temperature sensing systems [19].

Arduino-UNO microcontroller-based data acquisition strategy is implemented along with the interfaces of both analog and digital sensors in the agricultural sectors. Various parameters such as temperature, humidity, light intensity, and concentrations of gases have been monitored and calculated simultaneously. LabView helps to provide a better user interface and also acts as a system to understand situations and circumstances faster and efficiently. This system can be used in the agricultural field, industrial, as well as live environment monitoring sectors. With the help of the extracted data, one could process this data into information by using this system. As a means to store this data in cloud servers, Wi-Fi systems can be used. Due to a large number of applications and usage, this data collection system can be used and accessed anywhere in the world for various applications. This system can be manufactured easily and also it can be implemented easily as it is of low cost due to the usage of an open-source platform named Arduino-UNO. The graphical user interface that is used in this system can be understood by even the persons who do not know about the technologies. The data stored in the cloud server can be used later for further processing and analysis for a longer period. Various warning systems can be deployed along with this device to make an intimation to the users like farmers and landowners. This can be made automatic by coupling with the machine learning algorithms and modules to fasten up the process and thereby reduce the time [20].

Data acquisition systems and sensors play a major role in monitoring and controlling sectors of the agricultural fields, especially plants. Microchip PIC16C71 and the Intel 87C592 microcontrollers have been implemented in the wireless data acquisition device for the applications of agriculture. Along with these microcontrollers and microchips, radiofrequency linked to a base station has been used in a set of SPWAS. The base station controls the data acquisition stations and the storage of the collected data. The developed system can be applied to collect and procure outdoor and indoor climate data. Therefore, this system is used in Portugal to monitor the greenhouse effect inside and outside climate. Considering the low cost of SPWAS stations, it can be used as a flexible system for data procuring processes [21].

9.3 Proposed Methodology

The proposed system developed for the precision farming makes use of the DAQ device that will act as a portable device that is capable of placing in different agricultural fields. The circuit diagram of the proposed portable data acquisition system for precision farming is shown in Figure 9.1. The main goal of the proposed model assures the data transfer of the physical parameters of the agricultural lands. The DAQ has different sensors that are associated with the microcontroller and from the various sensor's, data are extracted; the first and foremost sensor used is a volumetric moisture sensor.

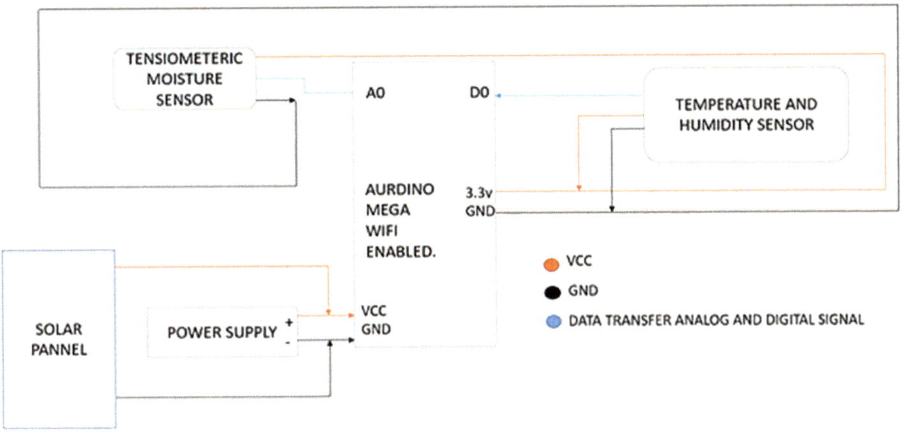

Figure 9.1 Circuit diagram of the portable data acquisition system for precision farming.

The volumetric moisture sensor is used to measure the moisture level through which the volumetric content of water in the soil, and the moisture measurements of soil drought, drying, and wet state can be measured. By utilizing a few other soil parameters, such as dielectric constant, electrical resistance, neutron interaction, and substitute dampness content, the sensors deployed in the system are capable of determining the volumetric water content. Here the different sensors are established among the unique sensors used in the device. Thus, each sensor will concentrate on unique parameters of the soil health and the data acquisition of the parameters that are required to be measured with respect to the time. The volumetric sensors are also known as the soil hygrometer detection modules. These sensors are frequently used to measure volumetric water content, while another group of sensors estimates water potential, a new feature of dampness in soils. Because they use tension meters, these sensors are commonly referred to as soil water potential sensors. This sensor uses capacitance to measure the water content in the soil. The water content in the soil can be shown in percentages when this module is initialized. When this sensor is inserted inside the soil, it aids the execution of tests to detect unknown moisture levels.

The preliminary parameters such as temperature and humidity decide the growth and the possibility for the yield of the crops. Different locations will have different scales for this device's DHT sensor. The DHT 22 could be an affordable temperature and humidity measuring sensor in digital form. These sensors are basically transducers that are compatible to connect with Raspberry Pi and Arduino microcontrollers. This sensor uses temperature sensitivity and capacitive wet detection to find the relative humidity. There are two anodes with a dampness holding dielectric between the anodes in the stickiness detecting sensors. The integrated circuit degree prepares these varied resistance values and converts them to digital data. The sensor is used to measure the negative temperature coefficient through thermistor, which causes resistance and will decrease as temperature rises. To acquire higher resistance values even for modest temperature changes, this sensor is often made of semiconductor ceramics or polymers. The sensor can be used to monitor humidity and temperature in heating, ventilation, and air conditioning systems, among other things; hence, it will be suitable for measuring the data in the agricultural field based on the size of the transducer used for the parameters like temperature. Climate stations also employ these sensors to forecast weather conditions; hence, these sensors are much suitable for monitoring the agricultural field. These sensors will provide different parameter values to the microcontroller in terms of analog or digital readings. The proposed method flow diagram is shown in Figure 9.2.

The device will be placed in the agricultural land where there is a need to acquire data to analyze or extract the current status and condition of the soil and the agricultural field. The proposed device first connects to the Wi-Fi to establish a connection that could access the gateway to transfer the data wirelessly, which will make this to use the MQTT protocol that allows us to transfer a large quantity of the data in real time with less latency to establish this there is need of an interface. A broker makes a difference in dealing with clients in MQTT innovation. It can oversee hundreds, thousands, or millions of associated MQTT clients at once, depending on the usage. Its fundamental capacities are: receiving information, decoding and sifting the messages received, determining which client will be inquisitive about which message, and transmitting these messages to clients depending on their interests. The MQTT door acts as a bridge for all MQTT gadgets and sensors or the web things. Moreover, cloud IoT centers can communicate with devices using MQTT as a bridge. The MQTT portal can act as a bridge by uniting information in a central area. The figure underneath appears as the MQTT portal capacities as a bridge. Here, a centralized cloud-based data collecting platform is established to save the data being collected from different agricultural fields. These storage cloud capacities employ information centers with gigantic computer servers that physically store information and make it accessible to customers online via the internet. It offers users to remotely transfer and store data and retrieve information as needed.

As the proposed DAQ device makes use of cloud technology, there is no need to buy a server, external drive, or memory stick to transfer information from one location to another. Cloud capacity works basically and simply

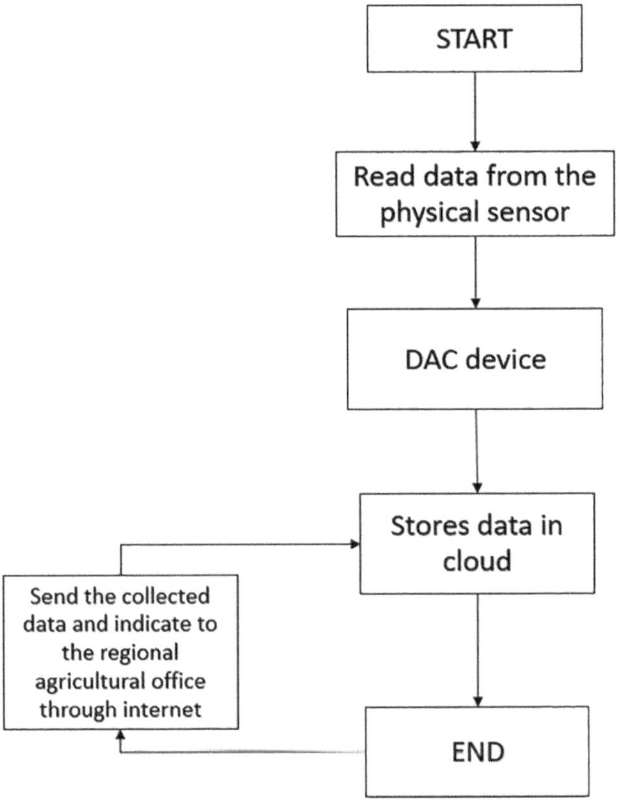

Figure 9.2 Proposed methodology flowchart.

where data is sent to cloud storage and information is stored in information centers, distributed around the world, and maintained by third parties. The information resides on a supported server and is easily accessible from the web interface. Cloud capacity uses a chain of servers that includes both control servers and other capacity servers. Then from there, pass the data for visualization and do real-time analysis. Even the data can be send to other apps like JSON and is capable of extracting it in CSV format. For XML data forms, the data is sent to the authorities using the SMTP protocol. SMTP or Basic Mail Exchange Convention is an application used to send, receive, and pass active emails between a sender and a collector. When an email is sent, it is exchanged over the web from one server to another utilizing straightforward information transmission. Thus, the working of data acquisition system and the data transfer wirelessly through the cloud needs more concentration. To make it sustainable and make it more efficient, an additional mini solar

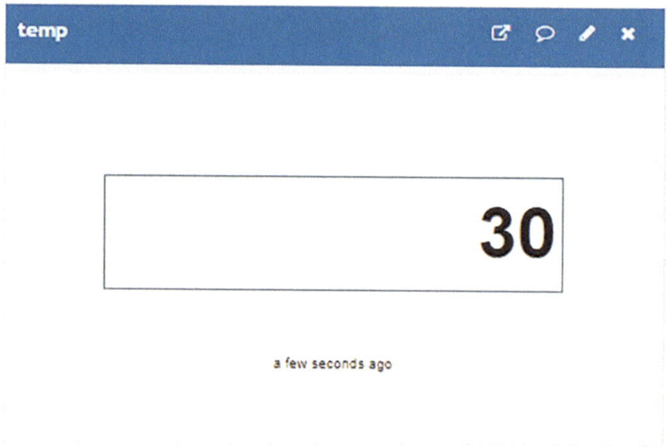

Figure 9.3 Temperature display.

panel is also embedded with the existing device to make it an efficient and eco-friendly device. Due to this feature, the device will be powered automatically by storing the current that is generated even when it is operating and it will recharge in the battery.

9.4 Results and Discussion

The proposed system has been implemented and the results have been extracted from the portable DAQ device where the different parameters are observed in different time intervals. Figures 9.3–9.5 show the data extracted through the respective sensor deployed with the various sensing device. The parameters measured are temperature, humidity, and moisture which are the main parameters that basically define the nature of the soil, its characteristics, and their availability to perform on the crop which is grown on it. The different plots like spline, bar, and column graphs are depicting the real-time representation of the data that are collected by the device for the dedicated parameters shown in Figures 9.6–9.8, where the data for a specific time of the data can also be viewed by clicking the graph where a legend will appear and it will show the data at that specific time and for a particular date and duration.

Figure 9.9 shows the data collected from the device and it has been stored in the CSV format. The data can also be stored in the cloud platform. This data can be directly used for ML deployment where it will increase the efficiency of the precision of the yield in the farming to be done on a specific soil type.

The data can also be extracted in different formats like XML and JSON as shown in Figures 9.10 and 9.11, where these two data will be more useful

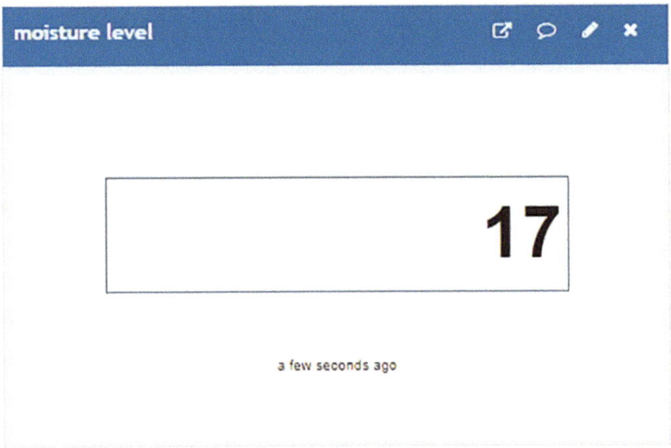

Figure 9.4 Moisture level display.

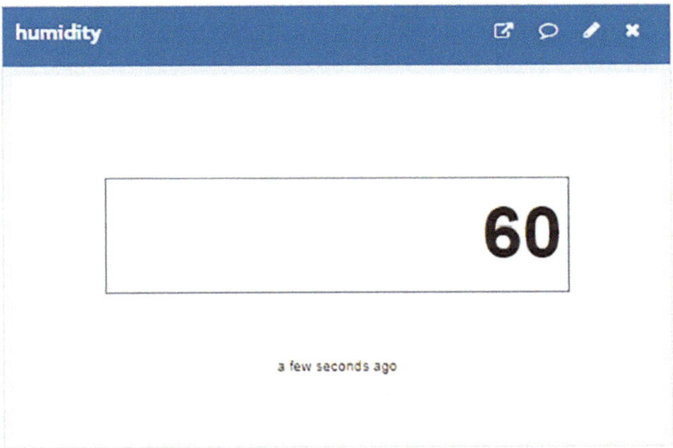

Figure 9.5 Humidity level display.

and efficient when an app or a web-based application is to be created, and it needs to be updated with live data. This system will support it as it is capable of parsing the data in a different type of formats. And it will be compatible to download the data in these formats to use them in real time with correlation with other applications if required.

9.5 Conclusion

The developed DAQ device for precision farming is capable of analyzing and extracting the physical data of the agricultural field. Thus, this system

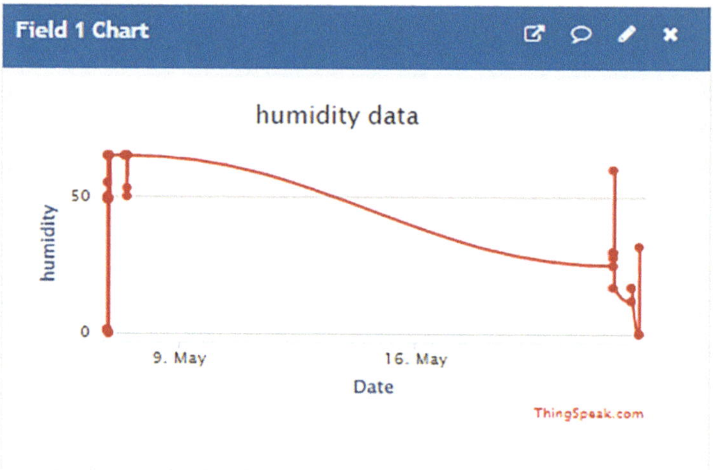

Figure 9.6 Graph representing the data of parameter humidity.

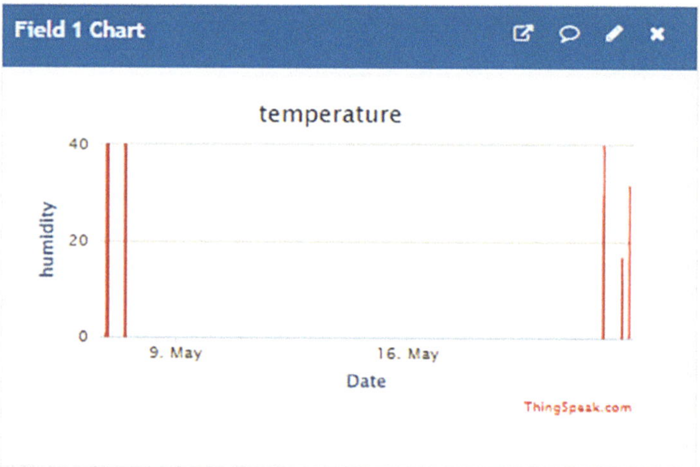

Figure 9.7 Graph representing the data of parameter temperature.

is available with different character analyzing capabilities and is capable of integrating it with different applications; it is completely compatible and portable and can be activated and deactivated whenever required. This device can be used in different places or objectives like precision farming where the data is much more important and used for analysis using artificial intelligence. For artificial intelligence the data plays a major role in smart farming to detect the abnormality of the parameters in the farming

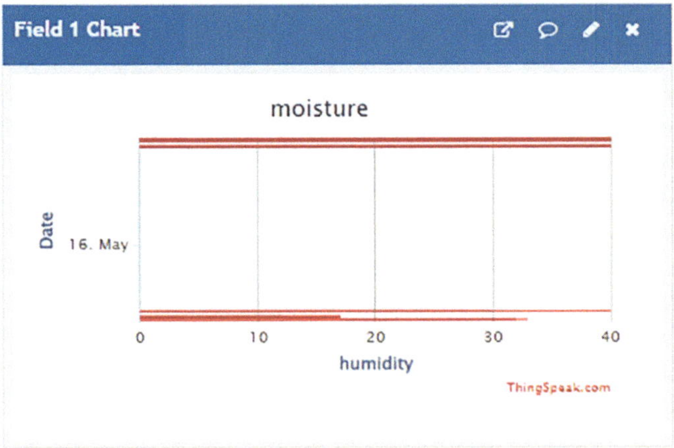

Figure 9.8 Graph representing the data of parameter moisture.

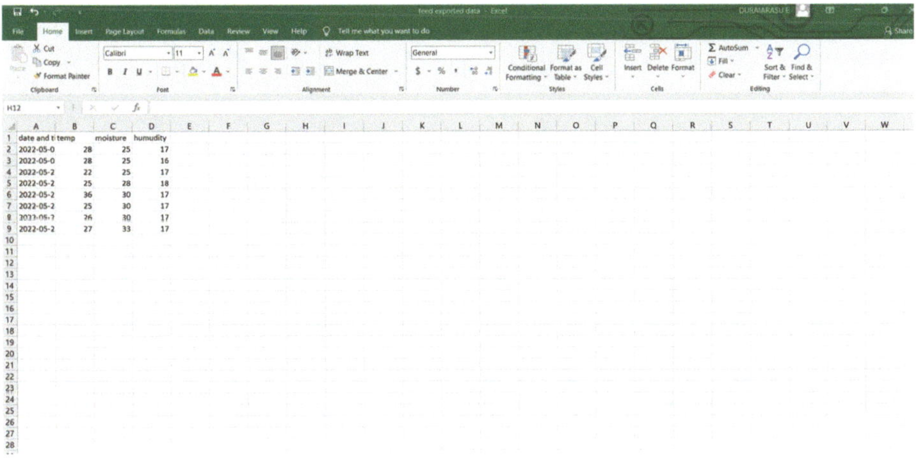

Figure 9.9 Extracted data in CSV format.

land or soil. This device, if recharged in the meantime, is even capable of acting as a smart plant or crop monitoring system for the terrace garden as well as the plants grown in bags. Then it can also be used in a hydroponic system way of growing plants by also changing the parameter with different thresholds.

The data for different land is stored in cloud and these data vary in real-time. Similarly multiple portable devices are used to collect different data and is used to feed the data of for various monitoring field or the ministry of agriculture to look into the future development of the agriculture and

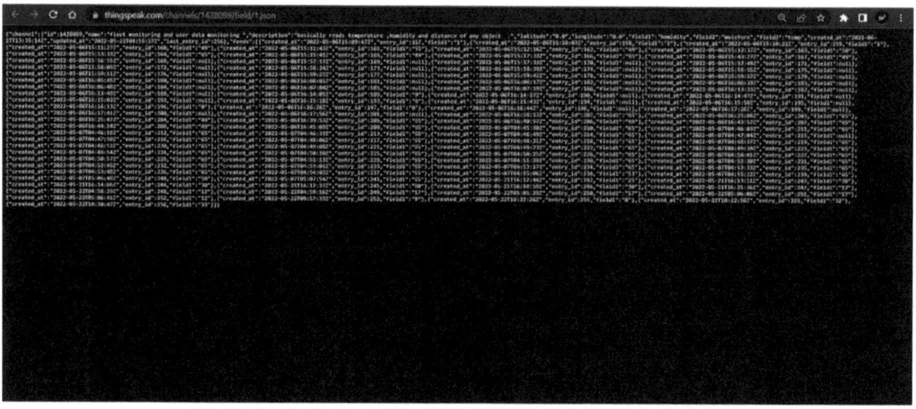

Figure 9.10 Extracted data in XML format.

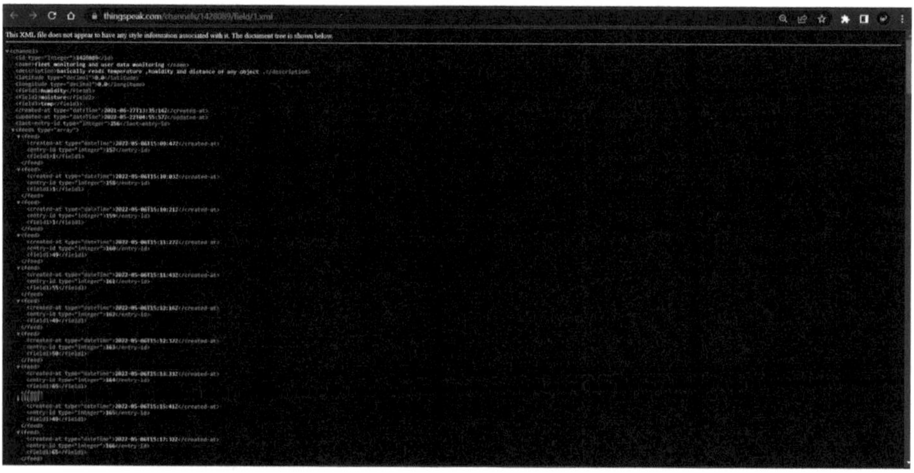

Figure 9.11 Extracted data in JSON format.

their productivity of the crops in the specific region. This data will be an asset for the government and the farm because even if the ML algorithms are being used in the precision farming technologies and the AI, the data should either be bought for the more precise data for the farming lands and there is no surety for the different data sample or collection of data for different soil types and crop variety; thus, this sorting device is in demand where it established all these facilities with these functions itself and it is compatible to use for different cases. The main advantage of this system is that this can be monitored from the same place; the data can be collected from different places and can also be viewed in the same place as a centralized monitoring system for the regional agricultural field testing and enhancement centers. As the data are extracted and converted to different formats, it has a wide opportunity in the future to use them for various applications like a dedicated app for VAOs (village administrative officers) and regional officers of the agricultural department to have a mini data center and data analyzed report for the agriculture fields which are responsible for overviewing. Then it is even convenient if the field inspector from the agricultural department is required to transfer the data and immediately report to the agricultural field. The data will be collected if he or she places the portable device in the fields; after that, the data will be collected and will be transferred to the higher authorities. Hence, this portable data acquisition system will be more efficient and cost-effective when compared to the other systems available in the market and will be more useful for the government and the farming industries.

References

[1] R. Dhanalakshmi, J. Anand, Arun Kumar Sivaraman, Sita Rani, 'IoT-based Water Quality Monitoring System using Cloud for Agriculture Use', Cloud and Fog Computing Platforms for Internet of Things, Edited by PankajBhambri, Sita Rani, Gaurav Gupta, Alex Khang, Routledge Taylor & Francis Group, 2022.

[2] J. Anand, J. Raja Paul Perinbam, 'Automatic Irrigation System using Fuzzy Logic', AE International Journal of Multidisciplinary Research, vol. 2(8), pp. 1–9, 2014.

[3] T. Manikandan, V. Nandalal, J.L. Mazher Iqbal, S. Joshua Kumaresan, S. Karthikeyan, 'Protection of plantation from wild animals usingwireless sensor network', Int.J. of psychosocial rehabilitation, vol. 24(5), pp. 2719–2723, 2020.

[4] Saurabh Kumar, Nagesh C. H, 'IoT-Based Multi-sensor Data Acquisition System for the Application of Smart Agriculture', Springer Advances in Computational Intelligence, Security and Internet of Things, ICCISIoT 2019, Communications in Computer and Information Science, vol. 1192, pp. 329–342, 2020.

[5] Daniela de Carvalho Lopes, José Helvecio Martins, Adílio Flauzino Lacerda Filho, Evandro de Castro Melo, Paulo Marcos de Barros Monteiro, Daniel Marçal de Queiroz, 'Aeration strategy for controlling grain storage based on simulation and on real data acquisition', Elsevier, vol. 63(2),pp. 140–146, 2008.

[6] Gsangaya K.R, Hajjaj S.S.H, Sultan, 'Portable, wireless, and effective internet of things-based sensors for precision agriculture, IJERT, vol. 17(9), pp. 3901–3916, 2020.
[7] Sean Tagle, Rica Pena, Frances Oblea, Hans Benoza, Nadine Ledesma, Jeremias Gonzaga, Laurence Lim, Gan A, 'Development of an Automated Data Acquisition System for Hydroponic Farming', Proc. In 10th International Conference on Humanoid, Nanotechnology, Information Technology,Communication and Control, Environment and Management (HNICEM) - Baguio City, Philippines, pp. 1–5, 2018,
[8] S. Castineira, T. Delwar, R. Duran and N. Pala, 'UAV-based agricultural monitoring and data acquisition system for precision farming', Proc. In SPIE, vol. 11754, 2021.
[9] SulaimaLebbe Abdul Haleem, Pravin R. Kshirsagar, Hariprasath Manoharan, Boppuru Rudra Prathap, Hemalatha S, Kukatlapalli Pradeep Kumar, Vineet Tirth, Saiful Islam, Raghuveer Katragadda, Temesgen Abeto Amibo, 'Wireless Sensor Data Acquisition and Control Monitoring Model for Internet of Things Applications', Hindawi, Scientific Programming, vol. 20, 2022.
[10] Ma, Qingyuan, Chen, Qiang, Shang, Qingsheng, Zhang, Chao, 'The Data Acquisition for Precision Agriculture Based on Remote Sensing, Proc. In Int. Conf. on Geoscience and Remote Sensing Symposium, pp. 888–891, 2006.
[11] Ajao.L.A, Agajo.J, Kolo.J.G, Maliki.D&Adegboye.M.A, 'Wireless Sensor Networks Based-Internet of Things for Agro-Climatic Parameters Monitoring and Real-Time Data Acquisition, *JASR*, vol. 7(6), 240–252, 2017.
[12] Reiser. D, Paraforos. D.S, Khan. M.T , Griepentrog. H.W, Vázquez-Arellano. M , 'Autonomous field navigation, data acquisition, and node location in wireless sensor networks', Springer, Precision Agriculture, vol. 18(3), 279–292, 2017.
[13] Dan Popescu, Florin Stoican, Loretta Ichim, GrigoreStamatescu, Dragana, Cristian, 'Collaborative UAV-WSN System for Data Acquisition and Processing in Agriculture', Proc. In 10th Int. Conf. on Intelligent Data Acquisition and Advanced Computing Systems: Technology andApplic. (IDAACS) - Metz, Franc, pp. 519–514, 2019.
[14] Ji-chun Zhao, Jun-fengZhang ,YuFeng, Jian-xinGuo, 'The study and application of the IOT technology in agriculture', Proc. In 3rdInt. Conf. on Computer Science and Information Technology (ICCSIT) - Chengdu, China, pp. 462–465, 2010.

[15] Chwan-Lu Tseng, Joe-Air Jiang, Ren-Guey Lee, Fu-Ming Lu, Cheng-Shiou Ouyang, Yih-Shaing Chen, Chih-Hsiang Chang, 'Feasibility study on the application of GSM-SMS technology to field data acquisition',*Elsevier*, vol. 53(1), pp. 45–59, 2006.
[16] N. Sharma, N. Sharma, P. Verma and R. Sharma, 'AgriTech: Data Acquisition in Agriculture', Proc. In Int. Conf. on Industrial Electronics Research and Applications (ICIERA), pp. 1–6, 2021.
[17] Li, Daoliang; Zhao, Chunjiang, 'Zigbee Wireless Sensor Network Nodes Deployment Strategy for Digital Agricultural Data Acquisition, Springer, IFIP Advances in Information and Communication Technology Computer and Computing Technologies in Agriculture, vol. 317, 2010.
[18] Tapankumar Basu, Vijaya Ravindra Thool, Ravindra C. Thool, Anjali C. Birajdar, 'Computer-Based Drip Irrigation Control System with Remote Data Acquisition System', Computers in Agriculture and Natural Resources, pp. 1–10, 2006.
[19] Seidel, Darren S. Field, Tyler C, Schinckel, Allan P, Stwalley, Carol S, Stwalley, Robert M, 'Effects of temperature probe orientation on the Purdue hog cooling pad data acquisition', Elsevier Computers and Electronics in Agriculture, vol. 175, pp. 105609, 2020.
[20] Mutha V. R, Kumar N, & Pareek P, 'Real-time standalone data acquisition system for environmental data', Proc. In 1st International Conference on Power Electronics, Intelligent Control and Energy Systems (ICPEICES) Delhi, India, pp. 1–4, 2016.
[21] Morais. R, Boaventura Cunha. J, Cordeiro. M, Serodio. C, Salgado. P, Couto. C, 'Solar data acquisition wireless network for agricultural applications', Proc. In 19th Convention of Electrical and Electronics Engineers in Israel - Jerusalem, Israel, pp. 527–530, 1996.

10

Decision-Making Support in Smart Farming

A. Jose Anand[1], K. Nirmala Devi[1], S. Vijayakumar[2], and Shiju C. Chacko[3]

[1]Department of ECE, KCG College of Technology, India
[2]Department of ECE, Paavai Engineering College, India
[3]Department of Electrical & Electronics Engineering, Uxbridge College, London
Email: joseanandme@yahoo.co.in; nirmalnkl03@gmail.com; svijiece@gmail.com; shiju83jose@gmail.com

Abstract

Modern agriculture mainly consists of automated techniques that lead to wastage of vital resources such as water. Water plays a significant part in the growth of plants and it is also regarded as a degrading resource on our planet. This chapter presents a solution to the problem stated above. In this system, the user can be able to manage the implementation of the sprinkler system with the help of a user-friendly Android application, thereby reducing the loss of water. They can activate the sprinklers only when the soil moisture is less. The quantity of moisture present in the soil is restrained with the help of a soil moisture sensor twice every 24 hours and if the value is less than the threshold, then a notification triggered with a global system for mobile (GSM) communication module will be sent to the customer's handset. Then the user can connect to the internet and then switch on the sprinklers by simply pressing a button that is available in the online-based application. This button activates the sprinklers by invoking a web service that turns the sprinklers on. This methodology is helpful for the farmers to utilize the water resource effectively.

10.1 Introduction

Agriculture is the word used to depict the production of food, fiber, and other things through forestry and farming. The introduction of agriculture was a crucial advance in the emergence of civilization and the domestication of animals. Agricultural science is the study of agriculture. Horticulture studies gardening, a related activity. There are numerous subspecialties of agriculture. Agriculture continues to be based on the farming of crops on the arable ground and the countrified herding of farm animals on rangeland. Sustainable agriculture and intensive farming have been distinguished during the previous century [1]. The use of modern agronomy, plant breeding, herbicides, fertilizers, and technological advancements has significantly enhanced crop yields. A significant part in the development of human civilization was played by agriculture. Up to the Industrial Revolution, agriculture was a cruel profession for the vast majority of people. The spread of these practices over some time is frequently referred to as an agricultural revolution. As agricultural systems have developed, agricultural production has constantly increased. Over the past century, there has been a notable change in agricultural methods as a result of new technologies. Particularly, the Haber–Bosch scheme for producing ammonium nitrate reduced the need for the conventional approach of recycling nutrients through crop rotation and animal dung [2]. The world's water resources are severely stressed. Fresh water for agriculture, drinking, and sanitation is obtained from rivers, lakes, and underground aquifers, while a sizable portion of the world's food supply is found in the oceans. However, these unique resources are now in danger in many regions of the world due to pollution, agriculture growth, damming, diverting, overuse, and diversion [3].

In particular, next to the watercourse now recognized as the Shatt al-Arab, which runs from its Persian Gulf delta to the flowing together of the Tigris and Euphrates, the Sumerians had developed core agricultural methods by 5000 BC, together with large-scale concentrated farming of land, mono-cropping, organized irrigation, and utilization of a specialized labor power. The domestication of untamed aurochs and mouflons into livestock and sheep, correspondingly, brought to the widespread usage of animals as food sources, sources of fiber, and pack animals. For sedentary and semi-nomadic communities, the shepherd joined the farmer as a vital provider. As early as 5200 BC, maize, manioc, and arrowroot were cultivated in the United States [4]. Farmers in North Africa and the Near East created and spread farming technology during the Middle Ages, together with irrigation schemes based on hydraulic and hydrostatic doctrine, the employment of Noria machinery, and the construction of dams, reservoirs, and water-raising machinery. They contributed to the widespread use of farms like sugar cane, paddy, citrus fruit, apricots, cotton, etc., as well as the writing of site-specific agricultural guides. Muslims also brought to Spain subtropical crops including bananas, cotton, almonds, figs, lemons, oranges, and other fruits and vegetables. The introduction of the Chinese-invented moldboard plough and the establishment of a three-field crop rotation system throughout the Middle Ages significantly increased agricultural productivity. The detection and succeeding farming of fodder crops, which permitted the overwintering of animals, was another significant advance toward the end of this period [5]. Following 1492, previously indigenous food and livestock breeds were traded all over the world. The tomato, maize, potato, cocoa, and tobacco were important crops that were intertwined in this conflict, as were more than a few varieties of sugarcanes, wheat, etc., that were brought from the ancient planet to the modern one. The horse and dog were the two most significant animal exports from antiquity to the present. The horse and the dog, despite not typically being used as food animals, swiftly took on crucial productivity responsibilities on farms in the western hemisphere [6]. Agricultural methods, tools, seed stocks, and domesticated plants were chosen and given distinctive names by the early 1800s because their decorative or practical qualities had advanced to the point where the yield per unit of land was many times greater than thought to be possible in the Middle Ages. Farming jobs are created at a speed and scale that were previously unimaginable, thanks to the speedy expansion of technology in the late 19th and 20th centuries, predominantly in the shape of the tractor. Due to increased productivity, some contemporary farming in the United States, Argentina, and other countries are now able to produce large quantities of premium goods on a given amount of land [7].

The Haber–Bosch technology for making ammonium nitrate marked a change and authorized crop yields to surpass earlier limitations. Agriculture over the previous century was distinguished by increased productivity, the use of labor-saving synthetic fertilizers and pesticides in place of natural ones, discerning breeding, mechanization, water stain, and farm subsidy. The unprocessed movement was born out of a repercussion besides the negative outside ecological effects of conservative agriculture in recent years [8]. In the meantime, in many parts of the world, agricultural research trips have been organized in the late 19th century to discover novel technologies and agricultural norms. Frank N. Meyer's fruit-nut gathering trip to China and Japan and the Dorsett–Morse Oriental Agricultural Exploration Expedition to China, Japan, and Korea to gather soybean germ plasma to support the expansion of soybean farming in the USA are two early examples of expeditions [9]. According to the International Monetary Fund, China produced the highest agricultural output in 2005, making up over a sixth of global production. The EU, India, and the USA were the next top agricultural producers. According to economists' measurements of the total factor productivity of agriculture, it is currently 2.6 times more productive in the United States than it was in 1948. Agri Surveil helps to reduce wastage of vital resources, and active treatment of fertilizer and thus increase the crop yield by frequently monitoring the thirst of the plant by checking the moisture content level [10]. Water is very critical in the field of agriculture in almost all places having a formative consequence on the ultimate yield. So it is essential to utilize the available resource efficiency so that the future generation is not deprived of it. Soil moisture has been used, which will be routed down in the crop field and it is programmed to monitor twice a day. If the humidity level is found to be less, then the consumer will receive a mobile notification through a mobile android application. The consumer has authorized power to control the sprinkler, which is routed down in the crop field. The processed data is stored in the cloud and the communication between the crop field and the application will be done through this cloud. The major advantage of this project proposal is that consumers need not be physically present to irrigate their crop field [11]. For a very small cost, field monitoring offers a solution to the food production issue. The return on your investment will occur within a reasonable amount of time depending on the crop you are growing. Agriculture water technology adaptation affects yield and cost savings. Our most precious resource is water, and producing more food with less water is not just a pipe dream. Every day, this reality is demonstrated. The applications of Agri Surveil include [12] low-cost field irrigation systems, pest control systems, crop protection systems, and home-based gardening systems. The chapter is structured as follows: Section 10.2 stretches a brief depiction of the associated work. Section 10.3 details the system design and implementation. Section 10.4 illustrates the outputs. In Section 10.5, conclusions are provided.

10.2 Related Works

This section discusses the relevant works in the arena of smart agriculture. An Internet of Things (IoT) service-oriented system [13] for agriculture observing briefly discusses precision agriculture, low energy consumption [14], wireless mesh network [15], aerial image mapping sensors like augmented reality (AR) drones with high-definition (HD) cameras, and the total data sampling and integration by the rules engine. For Agri Surveil, precision agriculture has been adopted, which makes the process simple and it can be controlled either manually or periodically. The aerial image mapping using the drones method is not adopted due to cost constraints. The mesh network method of connecting devices is adopted due to its flexible way of connecting, adding, or removing devices [16]. An IoT in precision farming uses wireless moisture sensor networks that deal with how sensor data are periodically sampled and shared over networks to the IoT [17]. It briefly discusses how the data has generated either contact-based or contact-free based, which confines for a different set of parameters. Also, it defines how value is generated by various means and the following processed steps. The data is shared over by ZigBee protocol of wireless communication, which is not followed in this proposal since adopted to develop an android application [18]. An android app to monitor the soil moisture condition with the use of Raspberry Pi, which will be sending the data from the sensor and the light emitting diode (LED), acts according to the data retrieved back from the android app [19]. Here, the LED setup has been done in a way to glow green, red, and blue colored LED based on the percentage retrieved from android. The android app will notify the status once in a while, whereas, in the application that has been designed, it is done for 12 hours. But the user can know the status only when the app is running [20]. The irrigation scheme has become a major consideration to give the farmer a smarter apparatus. It will help them to yield a good quality product. In this system, diverse types of sensors are utilized to monitor the field situation by using Arduino. Here, a Wi-Fi module attached to the Arduino-UNO-based controller is engaged to communicate with the sensor from a long distance in no time, which makes the user more productive. Long-distance communication is possible and a low-cost microcontroller is utilized. The total cost considered will also be less. Because of using a low-cost microcontroller, there will be less reliability [21].

10.3 System Design and Implementation

The following system has been proposed by utilizing the above components by arranging them as in the following architecture and circuit diagram in Figures 10.1 and 10.2, respectively. The proposed scheme contains both hardware as well as software. The software part includes the android application which must be mounted on the user's android phone. This android solicitation is advanced using Visual Studio. What the android application does is it consists of a single button. On clicking, it is coded to call the web server that includes the website. A web server is a program that has HTTP to provide files that make up web contents to clients in reaction to their needs, which are transmitted by HTTP users on computers. This website is one among the network that is created in IIS (Internet Information Services). This web server, which is shown in Figure 10.1, will act as an intermediate between the android application and the structured query language (SQL) database [22].

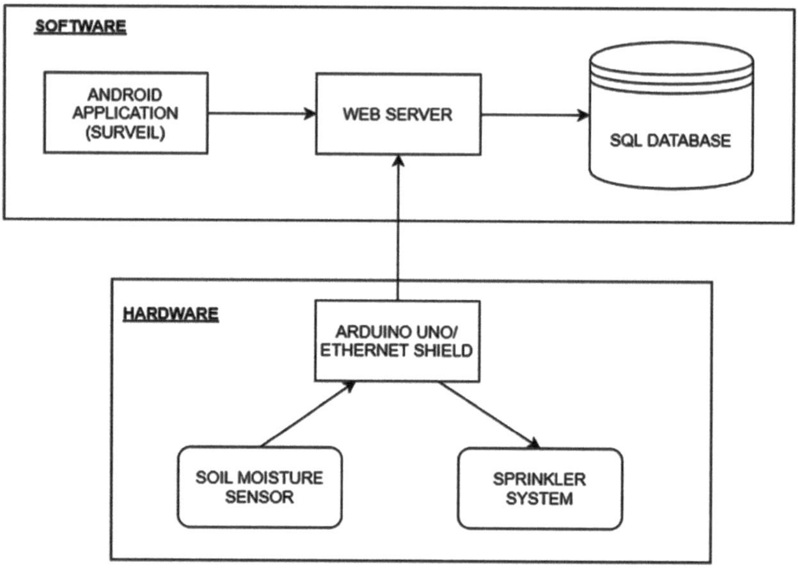

Figure 10.1 Architecture of the proposed system.

In the SQL database, a table has been created that holds the status of the sprinkler, whether it is on or off. It also includes a column for the time that it has been switched on or off. This is updated and the corresponding web server is invoked. The button in the android application is constantly monitored if it is clicked or not. Once clicked, the website in the web service is invoked, and also the table in the database present in the SQL server is updated to the current status of the sprinkler. The website that is invoked is based on how many times the button has been clicked. If it is an odd number of times, it sends an analog high signal to the sprinkler pin to initiate it. Otherwise, it is turned off. The hardware part includes the Arduino-UNO/Ethernet Shield and soil moisture and sprinkler system. The Ethernet shield is integrated with the UNO board and is assigned an IP address once it is given an Ethernet connection. Through this assigned IP address, the server can communicate with the board and sense from the sensor, and control the sprinkler actions. If the detected level from the soil moisture sensor is less than the inception and after the notification is sent and the button is pressed, then the sprinklers are activated. From the circuit diagram, the Ethernet cable is connected from the Arduino Ethernet shield to the Ethernet internet connection. The soil moisture sensor LM393 has been mounted to the third pin of Arduino [23].

Figure 10.2 shows the circuit diagram of the proposed architecture. There is an analogue input pin, and using the second pin, it has been

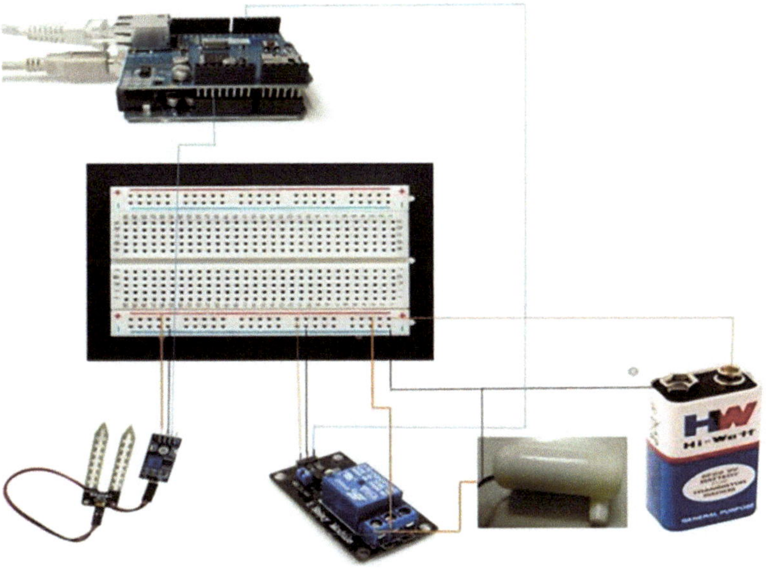

Figure 10.2 Circuit diagram of the proposed system.

connected. The sprinkler is connected to the third pin and in the coding is given a 0 or 255 as input to drive it. The sprinkler is allied via a relay system as shown above. The Ethernet shield is linked to the internet using an RJ 45 cable. Therefore, an IP address is allotted for the shield. The sprinkler is linked to the third pin of the Arduino, a free open-source platform with simple hardware components and software language. Boards can read inputs from things like a finger press on a button or light on a sensor, and they can transform those inputs to outputs like turning on or off an LED, starting a motor, or posting something online. Input/output (I/O) pins on the board, both digital and analogue, can be used to link to other circuits and expansion boards (shields). The board has 6 analogue pins and 14 digital pins. It may be programmed using the Arduino integrated development environment (IDE) with a type B USB connector. The Arduino Ethernet Shield enables an Arduino board to establish an internet connection. It uses the Wiznet W5100 Ethernet chip (datasheet). TCP and UDP are both supported by the IP stack that comes with the Wiznet W5100. It can simultaneously manage up to four-socket connections. Create sketches that connect to the internet via the shield by using the Ethernet library. With the help of lengthy wire-wrap headers that pass through the Ethernet shield, the Ethernet shield is connected to an Arduino board, as shown in Figure 10.2. This preserves the pin layout and allows for the stacking of another shield on top [24].

Sensors that detect soil moisture measure the volumetric water content of the soil. Soil moisture sensors using other sources quantify the water requirements by comparing the soil contaminant using the sensors and other electronic circuits. This is done because a direct gravimetric quantity of free soil moisture needs the removal, heating, and weighting of a sample. It is necessary to calibrate the relationship between the quantified property and soil moisture because it can alter depending on the kind of soil, the temperature, and the electric conductivity. Microwave radiation that is reflected and used for remote sensing in agriculture and hydrology is influenced by soil moisture. Portable probing instruments are useful for farmers and gardeners. Soil moisture sensors are widely used to describe sensors that estimate volumetric water content. Soil moisture measurement is critical in agricultural applications because it allows farmers to better manage their irrigation systems. Farmers can not only require minimum water to grow a crop when they know the accurate soil moisture situations on their farms, which can be measured with a capacitance as shown in Figure 10.3, but they can also improve yields and crop quality by better managing soil moisture during curious plant growth states.

Both the relational data stream management system (RDSMS) and relational data base management system (RDBMS) use the programming language SQL to handle data. SQL provides the idea of connecting multiple records with a single command, and it also does away with the requirement to mention the record fields. These two features make SQL superior to prior read/write APIs. Data query language (DQL), data definition language (DDL), data control language (DCL), and data manipulation language (DML) are just a few examples of the various statement types that make up SQL. Although SQL is sometimes referred to be a declarative language, which is to a large part, it also has procedural components. It became the most used database language despite not completely adhering to Codd's relational paradigm. In 1986, both the American National Standards Institute (ANSI) and the International Standardization Organization (ISO) recognized SQL as a standard. Since then, a wider range of functionality has been added to the standard. Despite the availability of such standards, the majority of SQL code requires modifications to work across several database schemes [25].

Donald D. Chamberlin and Raymond F. Boyce created SQL at IBM in the early 1970s after learning about the relational paradigm from Ted Code. This version, originally known as SEQUEL (Structured English Query Language), was created to access and handle data held in System R, which is a quasi RDBMS from IBM that is a team at the company's San Jose Research Laboratory. Square was Chamberlin and Boyce's initial trial at an

RDBMS, but its subscript notation made it challenging to use and developed the SEQUEL. Because "SEQUEL" was a trademark of the UK-based aircraft manufacturer Hawker Siddeley, the abbreviation SEQUEL was later altered to SQL. Seeing the performance by Codd, Chamberlin, and Boyce in the late 1970s, Relational Software, Inc. (now Oracle Corporation) created its own SQL-based RDBMS intending to sell it to the U.S. Navy, Central Intelligence Agency, and other U.S. government agencies. Oracle V2 (Version2) for VAX systems, the first commercially available SQL implementation, was released in June 1979 by Relational Software, Inc. The standard "Database Language SQL" language definition was formally adopted by ANSI and ISO standard bodies in 1986. In 1989, 1992, 1996, 1999, 2003, 2006, 2008, 2011, and most recently (2016), the standard was updated.

Data retrieval based on specific criteria is done via queries. This is a key component of SQL. Statements that have the potential to govern transactions, program flow, connections, sessions, and diagnostics as well as schemata and data are in the long term. SQL statements can also be terminated with a semicolon (";"). It is regarded as a standard component of the SQL syntax even if it is not necessary on all platforms. SQL statements and queries typically ignore small amounts of whitespace, which makes it simpler to prepare SQL code for readability. SQL helps in data manipulation, which is essential to generate the SQL table by eliminating all possible threats. Figure 10.3 shows that the whole process is cloud- and web-server-based, which works and provides services only on the real-time data from the application.

A tool used to water crops, lawns, landscapes, golf courses, and other locations is an irrigation sprinkler. They also serve to keep dust out of the air and to keep things cool. Sprinkler irrigation is a sort of irrigation technique that simulates rainfall in nature. Water is distributed through a network of pipelines using pumps. Sprinklers are used to spray the water into the air, where it fragments into tiny water drops that fall to the earth. Sprays or spray heads are the common names for sprinklers with a set pattern of spray. Sprays are often not intended to operate under pressure because misting issues could arise. A ball drive, gear drive, or impact mechanism propels higher pressure sprinklers that revolve around themselves (impact sprinklers). These can be made to circle completely or partially. Watering residential lawns often involve moving an oscillating sprinkler as needed. Home lawn sprinklers come in a wide range of sizes, prices, and levels of complexity. They consist of subterranean sprinkler systems, impact sprinklers, oscillating sprinklers, drip sprinklers, and others. Hardware stores and home and garden retailers sell tiny sprinklers for affordable prices. These are often placed momentarily and connected to an outdoor water faucet. Sprinklers installed underground

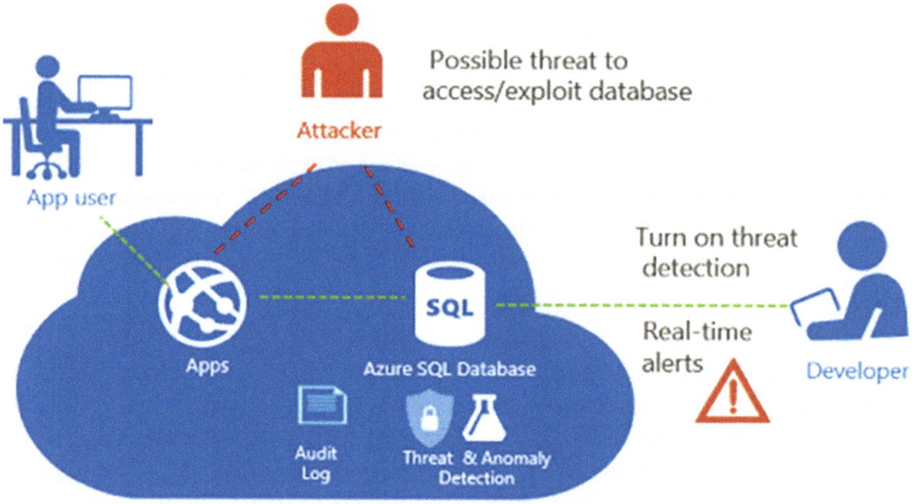

Figure 10.3 A real-time use of SQL database.

operate using simple electronic and hydraulic systems. A zone is a collective name for this valve and all of the sprinklers it will trigger. When the valve is opened, a small stainless steel plunger in the center is lifted by a solenoid that is magnetic and sits on top of the valve. This allows air to escape from the top of a rubber diaphragm that is situated in the middle of the valve due to the triggered (or elevated) plunger. The diaphragm is raised by water that has been charged and is currently sitting at the bottom of the same diaphragm. Farmers used sprinklers first in the form of residential and golf course sprinklers. While replacing the subterranean pipes and fixed sprinkler heads, these *ad hoc* systems interfered with farming and were expensive to operate. The rolling pipe type irrigation system, which has grown to be the most popular form for farmers irrigating big fields, was created for farms in the 1950s by a company based in Portland, Oregon, called Stout-Wyss Irrigation System. With this method, the big pipes with sprinkler heads attached to them travel slowly across the field.

A Microsoft IDE is Microsoft Visual Studio. It is used, among other things, to create websites, web applications, online services, and mobile applications. Among the Microsoft software development tracks utilized by Visual Studio are Windows API, Windows Forms, Windows Presentation Foundation, Windows Store, and Microsoft Silverlight. Both native and managed code can be produced by it. The code editor that comes with Visual Studio allows code refactoring and IntelliSense (code completion). Both source-level and machine-level debugging is possible with the integrated debugger. The additional off-the-shelf tools, such as code profiler, to design form through GUI applications, a web design tool, a class design tool, and a database design application. It receives plug-ins that motivate functionality almost at all stages, such as providing support for source control programs and new toolkits for programming languages with specialized functions, like editors and visual design tools, or toolkits for different stages of the software development lifecycle. Through plug-ins, support for additional languages including M, Node.js, Ruby, Python, and others is possible. In the past, Java and J# were supported.

Microsoft's extensible web server for the Windows NT operating system is called Internet Information Services (IIS), originally known as Internet Information Server. IIS supports the following protocols: FTP, HTTP, HTTP/2, HTTPS, etc. Despite not being active by default and possibly being absent in some editions, it has been a member of the Windows NT family. Although the security flaw CA-2001-13 that resulted in the infamous Code Red attack affected IIS 4 and IIS 5, versions 6.0 and 7.0 have no known concerns with this particular flaw. Microsoft decided to alter the pre-installed ISAPI handlers' behavior in IIS 6.0 to decrease the attack surface of IIS. Many of these handlers were responsible for the vulnerabilities in versions 4.0 and 5.0. Figure 10.4 depicts a typical screenshot of the IIS Manager.

IIS 5.1 and earlier, by default, operate the websites depicted in Figure 10.4, which shows the scenario of the system specifications, a Windows login with administrative privileges. Provided an extraordinary circumstance the work process operation is in under IIS 6.0, the vulnerability feature may not completely give the overall system performance because every action handling process is executed using Network Service account, which has a lot less access. Additionally, IIS 6.0 included a new HTTP stack in the kernel that included a response cache for both static and dynamic content and a tighter HTTP request parser. Furthermore, IIS 6.0 introduced a function known as "Web Service Extensions" that stops IIS from starting any program without specific authorization from an administrator.

Android applications are created using the Android Studio IDE for Google's Android operating system and are built on IntelliJ IDEA software from JetBrains. Users of Windows, macOS, and Linux can get it. As the main IDE for creating native Android apps, it replaces the Eclipse Android Development Tools (ADT). Android Studio was introduced on 16 May 2013, during the Google I/O conference. Beginning with version 0.1 in May 2013,

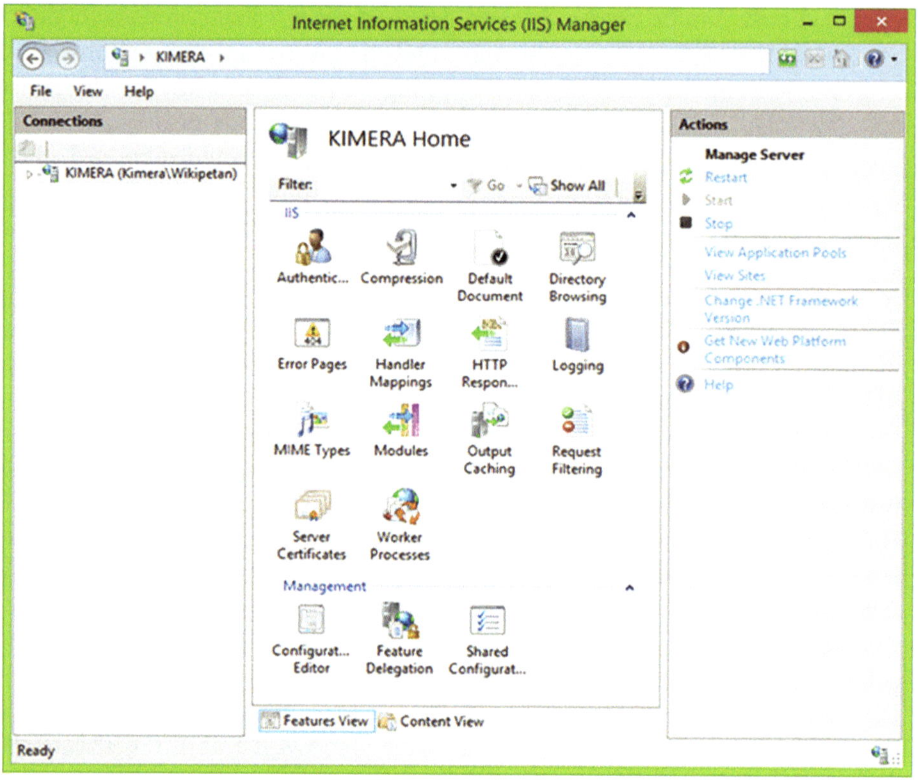

Figure 10.4 Typical IIS Manager.

It was in an early access preview stage until transitioning into the beta stage with version 0.8 in June 2014.

Figure 10.5 describes and shows as a list the various available emulators that are connected with the promised prototype. To run a particular emulator, the small green button is pressed. The same programming languages supported by PyCharm and IntelliJ, such as Python and Kotlin, are also supported by Android Studio. Version 3.0 of Android Studio also supports "Java 7 language facilities and a part of Java 8 language geographies that differ SQL by policy definition." Some Java 9 features are backported by external projects.

10.4 Execution Outputs

Figure 10.6 shows the flow diagram for the testing process. First, the moisture level is monitored from the farmer's field by using an LM393 soil moisture

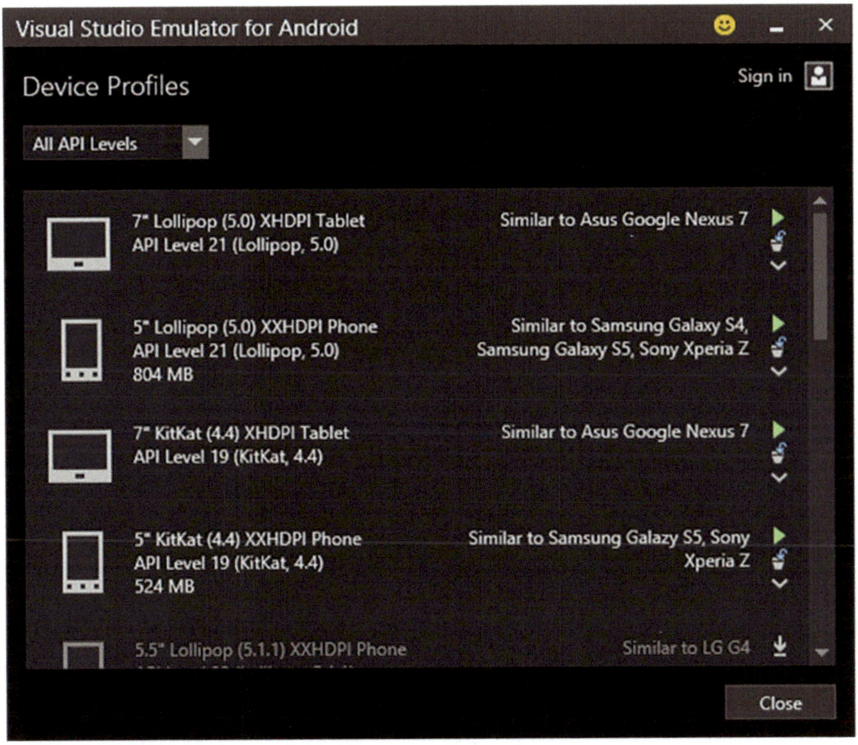

Figure 10.5 List of virtual simulators in Android emulator.

sensor. This is done when the sensor detects a value, which is greater than the threshold; no further action will take place. It starts to monitor the field again. If it is not greater than the threshold, then the notification is sent to the user's android mobile. After that, the user has to authenticate the further steps. When the button is pressed by the user, then check the web server, whether it is A.B.C/$1 or A.B.C/$2. The website that is invoked is based on how many times the button has been clicked. If it is an odd number of times, it sends an analog high signal to the sprinkler pin to initiate it. Otherwise, it is turned off. The hardware part includes the Arduino-UNO/Ethernet Shield and soil moisture and sprinkler system. The Ethernet shield is integrated with the UNO board and is assigned an IP address once it is given an Ethernet connection. Through this assigned IP address, the server can communicate with the board and sense from the sensor, and control the sprinkler actions. If the detected level from the soil moisture sensor is less than the threshold and after the notification is sent and the button is pressed, then the sprinklers are activated. Table 10.1 enlists the web service that is invoked and the status of

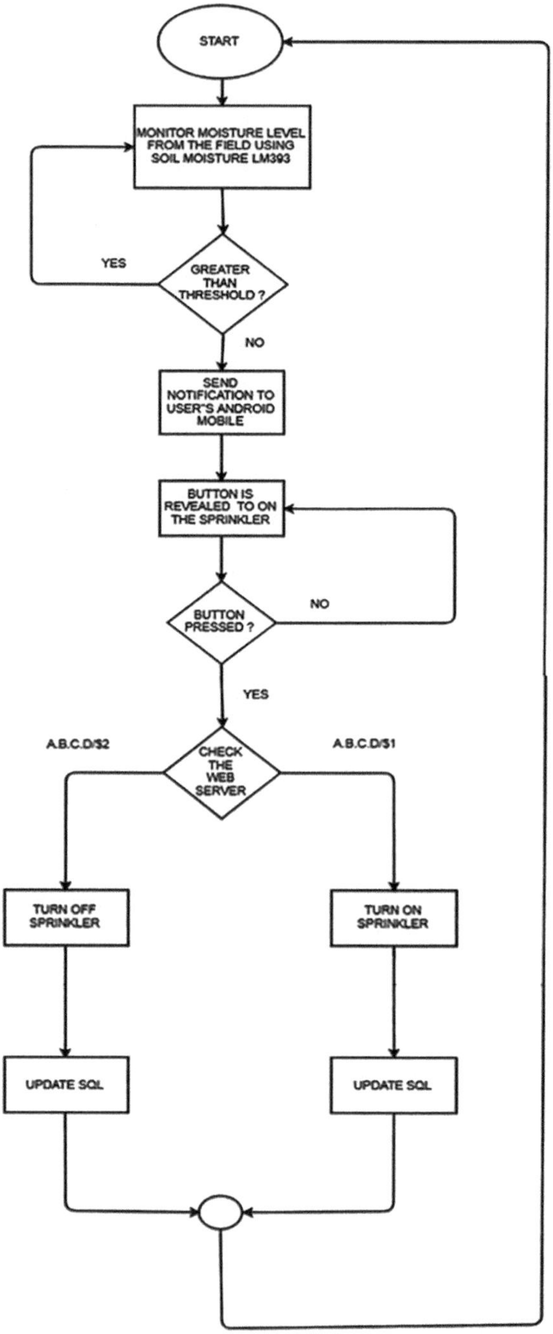

Figure 10.6 Testing flow diagram.

Table 10.1 Invoked web service and status of the sprinkler.

Status of sensor	Invoked web service	Action of the sprinkler
HIGH	A.B.C.D/$1	ON
LOW	A.B.C.D/$2	OFF

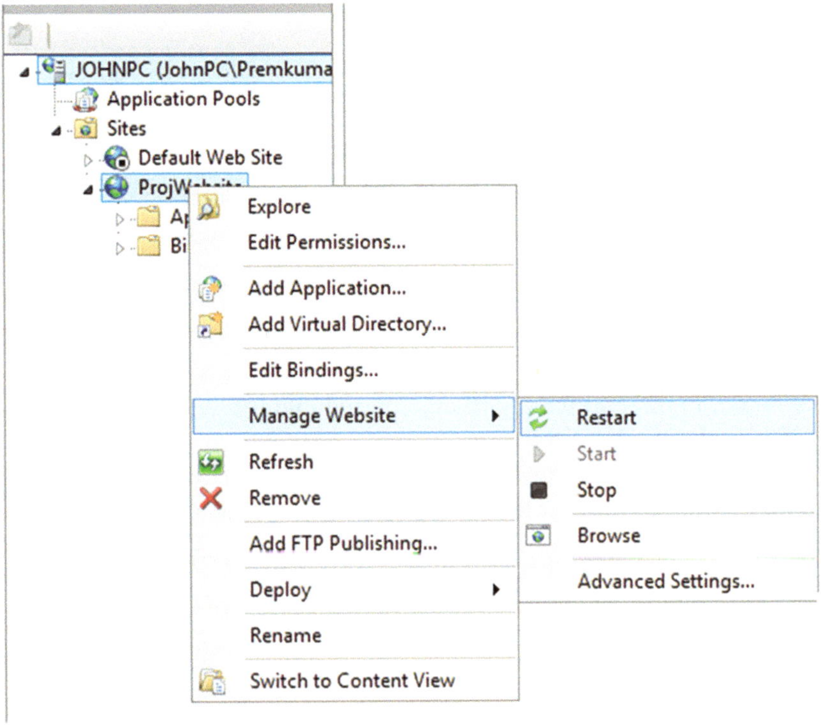

Figure 10.7 Refreshing the local network.

the sprinkler that has been initiated in the Arduino code. When the detected digital signal from the sensor is HIGH, then automatically the web service A.B.C.D/$1 is invoked. Otherwise, the sprinkler will be in the off state by calling A.B.C.D/$2. After the implementation of the circuit diagram, the following results have been recorded for the implementation.

The refreshing of the local network is shown in Figure 10.7. The local network that includes the web server being used is restarted to refresh the website. This local network is the network to which the PC is connected.

The IP address allocated to the Arduino Ethernet shield is 192.168.1.4 as shown in Figure 10.8. The Arduino serial monitor displays this information along with the switching on and off of the button in the application.

```
LocalIP address of the arduino:192.168.1.4
Server Started !!!
ON
OFF
ON
```

Figure 10.8 IP address allocation in the serial monitor of the Arduino.

Figure 10.9 shows the button in the virtual device that is obtained using the emulator and also the screenshot of the application through the emulator. Here, the button displayed has been clicked three times, which turns on, then off, and then on the sprinklers only if the water level below the threshold has been sensed through the water sensor connected to the Arduino.

SQL Management Studio 2014 is initiated as shown in Figure 10.10 to display the login details as shown above. Login details such as server type, server name, login, and password are entered and the server is connected.

Whenever the button is pressed in the application, the database will be updated as shown in Figure 10.11 for both "ON" and "OFF," along with the accurate time of clicking. The database can be opened in the SQL server by finding the table under the database on the left side. This table is updated three times and is as shown above, on the left side.

10.5 Conclusion

Irrigation is the application of a measured quantity of water to plants at desired intermissions. This helps nurture crops, uphold landscapes, and re-vegetate bothered soils in dry regions and throughout periods of not as much of as average rainfall. This project is related to the above-mentioned issue. When the land becomes dry, the user clicks on the button in the app and it will fill water in the land after the set time, and then the water will stop filling the land. Agriculture is the farming and breeding of animals, plants, and humans, but a recent survey says that between 2001 and 2010, over 2600 farmers committed suicide in the state due to no water on the land. The proposed project will surely help the farming lands and agriculture.

In the upcoming generations, the cloud can also be used instead of the database for faster communication. Furthermore, many sensors, such as NPK sensor, auto crop chopper, and pesticide level sensor, can be integrated using Arduino Mega. The NPK sensor can be used to detect the degrading levels of nutrients. Unfortunately, there is no single sensor on the market that can be

Figure 10.9 Image of the button in the virtual device using emulator.

used to test for the levels of NPK in the soil. These give accurate readings and are very expensive. Other levels such as herbicides, insecticides, and pesticides help to improve the environment for plant growth and to destroy plants. Those levels are measured and are made to indicate in the application as a warning. In future, advanced miniature sensors can be used for power savings.

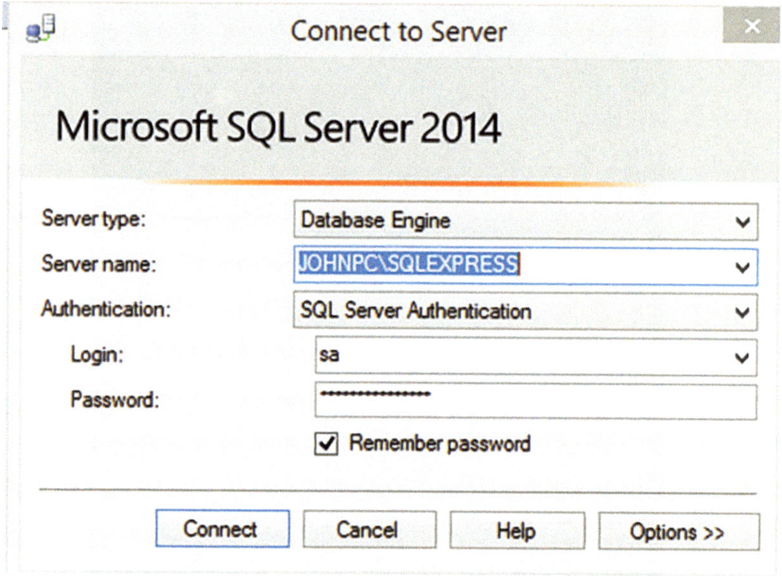

Figure 10.10 Connection to the SQL server by entering details.

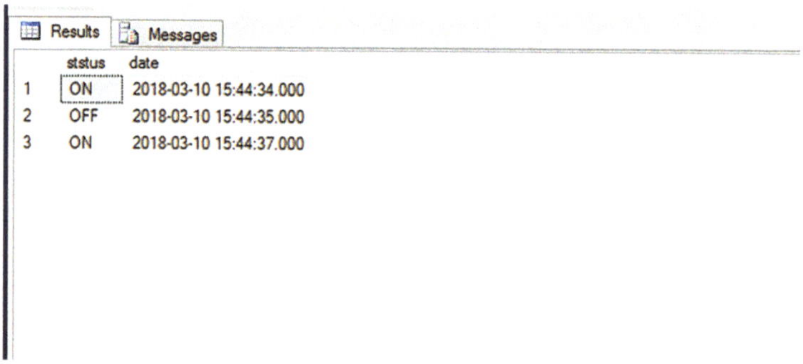

Figure 10.11 Status of the sprinkler updated in the SQL database.

References

[1] S. Ssushitha and H. Hegde, "Smart Farming-a key to Sustainable Agriculture Development in India" -A Study," 2022 4th International Conference on Smart Systems and Inventive Technology (ICSSIT), pp. 63–71, 2022.

[2] A. Khomenko, Z. Shen and S. O. Macheret, "Production of Liquid Nitrate Using Atmospheric Air Plasma Driven by Short Repetitive Pulses," 2018 IEEE International Conference on Plasma Science (ICOPS), pp. 1–1, 2018.

[3] Dhanalakshmi R., J. Anand, A. K. Sivaraman, and S. Rani, "IoT-based Water Quality Monitoring System using Cloud for Agriculture Use", Cloud and Fog Computing Platforms for Internet of Things, Edited by Pankaj Bhambri, Sita Rani, Gaurav Gupta, Alex Khang, Routledge Taylor & Francis Group, May 2022.

[4] Simon F., Bruce T., and Emma J., "Digitalization of Agricultural Knowledge and Advice Networks: A State-of-the-art Review", Agricultural Systems, Vol. 108, pp. 102763, April 2020.

[5] Evans A., Luyong Z., Lingmao T., Fuseini I., Kai T., Jiuhao L., and Lian H., "Moldboard Plowing with Direct Seeding Improves Soil Properties and Sustainable Productivity in Ratoon Rice Farmland in Southern China", Sustainability, Vol. 11, Issue 22, 2019.

[6] Evangelos D. Lioutas, Chrysanthi C., Giuseppe L. R., and Marcello D. R., "Key Questions on the Use of Big Data in Farming: An Activity Theory Approach", NJAS – Wageningen Journal of Life Sciences, Elsevier, Vol. 90–91, Dec. 2019.

[7] R. Eaton, J. Katupitiya, K. W. Siew and K. S. Dang, "Precision Guidance of Agricultural Tractors for Autonomous Farming," 2008 2nd Annual IEEE Systems Conference, pp. 1–8, 2008.

[8] E. Ramya, J. Anand, R. Renugha Devi, N. A. Issac, and K. R. Prasenth, "Solar Grass Cutter with Water Spraying Vehicle," 2021 International Conference on Advancements in Electrical, Electronics, Communication, Computing and Automation (ICAECA), pp. 1–6, 2021.

[9] Sushant W., Nishtha T., Ruturaj H., and Pragati T., "Farmer Friendly Application for Resource Mapping of Village with Government Aided Schemes", International Journal of Engineering Science and Computing, Vol. 7, Issue 12, pp. 15736–15639, 2017.

[10] T. A. Khare and A. C. Phadke, "Automated Crop Field Surveillance Using Computer Vision," 2020 IEEE International Conference on Distributed Computing, VLSI, Electrical Circuits and Robotics (DISCOVER), pp. 105–110, 2020.

[11] J. Anand, and J. Raja Paul Perinbam, "Automatic Irrigation System using Fuzzy Logic", AE International Journal of Multidisciplinary Research, Vol. 2, Issue 8, pp. 1–9, August 2014.

[12] M. Ayaz, M. Ammad-Uddin, Z. Sharif, A. Mansour, and E. -H. M. Aggoune, "Internet-of-Things (IoT)-Based Smart Agriculture: Toward Making the Fields Talk," in IEEE Access, Vol. 7, pp. 129551–129583, 2019.

[13] Jose A., Gowtham H., Lingeshwaran R., Ajin J., and Karthikeyan J., "IoT Based Smart Electrolytic Bottle Monitoring", Advances in Parallel

Computing Technologies and Applications, IOS Press, Vol. 40, pp. 391–399, November 2021.

[14] J. Anand, and J. Raja Paul Perinbam, "A Survey on Energy Efficient Biomedical Wireless Sensor Networks", American International Journal of Research in Science, Technology, Engineering & Mathematics, Vol. 3, Issue 7, pp. 212–216, June-August 2014.

[15] Helen Samuel, Sharnee Kaul, and J. Anand, "A Secure Routing Technique for Wireless Sensor Networks", International Journal of Engineering Research & Technology, Vol. 3, Issue 3, pp. 275–279, March 2014.

[16] C. Cambra, S. Sendra, J. Lloret, and L. Garcia, "An IoT Service-oriented System for Agriculture Monitoring," 2017 IEEE International Conference on Communications (ICC), pp. 1–6, 2017.

[17] J. Anand, A. Jones, T. K. Sandhya, and K. Besna, "Preserving national animal using wireless sensor network based hotspot algorithm," 2013 International Conference on Green High-Performance Computing (ICGHPC), pp. 1–6, 2013.

[18] I. Mat, M. R. Mohd Kassim, A. N. Harun, and I. Mat Yusoff, "IoT in Precision Agriculture Applications using Wireless Moisture Sensor Network," 2016 IEEE Conference on Open Systems (ICOS), pp. 24–29, 2016.

[19] J. Anand, T. G. A. Flora, and A. S. Philip, "Finger-Vein based Biometric Security System" International Journal of Research in Engineering and Technology, Vol. 2, Issue 12, pp. 197–200, December 2013.

[20] L. P. Dewi, J. Andjarwirawan, and R. P. Wardojo, "Android Application for Monitoring Soil Moisture Using Raspberry Pi," 2017 International Conference on Soft Computing, Intelligent System and Information Technology (ICSIIT), pp. 178–184, 2017.

[21] P. Singh and S. Saikia, "Arduino-based Smart Irrigation using Water Flow Sensor, Soil Moisture Sensor, Temperature Sensor, and ESP8266 WiFi Module," 2016 IEEE Region 10 Humanitarian Technology Conference (R10-HTC), pp. 1–4, 2016.

[22] O. Friha, M. A. Ferrag, L. Shu, L. Maglaras, and X. Wang, "Internet of Things for the Future of Smart Agriculture: A Comprehensive Survey of Emerging Technologies," in IEEE/CAA Journal of Automatica Sinica, Vol. 8, no. 4, pp. 718–752, April 2021.

[23] M. Rohith, R. Sainivedhana and N. Sabiyath Fatima, "IoT Enabled Smart Farming and Irrigation System," 2021 5th International Conference on Intelligent Computing and Control Systems (ICICCS), pp. 434–439, 2021.

[24] Ajesh, Mubeena Perveen T., Prashanth K. H. R., Abdul Jabbar, "IoT Based Intelligent Farming", Journal of Emerging Technologies and Innovative Research (JETIR), Vol. 6, Issue 5, pp. 75–79, 2019.

[25] Xing Yang, Lei Shu, Jianing Chen, Mohamed A. F., Jun Wu. Edmond Nurellari, and Kai Huang, "A Survey on Smart Agriculture: Development Modes, Technologies, and Security and Privacy Challenges", IEEE/CAA Journal of Automatica Sinica, Vol. 8, Issue 2, pp. 273–302, Feb 2021.

11

Indigenous Knowledge in Smart Agriculture

Pritam R. Ahire[1], Rohini Hanchate[1], and Vijayakumar Varadarajan[2]

[1]Department of Computer Engineering, Nutan Maharashtra Institute of Engineering Technology, Savitribai Phule Pune University (SPPU), India
[2]School of Computer Science and Engineering, The University of New South Wales, Sydney, Australia (UNSW Ranks 43rd in the 2022 QS World University Rankings, 1st in Australia for Research Excellence and Impact)
Email: pritamahire33@gmail.com; rohini.shanchate@gmail.com; v.varadarajan@unsw.edu.au

Abstract

Farming is not only affected by global warming, but it also contributes around 24% of greenhouse gas emissions. Another sobering truth is that the compositions of greenhouse gases have risen dramatically since the industrial revolution. Some modifications occurred as a result of human activity and/or overuse of natural resources. The type and extent of climatic change's effect(s) or danger differ depending on where you live. On the other hand, agriculture is itself affected by other aspects of society, including its cultural and economic values. Ages of humanity have built strong ties with the ecosystem, identifying natural patterns and inventing techniques to access nature's advantages before the advent of "Western Science," and even before science had a name. Forestry, farming, and aquaculture practices evolved over time as a result of trial and error in determining the optimum ways to practice agriculture in order to extract resources for survival, with some methods being more sustained than others.

Bihar is one of the Indo-Gangetic Plain states that have been negatively affected by climate change due to modest and distributed land holdings, i.e., around 16.1 million farm assets, with 91% being negligible. As a result, there is an urgent need to create location-specific sustainable agriculture methods to mitigate the effects of climate change.

As a process of modern agriculture and global demand of different agro-products, an abrupt loss of traditional agricultural knowledge related to rapid transformations and intensification of agricultural systems have been observed. More importantly, the culture that promoted such farmer experimentation still exists today, which, in turn, leads the sustainability in agriculture. Farmers not dependent on technology solutions usually solve using traditional methods. There is an urgent need to address. First, indigenous knowledge practices for sustainable agriculture can be done; second, how smart agriculture using wireless sensors network & IoT is helpful for agriculture; third, impacts of climate changes on smart agriculture; last, how machine learning concepts can be used to deal with indigenous knowledge (IK) and relate to current science technology solutions for handling change in climate.

11.1 Introduction

There are various key definitions for indigenous knowledge (IK). It is defined as knowledge passed down through the ages for more than a substantial amount of time from our ancestors to upcoming future generations. IK is "the ability to solve or to deal with problems in life by summing up the knowledge and practices that are based on best practices or experience of the people." Indigenous traditional knowledge [1] focuses on the unique, traditional, and local knowledge that exists within any culture or society and is established around the unambiguous circumstances of indigenous women and men in a certain geographic area and expressed in the form of myths, cultural stories, older methodologies, laws, agricultural practices such as bed preparation, sowing, plantation, management practices, and so on, which are then passed down to future generations.

The fundamental nature of IKs prevents over-exploitation of natural resources [2], proper use and conservation of existing natural resources, and application for current resources, among other things. ITKs are typically location-specific, based on the availability of resources and localized materials, and are the result of long-term familiar methods of research.

11.1.1 Indigenous knowledge is found to be:

- Sociologically desired
- Affordable and sustainable
- Long term

- Low danger to end user
- Conservation of resources
- Environmentally friendly
- Less capital intensive
- More labor intensive
- Profitable
- Proficient by-product usage
- Waste reprocessing

11.1.2 Why is indigenous knowledge needed for sustainable agriculture?

The use of IKs in agriculture was investigated; it was discovered that the majority of them are used in complex, risky locations and that they are generally used by minor and fringe farmers in growing cities [3]. IKs have scientific performance measures, and the most recent strategy in the age of modern agricultural technology includes integrating IKs into the qualitative research and assessing their scientific credibility in order to deliver quality and highly efficient location-specific alternatives in agro.

Farmers' engagement at the highest levels is expected whenever collective action is planned to promote developmental programs and provide a suitable climate for speedy information exchange and technology transfer. Farmers are more likely to participate in group action when they have access to IKs and local expertise. It is also worth noting that sustaining IKs is critical for aboriginal communities and farmers' existence because they are deeply established in their culture and customs.

The concept of sustainable development refers to the long-term management of agricultural resources, such as cultivation, both regular and periodic, agroforestry, and livestock, as well as the conservation measures required to ensure their long-term availability.

Erosion, overgrazing, desertification, excessive agricultural intensification, overfishing, and cultivation on marginal areas all contribute to the depletion of non-renewable resource bases, lowering agricultural yield potential and lowering/maintaining available capacity of farming [4]. To avoid the degradation of renewable or non-renewable resources, better management is required to build agricultural production systems that support our ecosystem.

11.1.3 Major IK practices for sustainable agriculture

- Water management
- Nutrient management
- Pest management
- Soil conservation
- Crop management

11.1.3.1 Water management

Techniques for predicting rain and preserving water are classic instances of information that will be handed down across generations in many conventional farming systems [5]. It is difficult to save water resources along with knowledge of rain and its different patterns, different time zones, and quantity in which it falls. Farmers might choose which crops and crop varieties to grow if they could predict how much rain will fall in the future season.

Major water management practices are:

- Inter-row and inter-plot water harvesting: Plants are typically planted in ridges in heavy soil, and ridges and furrows are usually constructed on cultivated fields. Such activities aid in the collection of rainwater in furrows rather than surface runoff. Inter-plot harvesting rainwater also entails the building of a cultivating plot on one side and a fallow plot on the other, with the fallow plot sloped to induce runoff to the next cropping plot.
- Rainwater harvest: During dry weather, harvesting periodical rain in reservoirs or canals can make water available for the rest of the season [6]. Large canals or water breaks, ponds in mountain terrains are useful for collecting and storing rainwater.
- Trenches for water conservation: The construction of several trenches along the slopes of hills aids in the collection and storage of excess rainwater, which later percolates as underground water and aids in the conservation of moisture in such places.
- Land leveling and bund making: The leveling up and construction of bunds on the field minimizes excessive water loss through runoff.
- Board beds and furrows to minimize runoff: Excess water from the beds is collected in the furrows between the beds and routed to neighboring ponds or streams, where it can be utilized for future agricultural production.

11.1.3.2 Nutrient management

- Organic manure: For example, in several parts of Nepal, farm yard manure was commonly used as a source of nitrogen. Farm yard manure was typically the primary source of nutrition adapted to the field prior to the advent of artificial fertilizer [7]. Cow urine also includes potassium, which could also help crops meet their potassium needs.
- Green and brown manures: Sun hemp (*Crotolaria juncea*), dhaincha (*Sesbania rostrata*), mung bean, and cowpea are also employed as fertilizers for the harvests.

11.1.3.3 Bio-pesticide for pest management

For several decades, pest management has included the use of neem leaves as a bio-pesticide and storage [8]. Bio-pesticide is made from neem leaves, as well as other locally accessible herbs and animal manure, and it is used to control insect infestation in the field. Neem leaves, titepati, aloe vera, aakh leaves, ginger, garlic, kapur, turmeric, cow dung, and cow urine are blended in suitable proportions to make a simple DIY bio-pesticide that is resistant to a variety of pests in the field. Similarly, neem leaves and kapur are employed to combat pests in storage.

11.1.3.4 Soil conservation

Low-tech instruments, such as the indigenous plough, or halo, minimize the erosion of soil and are environmentally friendly. Cultivation of leguminous crops in the field's bunds also aids in soil erosion reduction and nitrogen fixation [9], increasing the microbial activity of the soil microorganisms. Crop rotation is also conducted in the traditional way, which aids in soil protection.

Other indigenous traditional practices include the following:

- Pests are controlled by burning stubbles in the field.
- Termite control is accomplished by burying neem leaves in soil [10].
- Ash is applied for seedbeds and fields of onion before sowing and planting to promote bulb growth and quality [11].
- Tobacco residues are integrated into soil to prevent organisms.
- During summer season, deep ploughing is done in drylands to prevent hardpan in the soil, increase water absorption, and manage pests.
- Mixing up the seeds like coriander seeds are mixed with sorghum seed before sowing to completely control Striga (parasitic weed).

- To promote the alkalization of water and fish growth, banana stems are placed in ponds.
- To improve germination, using cow dung is one of the most effective methods for sowing.
- Intercropping Sesbania with tomato during the summer has been shown to increase the tomato crop yield.
- Cotton pest and disease incidence was reduced by sowing methods.
- To avoid pest damage, neem leafs and Pungam are used.
- To control pests, cow dung is one of the effective ways.

11.1.4 Smart agriculture using WSN and IoT

The understanding that precision farming methods might help to resuscitate the failing traditional agricultural sector like water management, moisture & humidity measurements Precision farming, are all sure-fire ways to enhance production per acre of land [12]. Precision agriculture reduces the overuse of pesticides and fertilizers by allowing farmers to use land according to its quality and natural characteristics. Precision farming could be a lifesaver at a time when India's water levels are rapidly dwindling due to enormous demand from the agricultural and industrial sectors. Farmers are still delaying or refusing to change their traditional techniques, which might further depress India's GDP. Migrants from all across the world have recently gained new skills.

In the current pandemic situations, majorly skilled migrants across India selected agriculture as a profession and had no plans to return [13]. These migrants can now move closer to smart agricultural systems since it saves a lot of time to persuade them to use modern agricultural systems than it does for traditional farmers.

Wireless sensor network is used for IoT smart agriculture to monitor the weather conditions like temperature, humidity, rain dampness of soil, etc. [14]. WSN can also work as controller for providing inputs for seeds, harvesting, fertilizing, pesticides, etc.

The WSN based application development farmers in need of information for cultivation, as well as an input feeder control system on agricultural machinery. Inadequacies and breakdowns [15], such as sensor devices and power supply problems, as well as data security, can all be a serious challenge in WSN systems.

Based on the soil moisture in the sensors, there is the need of maintaining water level and flow of water so that pump can be switched on/off. To identify moisture in the soil ,soil sensor is installed beneath the ground in the field. Initially, a reading of the water level is taken and decisions are made based on it.

Using temperature, the sensor is fixed to get the reading of the temperature of soil [16]. To gather data, several sensors are linked to an Arduino UNO, and, further, it is forwarded to WSN depend on the crop cultivation threshold values will be marked once threshold value reached alarm system gives alert & corresponding action will be taken according to it.

Indigenous smart agriculture uses various smart solutions like smart agriculture application using IoT. There are various apps that are designed to provide the agricultural solutions by monitoring soil conditions, water levels, temperature, and humidity and to performs analysis of the level of fertilizers, equal proportions of pesticides needed for land area, etc. They remotely monitor and measure soil water humidity levels in order to make the water supply to crops [17] and to avoid crop damage or loss by automated sprinkling systems. This concept will enhance crop yield and management.

11.1.5 Applications of IoT in smart farming with the adoption of IoT

With the adoption of multiple areas such as industry, homes, vehicles, and even cities, there is a large potential to make everything intelligent and smart. The agricultural industry is also adopting the IoT technology, which has resulted in the development of smart farming using Internet of Things (IoT) or precision farming [18] (see Figure 11.1). The diagram depicts the various application areas where IoT is used to create smart agriculture. Crop management with an IoT: Multiple cropping system allows for disease detection in the leaf, weed, and water level detection, pest detection, and animal intrusion into the field of crop growth in an agricultural field [19]. Farmers can effectively predict crop growth by understanding nutrient requirement.

Soil characteristics, weather conditions, moisture, temperature, and other parameters must be monitored in farmlands. To identify the crop health and to identify diseases, various sensors and RFID chips are attached to plants. Intelligent irrigation system or smart irrigation, also known as automated irrigation, recognizes the water requirements of a plant or field [20]. It has an efficient and smart way to determine the necessity of water in each plant. The automated smart irrigation system in smart agriculture saves time, water, and effort of the farmer. Traditional irrigation techniques necessitate

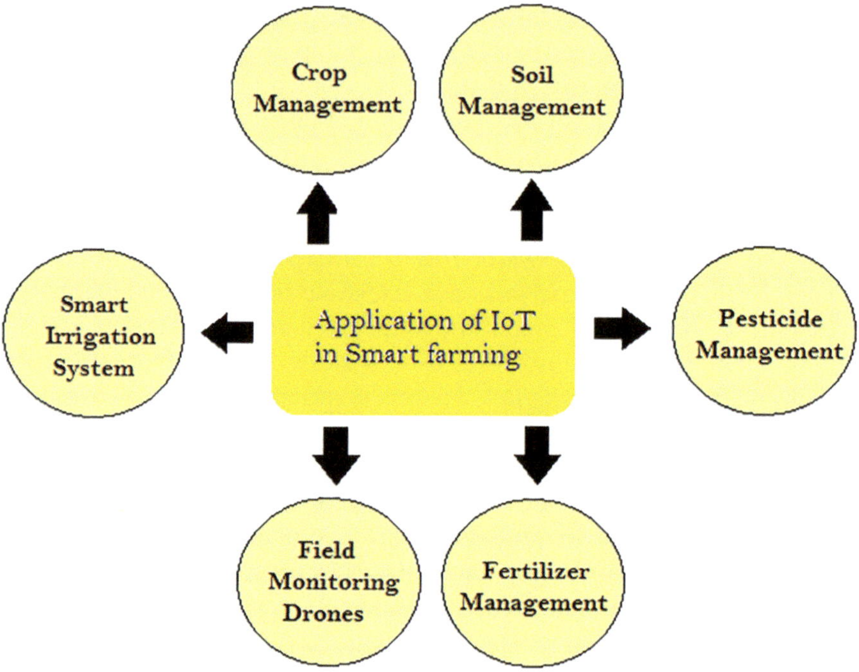

Figure 11.1 Applications of IoT in smart farming.

manual intervention. This automated irrigation technology can reduce the need for human intervention.

One of the major tasks is to perform soil management and monitoring with IoT, demonstrating how growing conditions and properties change over time [21]. To create suitable crop growing conditions, every farmer must explore the fundamental types of soil and soil mineral prerequisites. The temperature, pH values, and humidity content of the soil are the basic parameters that help to characterize the soil and thus make the best choices about fertilizer use as well as crop selection.

Drone monitoring has been a well-known technique for the field. Agricultural drones also known as UAVs (unmanned aerial vehicles) are utilized in smart agriculture, which are used to collect the agro data. Apart from monitoring capabilities, drones may also undertake jobs that traditionally required human labor, such as agricultural planting, pest and illness management, agricultural spraying, security aspects, crop monitoring, etc.

Cattle farm animals may be equipped with IoT agricultural sensors, similar to those used in other applications, to measure their overall health and fitness, which is done via the use of sensors.

Livestock tracking and monitoring assists in the collecting of information on cow health, well-being, and geographical location. Cattle tracking sensors can detect ill or damaged animals, allowing farmers to separate individuals from the swarm and prevent diseases from spreading. Utilizing drones for cow tracking saves farms money on labor.

Pesticide and fertilizer administration: Monitoring pesticide levels on crops can help farmers improve crop productivity. A farmer may need to apply pesticides more frequently when it rains; yet, the influence that a storm has on different portions of a field can result in excess and under-application of pesticides in different spots. Chemical levels can be measured using sensors placed in the soil or near plants above ground. Pests and their activity, location, and patterns can be understood through pesticide and fertilizer control.

11.1.6 Implementation challenges of IoT in smart farming

In this section, numerous obstacles must be handled. The challenges of deploying IoT in farming are classified depending on the technological challenges in farmland. When implementing cutting-edge technologies such as IoT, Big Data, and cloud computing in farming areas where basic internet connectivity is a barrier [22]. If a farmer decides to invest in smart farming, he or she must be aware of certain problems. The issues that the agricultural industry runs into while using IoT technology are like issues with interoperability standards in a use case for IoT in agriculture, where information sharing between all related IoT devices is required.

These devices can communicate using a common protocol, as well as the necessary connectivity and standards. As a result, the most difficult fact is interoperability. It is challenging to combine heterogeneous data from various sensors (such as moisture sensors, soil sensors, temperature sensors, and so on). We must spend a significant lot of time and cost to build standardized common protocols for all IoT devices. Storage issues for enormous agricultural data: There is a large volume of data generated by various IoT devices (such as sensors, cameras, weather stations, and so on) connected in a farmland. A vast repository is required to store such a large number of data.

Storing data in a database is insufficient for dealing with such a huge data. Technologies such as cloud computing and fog computing would provide greater storage and latency performance. Problems with connectivity: Because wireless communication is too difficult to achieve, providing connectivity with data interchange between an IoT device and a database or cloud is a key difficulty [23].

The fundamental issue is a lack of internet connectivity in farms. Most farms are located in distant places where internet access may not be adequate for fast transmission speed. Furthermore, crops, severe weather, and other physical obstructions may disrupt communication cables. In the future, 5G technologies that utilize space-based internet could be the solution. For constructing an agri-based solution, we must first identify the appropriate hardware and software tools and procedures, as well as maintain them. Sensor quality, improved data storage systems, and advanced data analytics tools.

Because the outcomes will be reliant on the correctness and dependability of the data. Because the sensors used to monitor the field may be impacted by animals, high winds, rain, and other factors, hardware maintenance is a difficult undertaking in agriculture. As a result, we must ensure that the hardware is long-lasting and easy to repair. Otherwise, we would have to replace sensors more regularly, which would incur additional costs [24]. Security concerns: When adopting IoT on farms, farmers must understand the security concepts and establish security regulations.

Smart agriculture and embedded systems need working with massive amounts of data from various sensors, to which increases the number of probable security flaws that offenders can use for data theft and hacker assaults. Providing safety in agriculture is an unknown and difficult undertaking. Drones that broadcast data to farm gadgets are used on many farms. These gadgets are linked to the internet but lack security features such as secure passwords or remote access authentication.

11.1.6.1 Impacts of climate challenges of IoT in smart farming

- Climate change impacts indigenous communities in SSA in a variety of ways, including adapting and mitigating the effects, as well as observing weather patterns and climate change. Some of the proven practical techniques used in developing climate resilience in smallholder farming communities include terracing, agroforestry, mulching, crop rotation, mixed farming, mixed crops, ridge and counter plowing, weather forecasting, water diversion ditches, and improved grazing [25].
- Intercropping and crop rotation: Crop rotation and intercropping are cultural techniques or systems based on mulch cover and biological tillage that need less mechanical ploughing; alternative pest and weed control methods must be developed. Crop rotation and intercropping offer various benefits, including reduced insect and weed infestations and improved water and nutrient distribution across the soil profile.

- The weather forecast is linked to the forecasted weather in the communities studied, observing wild animals and domestic animals, as well as sky signs, which is critical to understanding human behavior. Different types of forecasts are available: long-term forecasts for months when precipitation is forecasted, to short-term forecasts for current weather conditions, days ahead, or seasonal forecasts.
- Integrated agriculture: The practice of mixing crop production and livestock raising is known as diverse agriculture. Growing food, cash feed crops, etc. are cattle on the same agricultural site [26]. Livestock enterprises supplement agricultural output, resulting in a more balanced and productive farming system. Because the farm's crop and animal components support each other, it reduces dependency on external sources such as fertilizers.
- Gardening crops are planted on the slopes of hills or mountains on graded terraces built into the slope. Despite its labor-intensive nature, the method has been effectively employed to enhance arable land area on diverse terrains while minimizing soil erosion and water loss.
- In contour ploughing, the runoff water is regulated, moisture is absorbed and retained, and thus soil quality and composition are improved, thus reducing the effects of flooding, storms, and landslides on crops [27]. A contour farming approach can incorporate strip cropping concepts to maximize soil conservation.
- Ditch for water diversion: Water diversion ditches are a historic strategy used to decrease soil erosion caused by runoff. It is built along contour lines and across slopes to catch surface runoff and redirect it to appropriate outlets. Ditches are soil conservation for upland areas.
- Agriculture plays a very important role in local and global economy. In India, more than large population residing in rural areas as growing population pressures on agriculture methods like digital agriculture, smart agro to increase the productivity of yielding. Using smart agriculture, data is generated, which is a collection of various sensors, devices, actuators, etc., to understand the operational environment (an interaction of dynamic crop, soil, and weather conditions) and the operation itself (machinery data), to perform the accurate and efficient decision-making system for smart agriculture [28].
- Smart agriculture is a new model that uses high-precision computers to predict agricultural more efficient and reliable. Machine learning – the

scientific area that provides robots the ability to be smart learner system without rigorously programmed – is the mechanism that drives it. It has arisen with big data technology and high-performance computers to open up new avenues for unravelling, quantifying, and comprehending data-intensive processes in agricultural operational contexts.

11.1.7 Role of machine learning in smart agriculture

Machine learning is a popular technology that may be applied in the current agriculture business. The application of ML in agriculture aids in the development of healthy seeds. The approaches in machine learning agriculture are developed from the learning activity. To fulfill a certain task, these approaches must learn via experience. The ML is made up of data that is based on a set of examples. A single example is defined as a collection of characteristics. Attributes or features are the names given to these groups of qualities. A feature might be binary, numeric, or ordinal in nature. The performance measure is used to calculate the machine learning's effectiveness. The ML model's performance goes up as it gathers experience. Various mathematical and machine learning models used static evaluations in agriculture. After the learning process is complete, the model may be used to make assumptions, categories, and test data. This is accomplished after accumulating continuous training experience.

Machine learning is included throughout the growth and harvesting process. It all begins including the planting of a seed in the ground [29] – from soil preparation to seed hatching and water feeding assessment – and concludes with robots picking up the yield and judging ripeness using machine learning.

11.1.7.1 Species management

Species breeding: This technique is our preferred one since it is both sensible and unpredicted, considering what you usually hear about crop estimate or environmental circumstances managed at later phases. There is time consumption in species selection, which impacts the efficacy of water and nutrient utilization, adaptation to environmental change, disease resistance, and nutritional content or flavor [30]. Deep learning methods, in particular, use eras of field facts to examine crop performance in varied climates and novel features generated in the process. Based on this information, scientists may create a possibility model that predicts which genes are most likely to give a positive characteristic to a plant.

Species recognition: While the typical human techniques to seedlings categorization compare leaf color and shape, computer vision can deliver more precise and timely responses by studying leaf veins architecture, which gives more information regarding leaf properties.

11.1.7.2 Field conditions management

Soil management: Agricultural experts describe soil as a varied renewable resource with muddled methods and mechanisms. Its temperature can provide information on how climate change affects area output. To better comprehend environmental dynamics and agricultural implications, deep learning methods study evaporation processes, soil humidity, and weather [31].

Water management: Agriculture's water management has an influence on hydrologic, atmospheric, and agronomic balance. Till now, the application-developed ML-based applications are associated with various estimation evapotranspiration, allowing for more efficient irrigation system use and prediction of daily vapor pressure, which aids in identifying predicted weather conditions and estimating absorption and convective cooling.

11.1.7.3 Crop management

Crop predicting is one of the most important and often debated topics in smart farming since it impacts yield modeling and estimation, crop supplies and requirements matching, and cultivation strategies. Modern systems go well beyond basic prediction based on past data, including machine learning technologies.

Crop quality: Accurate identification and categorization of crop quality factors can boost product pricing while decreasing waste. In comparison to human specialists, robots can employ seemingly useless data and linkages to disclose and discover new attributes that have a major role in the qualitative production of crops.

Disease detection: A most extensive disease management practice in both open-air and greenhouse situations is to evenly spray insecticides across the cropping area. This method requires a large amount of pesticides to be successful, which comes with a considerable financial and environmental cost [32]. ML is used in generic precision agriculture control to target pesticide input terms of duration, area, and affected plants.

Weed detection: Grasses are the greatest serious hazard to agricultural yield, aside from diseases. The most difficult challenges in weed control are identifying and differentiating crops. Computerized solutions and machine learning techniques can enhance for identifying and classifying at low cost with no effects on environmental state. These technologies will power weed-killing robots in the future, minimizing the need for pesticides.

11.1.7.4 Livestock management

Livestock production: Computer vision, like agronomic practices, allows agricultural parameter predicting and estimate to improve and develop economical feasible systems, such as weight prediction systems that may project future weights 150 days before slaughter, letting farmers to change meals and circumstances accordingly.

Animal welfare: Currently trending with, cattle are regarded not only as food containers but also as creatures that can grieve and tire out on their farms. Analysts of animal behaviors may link the symptoms of chewing with the need for dietary modification, and their movement patterns, including posture, movement, eating, and drinking, can detect an animal's stress level and predict its pathogenesis, weight gain, and productivity.

Farmer's little helper: Consider this a supplementary application. Considering an agriculturalist, for him, it is really hard time figuring out the next stages in crop management. The very first question comes is whether he can sell to a local producer or attend a regional fair presently. He needs someone to walk him through the many possibilities before making a final selection. Companies are now developing specialized chatbots that can talk with farmers and offer them with vital statistics and analytics to assist him. Farmers' chatbots are projected to be wiser than consumer-oriented Alexa and similar assistants, since they will be able to not only provide numbers but also evaluate them and advice farmers on difficult issues.

Models behind: Though reading about the future is always exciting, the most significant aspect is the innovation that opens the way for it. Agricultural machine learning, for example, is a well-defined set of models that gathers relevant data and use certain methods to obtain desired outcomes. So far, the distribution of intelligence algorithm is uneven [33]. Machine learning approaches are most commonly utilized in agricultural management operations, followed by agricultural condition management and animal management.

According to the literature study, the most popular agricultural models are artificial intelligence, neural network, deep learning models like ANN and SVM, etc.

In the same way that ANNs are influenced by the features and functions of the human intelligence, they also stimulate the complexity of neural processes in the brain. They can create patterns, and learn and make decisions. Models of this type are commonly used in regression and classification tasks, showing their use in crop management, detecting weeds and illnesses, and identifying unique traits. Deep learning of ANN has recently increased the application of this type of model across a wide range of tasks.

SVMs are binary classifiers that categorize data instances by constructing a linear separation hyper-plane. SVMs are employed in the regression model, classification of algorithms [34], and clustering tasks. They are used in agriculture to forecast crop output and quality, as well as livestock production.

Using machine learning models like Bayes, Bayesian inference different complex tasks, such as evaluating animal welfare, involve the use of many classifier methods are combination in ensemble learning or Bayes models – graphical models in which the analysis is carried out using Bayesian inference.

Developing artificial intelligence enables farms, albeit being in their early stages. Currently, to provide the agricultural solutions using machine learning which is integrated with various interconnected systems like data analysis, decision-making systems, automated systems, etc., various farming practices would shift to knowledge-based agriculture, which increase the capability, levels of production, and product quality.

11.2 Conclusion

For the indigenous knowledge required for smart agriculture is defined in this article with their implementation issues needed to be addressed using the IoT or smart agriculture methods and by using machine learning or deep learning process considered for more effective and gives best results smart agriculture. Different smart agriculture methods are defined further to get improved smart agricultural analysis models with predictive models using machine learning algorithms.

References

[1] Hart, T. and Mouton, J. Indigenous knowledge and its relevance for agriculture: a case study in Uganda, Indilinga: African Journal of Indigenous Knowledge Systems, Vol. 4, No. 1: pp. 249–263,2005.

[2] UNFCCC (United Nations Framework Convention on Climate Change). United Nations Framework Convention on Climate Change (FCC/ INFORMAL/ 84/ Rev.1), Bonn, Germany,1992.

[3] FAO (Food and Agriculture Organization). Climate Smart Agriculture: Building Resilience to Climate Change. Natural Resource Management and Policy, Springer International Publishing AG, Cham, Vol. 52, 2018.

[4] Teklewold Haile Mariam, Menale Kassie, and Bekele Shiferaw. Adoption of multiple sustainable agricultural practices in rural Ethiopia, Journal of Agricultural Economics, Vol. 64, No. 3: pp. 597–623,2013.
[5] Orlove, B., Lazrus, H., Hovelsrud, G. K., and Giannini, A. Recognitions and responsibilities: On the origins of the uneven attention to climate change around the world. Current Anthropology, 55(3): 249–75,2014
[6] Nkonya, E., Jawoo, K., Edward, K., Timothy, J., et al. Climate risk management through sustainable land and water management in Sub-Saharan Africa. In Lipper, L., McCarthy, N., Zilberman, D., Asfaw, S. and Branca, G. (Eds). Climate Smart Agriculture, Natural Resource Management and Policy, Springer, Cham, Vol. 52, pp. 445–476, 2018.
[7] Odero, K The role of indigenous knowledge in responding to climate change: local-global perspectives. Panel 10: Roles of local and indigenous knowledge in addressing climate change. In Proceedings of the African Adapt, Climate Change Symposium, Addis Ababa, Ethiopia: pp. 9–11, 2011.
[8] Mafongoya, P., Jiri, O., Mubaya, C., and Mafongoya, O. Using indigenous knowledge for seasonal quality prediction in managing climate risk in Sub-Saharan Africa. Indigenous Knowledge Systems and Climate Change Management in Africa, Centre for Agricultural and Rural Cooperation (CTA), Wageningen: p. 43, 2017.
[9] K. Gunasekera, A. N. Borrero, F. Vasuian, and K. P. Bryceson, "Experiences in building an IoT infrastructure for agriculture education," in Proc. 3rd International Conference on Computer Science and Computational Intelligence, Alam Sutera, vol. 135, pp. 155–162, 2018
[10] Khanna and S. Kaur, "Evolution of Internet of Things (IOT) and its significant impact in the field of precision agriculture," Computers and Electronics in Agriculture, vol. 157, no. 3, pp. 218–231, 2019.
[11] J. Muangprathub, N. Boonnam, S. Kajornkasirat, N. Lekbangpong, A. Wanichsombat, and P. Nillaor, "IoT and agriculture data analysis for smart farm," Computers and Electronics in Agriculture, vol. 156, pp. 467–474, Jan. 2019
[12] Dr. N. Suma, Sandra Rhea Samson, S. Saranya, G. Shanmugapriya, R. Subhashri, IOT Based Smart Agriculture Monitoring System. International journal on recent and innovation trends in computing, energy efficiency and communication-IJRITCC volume: 5 issue, 2017.
[13] M.K.Gayatri, J. Jayasakthi, Dr. G.S. Anandha Mala, Providing Smart Agricultural Solutions to Farmers for better yielding using IoT. IEEE International Conference on Technological Innovations in ICT for Agriculture and Rural Development (TIAR 2015).

[14] Nikesh Gondchawar, Dr. R.Complexion.Kawitkar, "IoT based agriculture", all-embracing almanac consisting of contemporary analysis smart minicomputer additionally conversation planning (ijarcce), vol.5, Overall Journal on Recent and Innovation Trends in Computing and Communication ISSN: 2321- 8169 Volume: 5 Issue: 2 177–181 3, 2016.
[15] SoumilHeble, Ajay Kumar, K.V.V Durga Prasad, Soumya Samirana, P.Rajalakshmi, U. B. Desai. A Low Power IoT Network for Smart Agriculture [15] Rajesh M, Salmon S, Dr. Veena 2015.
[16] Dr. Vidya Devi, lockup. Meena Kumari, "continuous mechanization along with patrol process under the authority of most aerodynamic agriculture", universal newspaper made from appraisal furthermore probe contemporary scientific knowledge together with structures (ijrrase) vol3 no.1. pp 7–12, 2013.
[17] Belachew Agere, Mekuria Wuletaw, Nachimuthu K. (2020). Factors influencing adoption of soil and water conservation practices in the northwest Ethiopian highlands, Int Soil Water Conserv Res., 8: 80–89. https://doi.org/10.1016/j. iswcr.2020.01.005, 2020.
[18] Collier, P. and Dercon, S.African agriculture in 50 years: smallholders in a rapidly changing world? World development, 63: 92–101, 2014.
[19] Kagoya, S., Paudel, K. P., Daniel, N. L. Awareness and adoption of soil and water conservation technologies in a developing country: a case of Nabajuzi Watershed in Central Uganda. Environ Manage, 61(2): 188–196, 2017.
[20] IPCC (Intergovernmental Panel on Climate Change). Climate change: impacts, adaptation, and vulnerability. Part A: global and sectoral aspects, Contribution of Working Group 2 to the Fifth Assessment Report of the IJCCSM 12, 2 284, Cambridge University Press, New York, NY: p. 1132, 2014.
[21] Mafongoya, P. and Ajayi, O. C. Indigenous Knowledge Systems and Climate Change Management in Africa, CTA, Wageningen, 2017.
[22] Nchu, I. N., Kimengsi, J. N., and Kapp, G. Diagnosing climate adaptation constraints in rural subsistence farming systems in Cameroon: gender and institutional perspectives. Sustainability, 11(14): p. 3767, 2019.
[23] Boansi, D., Tambo, J. A., and Müller, M. (2017). Analysis of farmers' adaptation to weather extremes in West African Sudan Savanna. Weather and Climate Extremes, 16: 1–13, 2017.
[24] Tume, S. J. P., Jumbam, M. S., Nsoseka, N. A., Nyarka, N. D., Yenla, L. J., Njodzeka, N. G. (2018). Role of Media in Climate Change Communication in the Northwest Region of Cameroon. In Handbook of Climate Change Communication: Volume 2, Climate Change

Management; Leal Filho, W., Manola, E., Azul, A., Azeiteiro, U., McGhie, H., Eds, Springer, Cham, Switzerland, pp. 47–60, 2018.

[25] Kronik, J., Verner, D., Mearns, R., and Norton, A. (2010). The role of indigenous knowledge in crafting adaptation and mitigation strategies for climate change in Latin America. Social Dimensions of Climate Change. The World Bank, Washington DC: p. 145, 2010.

[26] Nyong, A., Adesina, F., and Elasha, B. O. The value of indigenous knowledge in climate change mitigation and adaptation strategies in the African Sahel. Mitigation and Adaptation Strategies for Global Change, Vol. 12 No. 5: pp. 787–797, 2007.

[27] Odero, K. The role of indigenous knowledge in responding to climate change: local-global perspectives. Panel 10: Roles of local and indigenous knowledge in addressing climate change. In Proceedings of the African Adapt, Climate Change Symposium, Addis Ababa, Ethiopia: pp. 9–11, 2011.

[28] Mafongoya, P., Jiri, O., Mubaya, C., and Mafongoya, O. Using indigenous knowledge for seasonal quality prediction in managing climate risk in Sub-Saharan Africa. Indigenous Knowledge Systems and Climate Change Management in Africa, Centre for Agricultural and Rural Cooperation (CTA), Wageningen: p. 43, 2017.

[29] Nkonya, E., Jawoo, K., Edward, K., Timothy, J., et al. Climate risk management through sustainable land and water management in Sub-Saharan Africa. In Lipper, L., McCarthy, N., Zilberman, D., Asfaw, S. and Branca, G. (Eds). Climate Smart Agriculture, Natural Resource Management and Policy, Springer, Cham, Vol. 52, pp. 445–476, 2018.

[30] FAO Drought-related food insecurity: A focus on the Horn of Africa. Drought emergency, Emergency Ministerial-Level Meeting, Rome, 2011.

[31] Lipper L Climate - Smart Agriculture – Policies, practices, and financing for food security, adaptation, and mitigation. Food and Agriculture Organization of the United Nations, 2010.

[32] Gyampoh BA, Idinoba M, Amisah S Water scarcity under a changing climate in Ghana: Options for livelihoods adaptation. Development 51: 415–417, 2008.

[33] Environmental Protection Agency-EPA Climate change in Ghana: Disaster risks are an increasing concern for all-climate change related disasters in Ghana", EPA Policy Brief No. 3: 1–4, 2010.

[34] Shridhar M, Chinmay P, Piyush W, Aniket P, Vaishali D, "A Survey on Predicting Suitable Crops for Cultivation Using IoT," International Journal of Innovative Research in Computer and Communication Engineering, January, 2017.

12

Climate Change and Its Impact on Agriculture

M. Gomathy[1] and K. Kalaiselvi[2]

[1]School of Computing Sciences, Vels Institute of Science, Technology and Advanced Studies (VISTAS), India
[2]Department of Computer Applications, Saveetha College of Liberal Arts and Sciences (SIMATS), India
Email: marimuthu.gomathy@gmail.com; kalaiselvik.sclas@saveetha.com

Abstract

Changes in weather patterns are currently the greatest global challenge. Climate change has a detrimental impact on agriculture. The decline in yield and nutritional quality of crops is owing to several factors causing climate change. It has created a negative impact on some food crops, which could lead to food shortages shortly. Climate change, precipitation, greenhouse gas emissions, and natural disasters are all to blame for the effects. All of these negative effects harm the environment, which is the root source of environmental problems. Climate change is having a major impact on all life forms on earth, especially in agriculture. IoT devices built using machine learning algorithms have accelerated smart farming in different climates. Machine learning has made a significant contribution to data prediction and analysis in a variety of fields. It is a field of science where machines can acquire knowledge without being programmed. It has emerged in unison with big data technologies and high-performance machines to generate new avenues for unwinding, analyzing, and comprehending data-intensive farming activities. Machine learning is used in agro at all stages, from planting to reaping. The procedure includes land preparation, seed prognosis, weather forecasting, irrigation feed monitoring, and assisting robots in appraising the growth or ripening result. This chapter describes a global scenario for how climate change would affect agriculture, as well as a number of IoT tools used for smart farming including its challenges. The chapter also briefly explains how machine learning can be implemented to increase the production level in agriculture to changes in weather conditions.

12.1 Introduction

A change in the weather pattern is a prolonged change in the temperature of the environment and climate variability. This might happen as a result of natural disasters or human behavior. Some of the repercussions of climate change include glacier melting, which raises water levels, a rise in heat waves, which leads to a long summer, and a change in precipitation pattern from region to region. People have been forced to migrate as a consequence of elevated sea levels and saline intrusion, and communities are fearful of famine as a result of repeated droughts.

Environmental problems are having an adverse effect on agricultural production. Changes in climate patterns have an effect on the productivity of soil and crops. The preponderance of natural vegetation is impacted by environmental variations. However, there is a worry that crops may lose their nutrient quality as a result of the usage of fertilizers and manures, which help to increase productivity.

Conservation agriculture strives to increase crop output while simultaneously enhancing economic and environmental benefits. Fuel efficiency, elevated soil, and crop moisture have all increased as a result of using conservation farming approaches using IoT. As a result, the impact of anthropogenic global warming has been lessened. Although infrastructural and financial obstacles remain, the dissemination of sensors and network devices could encourage the adoption of innovative methodologies such as smart farming, which uses the least chemical pesticides, fertilizer, and moisture potential, along with weather prediction systems, which are especially important, given the devastating impact of extreme weather events, such as droughts, on local communities well-being and their economic development. As a result, the impact of natural and anthropogenic global warming has been lessened.

12.2 Global Scenario and Evolving Context

12.2.1 Climate change

The physical environment, biodiversity, and human society are all impacted by human behavior's contribution to climate change. Future effects of climate change in various nations can be predicted based on reductions in greenhouse gas emissions. Today, as predicted by experts earlier, sea ice loss, fast sea level rise, and longer and more intense heat waves are all occurring. It is claimed that environmental challenges are not uniformly prevalent around the world. For example, oceans change more slowly than land regions do, and northern high latitudes change more quickly than equatorial. Three main ways that regional climate will be impacted by global warming are the melting of glaciers, alteration of the water regime (moisture and rainfall), and disruption of ocean currents.

Climate change affecting the Philippines:
The Philippines has a long history of being particularly susceptible to extreme weather. However, in recent years, the country has been hit by even more powerful storms, such as Typhoon Haiyan. Around 20 tropical cyclones make landfall to submerge in Philippine waters, with eight or nine of them due to the impact of climate change. However, as the temperature of the ocean's surface rises as a result of climate change, more heat is emitted into the atmosphere, increasing the Philippines' annual debt by at least 4%.

Climate change affecting Canada:
Since 1948, average global temperatures in Canada have risen by 1.7 °C. Sea levels are rising in tandem with rising temperatures and greater rainfall. The Arctic is warming far more quickly than the rest of the world. Scientists have already noticed significant reductions in the Arctic Sea ice cover, especially during the summer. The melting sea ice disturbs regular ocean circulation, causing temperature and weather changes all across the world. Wildfires in the Canadian West are raging faster and more fiercely than they have ever been. The people's and country's economies have suffered as a result of the changing circumstances brought on by climate change.

Effect of climate change on Mexico:
Climate change has emerged as Mexico City's greatest long-term threat. It is linked to groundwater, hygiene, air quality, flood-related transportation disturbance, and landslide-prone homes. Huge numbers of people in Mexico are threatened by a shortage of enough water as a result of climate change. Food insecurity is a result of water scarcity. Warmer temperatures increase evaporation in the soil, affecting plant life and further reducing rainfall. Mexico's soil is becoming significantly less suited for raising food and crops as a result of climate change, which is affecting Mexican households today.

Effect of climate change on Germany:
Over the past several decades, Germany has experienced an increase in the frequency of days with record heat. When buildings that absorb and store more heat are built in place of natural landscapes, cities become warmer than

their surroundings. German cities can experience midsummer temperatures that are up to 10 °C greater than those found in the nation's remote regions. River basins are becoming smaller, which has significant consequences for German agriculture and energy production. Wildfire risk has increased in the Alps as a result of the drought brought on by climate change and everything that goes along with it.

Climate change affecting the UK:
The repercussions of the climate crisis are being felt throughout the UK, including harsher seasons, rising sea levels, and altered patterns of precipitation. New and emerging pests and illnesses, as well as invasive non-native species, pose serious threats to people, animals, and the natural environment in the United Kingdom.

12.2.2 Agriculture

Climate change has altered nature by elevating temperatures, drying soils, and increasing the risk of wildfires. Due to rising temperatures, agricultural production and the availability of freshwater resources are in jeopardy. Climate change has a significant impact on human health, both intrinsically and extrinsically through heat stress and the development of infectious diseases.

The consequences of increased carbon dioxide in the air, hot temperatures, altered monsoon, and transpiration patterns, increased rate of catastrophic weather, and enhanced weed, bug, and pathogenic pressure will all have an impact on agricultural and food production around the world. Droughts cause agricultural failures and the loss of cattle grazing. Soil erosion occurs 10–20 times more frequently than soil accretion in farmlands where no-till farming is practiced. Due to global climate change, soil erosion and desertification are becoming increasingly widespread.

Agriculture in Brazil:
One of Brazil's most significant economic pillars is agriculture. Brazilian agriculture and food production are first and fourth, respectively, in the world (such as soy, sugarcane, and maize). Brazil exports the most coffee, cattle, ethanol, and soybeans in the world. Approximately 7% of Brazil's geographical area is dedicated to crop production, including soybeans. Brazil produces 13% of the world's oranges.

Agriculture in Mexico:
Despite its small share of the overall economy, agriculture plays a significant role in Mexico's economy. It is essential for fostering business ties with the

United States and creating jobs in Mexico. Some of Mexico's most important crops include grains, sugarcane, peppercorns, corn, bananas, sorghum, blue agave, avocados, beans, other tropical fruits, and also more. In Mexico, agriculture occupies around 15% of the land, with cattle accounting for the remaining 50%.

Agriculture in France:
One-third of France's economy is based on agricultural land. It is one of the most prosperous industries in the country. Their main products are wheat, barley, and grapes. Agriculture's actual income increased by 4% over the previous year. France is the top European producer of meat, milk, wine, sugar beets, oilseeds, and grains. Sugar beets are produced in France by about 29 million metric tons.

Agriculture in Canada:
The agriculture and agri-food system in Canada is a major economic driver. In 2018, the system generated $143 billion, or 7.4% of GDP, and employed one out of every eight Canadians. Agriculture is a robust and prosperous industry that has seen significant expansion over the last decade.

Agriculture in Germany:
Agriculture and forestry cover more than 80% of Germany's territory. Organic farming is practiced by roughly 10% of Germans. The third-largest supplier of farm commodities worldwide was Germany.

12.2.3 Impact of agriculture on climate change

Canada:
As a result of anthropogenic climate change, Canada and other Arctic countries are predicted to warm faster than any other on Earth in the future decades. Warming may give new chances for agriculture in some parts of Canada where temperatures are low and the growing season is short, as the growing season lengthens. Perennial crops and grazing pastures may surpass annual crop output, altering current carbon sinks. Climate change is one of the major challenges for Canada's agriculture. The researchers claim that droughts may grow more intense and frequent as a result of future climate change, while storms may become more violent and damaging. Crop output would suffer in Canada's semi-arid regions because of the change in weather patterns. Droughts and floods have wreaked havoc on numerous crops across Canada, resulting in a 50% loss in average yields owing to climate change.

Germany:

Agriculture is Germany's greatest land user and a major contributor to environmental hardship. Agriculture, on the other hand, is impacted by the consequences of climate change. Excessive nitrogen compound inputs have detrimental consequences on climate, biodiversity, and landscape quality. Excess nitrogen compounds can infiltrate surrounding rivers or the air if manure fertilizers (from animal husbandry or mineral fertilizers) apply more nitrogen to agricultural soils than is absorbed from cultivated plants. Critical loads have already been exceeded in places in northern Germany where extensive livestock rearing is practiced. Wheat production has been whittled down because of a lack of rain as a consequence of climate change. German farmers are quite concerned about livestock since grass and maize, which are used to feed animals, are already running out, and some immature wheat has had to be cut down in some parts of Germany. In the spring and summer, Northern and Eastern Germany experiences prolonged periods of high temperatures, which exacerbates drought conditions. Drought also wreaked havoc in the southeast, devastating a slew of crops. Droughts were observed on some farmland, while farmers just 2 km away struggled with the effects of excessive rains spurred by global warming.

Mexico:

Climate change has begun to disrupt Mexican agriculture, decreasing the yield of some key items in the local diet. Droughts, an unpredictable climate, insect proliferation, occasional but more heavy precipitation, hailstorm, and the impacts of human behaviors are all harming an area near the Mexican capital that is vital for food and water supply and climate regulation. Agriculture is extremely reliant on local meteorological conditions and is projected to be extremely vulnerable to climate change. Climate change in Mexico has resulted in increased frequency of fungi and insects, soil alteration, diminished area and groundwater availability for agriculture, and changes in agrobiodiversity.

Philippines:

The Philippines has a tropical marine climate due to its location on the equator. Due to changes in weather patterns, typhoons have become more regular in the Philippines in recent years. Farmers rely on the constancy of this pattern to predict the ideal time to plant for the best harvest despite the frequent occurrence of natural calamities. Climate warming has significantly altered local weather patterns, making it more challenging for farmers to choose when to plant their crops. Furthermore, the dry season has been extended, increasing the likelihood of drought. Due to shortage, this danger will directly affect the amount of water available for domestic and agricultural use. Obtaining water for crops will become more and more challenging for farmers with already meager incomes as expenses grow. Additionally, a lack of rain will restrict the range of crops that farmers can grow because the majority of the major crops in the nation are rain-fed.

12.3 An Overview of the Indian Scenario

12.3.1 Climate change in India

India is the world's fifth country that is extremely vulnerable to climatic changes. Climate change cost India almost 37 billion dollars in 2018 (almost twice what it lost between 1998 and 2017). According to the NITI Aayog (2018), over 600 million Indians would face "severe water scarcity" in the next years. If surface warming increases by 2 °C over preindustrial levels, major flooding is expected to increase in more than 89 urban regions of India, according to a report given by MIT. There have been more heat waves, which have led to violent cyclones and a rise in sea levels.

The capacity of younger generations to accept, change, and succeed in life is impacted by climate change. Severe weather conditions, including cyclones, endanger even their own lives and imperil their capacity to survive. Floods cause threats to cleanliness and access to sewerage systems and services, resulting in diseases like cholera, which children are especially susceptible to. Droughts result in crop failures, higher food prices, and nutritional deprivation for the poor, all of which can have long-term effects like food insecurity and malnutrition. Environmental resources such as topsoil and forestry are being destroyed as a result of climate change. Children's access to ecosystem services is under threat due to the increasing frequency and severity of natural disasters as well as broader ecological pollution.

In India, there are serious worries about ecological sustainability and poor air quality. Most of the highly polluted cities in the world are located in India (IQ Air Report, 2020), and a Lancet study from 2018 estimates that 1.24 million people in India died due to air pollution in the year 2017. Based on four performance indicators examined in 2018, India was placed 177th out of 180 countries by Yale's Environmental Performance Index. Our regular lives are highly impacted by climate change, and as such, adaptation and prevention strategies addressing its effects must be included in the present and future UNICEF projects.

12.3.2 Agriculture in India

Agriculture has been the backbone of India for nearly 1000 years, providing a livelihood for the majority of the population. It also plays an important part in the country's economy and provides numerous job opportunities. In comparison to the rest of the globe, India ranks second in agricultural product production. Punjab, Uttar Pradesh, Madhya Pradesh, Haryana, Bihar, Andhra Pradesh, Maharashtra, West Bengal, and Gujarat were among India's agriculturally developed states. These states are crucial in India's agrarian product production. India's total arable land area is 15,73,50,000 km^2, accounting for around 52.92% of the country's total land area. Because of the constant strain of an ever-increasing population and expanding urbanization, arable land in India is dwindling.

Some of the crops that India is top in producing are: Anise, Fresh fruit, Badian, Fennel, Tropical fresh fruit, Coriander, Pigeon peas Jute, Spices, Pulses, Castor oil seed, Millets, Safflower seeds, Sesame seeds, Limes, Lemons, Dry chilies and peppers, Cow's milk, Cashew nuts, Chickpeas, Ginger, Okra, Guavas, Turmeric, Goat, Milk, Mangoes, Meat, and Buffalo milk.

Furthermore, the country is the leading producer of cereal crops such as Bajra, Jowar, and Ragi. India is the world's second-largest producer of rice, after China. India produces approximately 10% of the world's fruits. In the production of wheat, rice, cotton, sugarcane, and groundnuts, India holds the second position. India is a major producer and consumer of silk.

12.4 Impact of Climate Change

12.4.1 On agriculture land

Two-third of India's agricultural land is dependent on rain, and the irrigated network is reliant on monsoon rain. Therefore, climate change issues, notably drought, pose a significant threat to farming practices in India. Due to increased air temperature, the frequency of severe climatic extremes has grown in recent years, resulting in greater hazards and a substantial reduction in crop cultivation.

Directly or indirectly, agriculture is impacted by the consequences of climate change on goods, land, animals, and insects. C_3 photosynthetically active crops are replaced by higher CO_2 concentrations, which stimulates their growth and output. The growing season is shortened by excessive heat, which also tends to increase agricultural metabolic activity, skews the synthesis of compounds with the aid of radiant energy, and has an impact on the survival and expansion of insect populations, resulting in a new balance between crop production and pests, facilitates rapid nutritional mineralization in soil properties, significantly improves the nutritional value of the soil despite eliminating the need for fertilizer's which were once employed to raise the productivity of the land. Global warming has a substantial and conflicting impact on India's farming. Soil organic matter changes are declining due to a lack of irrigation water, as well as the frequency and severity of droughts and floods. Because of soil degradation and coastal land submersion, changes in insect characteristics are also becoming more prevalent in agricultural areas.

The agriculture sector would confront significant hurdles as a result of climate changes, which are as follows:

(i) Changes in precipitation, fluctuations in stream flow, and a rise in crop water requirements all affect water availability.

(ii) Water quality degradation due to seawater intrusion, salt movement from bottom sediments because of improper irrigation practices, and overuse of aquifers.

(iii) Unpredictable weather conditions such as droughts, floods, and cyclones are becoming more common and intense, affecting production rates.

(iv) Heatwave is caused by a rise in temperature during an intense phase of agricultural production.

(v) Changes in the increase of insects and plant diseases are unpredictably variable. With shifting climatic conditions, it is also possible for minor pests to become major pests.

12.4.2 Effects on crops, water, livestock, fisheries, and pest diseases

Crops:
Unpredictable weather occurrences including flooding, famine, thunderstorms, and extremely warmer waves occur more frequently and last longer, in turn, reducing agricultural productivity. Enhanced crop capacity utilization and changes in weather patterns during the monsoon season result in lower yields in rain-fed areas. The quality of fruits, vegetables, tea, coffee, aromatic, and medicinal plants is also declining as a result of the change in weather conditions.

Water:
When the temperature rises, water loss climbs as well, resulting in increased irrigation demands. In some areas, this may result in a drop in the groundwater table. In the near term, melting glaciers in the Himalayas may improve the water supply in the Ganges, Brahmaputra, and their tributaries, but in the long term, water availability will decline significantly. Flooding may grow more frequently and stay longer as a result of it, as well as soil degradation. Because of the incursion of seawater, the aquifer features, along the coastline, will be deteriorated.

Livestock:
Weather pattern change has a considerable effect on feed production as well as the nutrition of cattle. Food and fodder output would be reduced if water scarcity increased. Climate change has a significant effect on vector-borne disorders of cattle in cooler places due to the increase in vector populations. Heat stress in ruminant animals is predicted to worsen as a result of climate change, impairing their reproductive function.

Fisheries:
Fish spawning, migration, and harvest are likely to be affected by rising sea and river water temperatures. Because of the rising sea surface temperature, coral bleaching is anticipated to worsen.

Pest disease:

- Insect pests and illnesses' geographic ranges are expanding.
- Variations in infection and arthropod rates of growth.
- Introduction of new ailment concerns, as well as a higher danger of migrant pests and diseases.
- Various systems of pathogen and whitefly prevention have a lower efficiency

12.5 Technology Used to Overcome Problems in Farming

In the majority of emerging economies, agriculture is the primary source of revenue. Agriculture provides almost 70% of employment in these countries. The country's fertile soil is one of the key reasons. It ensures the high availability of various crops due to the varied nature of climate change in different regions. Despite the abundance of resources, due to scarcity, poor technology use, agrarians' ignorance, and lack of knowledge, as well as the application of some outmoded practices, it was unable to meet the demand.

Furthermore, most crops are being afflicted by new pests and diseases, resulting in lower yields. Fertilizers used to manage the insect have a detrimental effect on the standard of grains grown. Changes in meteorological conditions have resulted in considerable losses in agricultural production owing to pests and illnesses. As a result, efficient resource management is required to design a viable technology that would benefit farmers.

Farmers should be able to use technology to forecast and prevent agricultural diseases in advance, as well as enhance the effectiveness of their harvests. With the help of some sensor devices, these smart technologies should be able to analyze the environment, soil, and growth patterns remotely. An

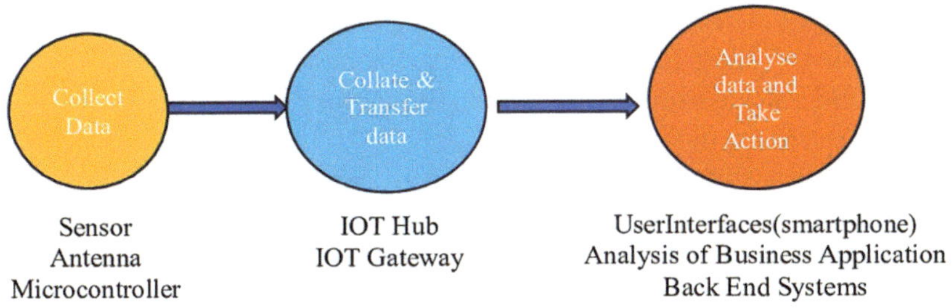

Figure 12.1 Working of IoT.

alert is sent to the farmers whenever any of the stored parameters change. This could assist them in taking the required steps to combat pests, illnesses, or other factors that impede growth. These technology and accurate forecasts help to limit the usage of chemicals, resulting in increased production and product quality, which in turn has decreased the percentage of loss.

12.5.1 Overview of IoT

Smart farming relies heavily on wireless technology and sensor-embedded objects. A huge network of connected gadgets that are equipped with sensors, algorithms, and other tools to interface with other gadgets via the internet is referred to as the "Internet of Things" (IoT). IoT devices are aware of their surroundings and assist in obtaining anything at any time and from any location. The working of IoT devices is explained in Figure 12.1.

Digital sensors that can connect to the internet make up an IoT ecosystem. Such integrated sensors use implanted devices to collect, disseminate, and respond to data from their environment. To send data from the sensors, IoT devices connect to an IoT network or perhaps another edge machine. Such uploaded pieces of information are either processed locally or sent to the cloud for analysis. These devices could periodically exchange messages and examine the data they obtain.

Wireless technologies are used in IoT to collect data from diverse sources and transmit it to other devices. Wireless sensor networks and radio frequency identification are the two most common technologies used in the Internet of Things for sensing and communication.

12.5.2 IoT's importance and benefits to businesses

The Internet of Things can enable human habitation more intelligently and gain total control over the situation. In addition to offering smart devices to regulate homes, IoT is essential in any organization. IoT provides information about everything from equipment performance to manufacturing to distribution and administrative processes, giving businesses a proper perspective on the operation of their systems.

The Internet of Things can be used by businesses to streamline procedures and save labor costs. Bringing down the cost of producing and shipping goods as well as bringing transparency to consumer interactions helps to minimize waste and improves the quality of service.

Because of this, the Internet of Things is among the most important technologies in day-to-day life, and it will only elevate high as more businesses see how important smart gadgets are in keeping them competitive.

Numerous opportunities exist for businesses to gain from the Internet of Things. Some benefits are sector-specific, whereas others apply to a wide range of industries.

- The Internet of Things (IoT) can help agriculturalists by rendering their work easier. Sensor gadgets can collect information on rain, moisture, heat, and soil composition, among other things, to facilitate the automation of agricultural techniques.
- The capability of monitoring equipment functioning is another feature that IoT shows its efficiency. To monitor the incidents or alterations in structural buildings, pipelines, and other infrastructures sensors can be utilized. Various advantages, including cost reductions, improved efficiency, reliability in workflow modifications, and a digital workflow are applicable.
- Accessibility of data on any device, at any moment, and from any region.
- Information exchange among electronic gadgets that are connected is enhanced.
- Reducing time and costs by sending data packets across a networked system.
- Reduces human intervention and improves the quality of services using automated machines.

12.5.3 Internet of Things (IoT) in farming

By gathering real-time data, IoT sensor nodes play a critical role in forecasting farming circumstances. These nodes can gather information from the fields, such as land or plant characteristics, to assist farmers in making rational

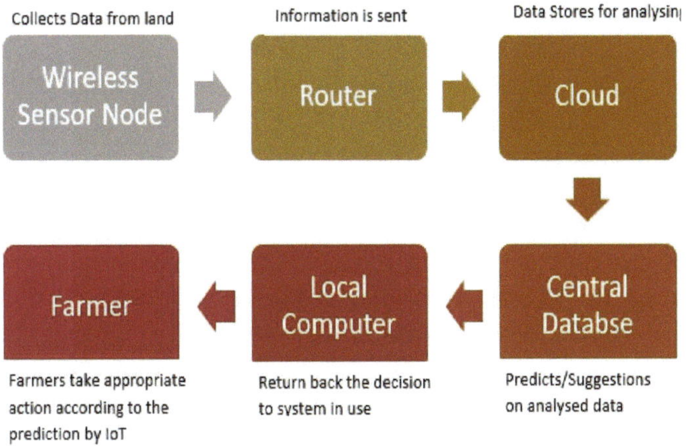

Figure 12.2 Model of different stages of smart agriculture.

decisions. Agribusiness has advanced with the advancement of machine learning and data analytics. IoT and machine learning are also exploding in other industries.

IoT devices are used in smart farming to keep farmers updated on the state of their crops at periodic intervals. The paradigm of precision agriculture contains different stages, which are explained in Figure 12.2.

In the beginning, wireless sensor hubs use sensors to identify and detect factors, such as soil conditions, vegetation, and nutrient levels. The acquired information is subsequently delivered to the cloud through the router in the second stage. Information in the cloud can be utilized for more complex computations as well as remote data monitoring. Analytical procedures are used in the third section to specify the precise nature of the crops and whether the computed values have been reduced to a threshold value. The analytical results are then communicated to the farmer, who is the end-user. As a result, the actuators would perform essential steps such as elevating the topsoil level of the water or sprinkling compost to improve crop yield. Smart sensor devices and actuators interact in real time to detect or predict critical circumstances.

12.5.4 Application of IoT and WSN in farming

1) Choosing the right soil:

The farmer's initial step is to identify and analyze the soil's characteristics. Soil testing can reveal information about the nature of the soil, such as the nutrients present, the fertilizers demanding the irrigation level, and the biomass plant to be planted. Farmers are able to make the best choices for their fields based on the obtained information. The urge to produce more from the minimum area of arable land is the main driving force behind precision farming. Sensor-based technology is extremely useful in determining the ideal soil for a healthy plant.

Any agriculturalist can utilize the Lab-in-a-Box soil testing toolkit to learn about the nature of the soil. The profundity and proximity for implanting seedlings are known using vision-based technology and sensors. IoT technologies used for seed planting include GPS (global positioning system), sensors, and an autonomous robot called Agribot. These technologies aided farmers in selecting the best acreage for a particular crop. Figure 12.3 explains the application of WSN and IoT in smart farming.

2) Irrigation:

Water is essential for the development of high-quality crops. Existing irrigation systems are insufficient to increase plant water demand due to water scarcity caused by climate change. By installing technologies such as IoT and WSN (wireless sensor network) in the farmland, farmers may immediately assess soil moisture and decide when to raise the water level. The IoT sensor device, the crop water stress index is employed to collect the moisture level of the soil. Following that, the data is transferred to the cloud or a central processor. The system compares the data from weather stations with the collected information and sends a notification to the farmers. The farmer will then determine whether or not to hydrate the soil.

3) Fertilizers:

Fertilizers aid in increasing the nutritional value of crops. Nitrogen, potassium, and phosphorus (NPK) are three fertilizers that contribute significantly to preserving the well-being of the flora and fauna. Excessive use of these fertilizers also causes plant development to become unbalanced. Emerging IoT technologies such as NDVI (normalized difference vegetation index), VRT (variable rate technology), geo-mapping, and GPS assist farmers in determining soil nutrient levels and accordingly the type and proportion of fertilizers to be used to boost the growth of crops and flora.

4) Crop diseases:

There are various crop diseases, like Black Spots, Other Leaf Spots, Powdery Mildew, Downy Mildew, Blight, and Cake, which cause damage to the entire crops. Farmers also lose money as a result of it. To capture the state of the crops, advanced technology such as wireless sensors and drones are used.

Figure 12.3 Application of IoT and WSN in smart farming.

The acquired information is forwarded to the backend processor. The decision support system in machine learning delivers a notion of the nature of the sickness as well as corrective strategies to combat it.

5) Controlling pests:
Agrochemicals, germicides, and insecticides used in excess will disrupt the ecology. These approaches may aid in the rapid growth of plants or flora. Consumption of these items, on the other hand, may cause fatal and chronic disorders in humans. As a result, pest management employing IoT devices is beneficial in gaining control over the use of these manures. Pesticide usage monitoring, establishing the type of agricultural disease, and disease prediction at an early stage can all be aided by modern IoT systems.

6) Yield surveillance:
The ultimate goal of farmers is to improve the crop's quality and quantity. Monitoring the field from inception to delivery, including harvesting times, storage locations for harvested crops, and dealing with agricultural diseases is a major work. Most of the obstacles cannot be handled solely by physical

labor. To monitor yields, farmers must use current technologies such as many optical sensors and IoT devices.

12.5.5 IoT technologies in predicting climate change

Climate change is a global concern in the present environment. Urbanization and rising populations both contribute to climate change. Health and agribusiness are negatively impacted by greenhouse gas emissions. Nobody has the authority to alter the current climatic situation. However, predicting and monitoring climate change could result in significant financial savings for agricultural firms.

Extreme weather can cause temperature and water level increases, which can lead to drought in some areas and flooding in others. The use of IoT technologies to predict weather forecasts could assist farmers in equipping themselves to deal with difficulties.

Weather monitoring technologies:
A vast amount of data is necessary for all smart weathering techniques. These data are used to monitor normal weather conditions that affect crop growth in addition to forecasting extreme weather events. Some of the technologies used in the prediction of weather conditions are IoT weather stations, meteorological data, and artificial intelligence weather forecasting. Agricultural businesses utilize these data to predict climate changes and react to them quickly and support necessary initiative measures to be taken.

IoT weather station:
It is a smart device that monitors the weather conditions in the area where they are positioned. Air pressure, air humidity, rainfall, wind, and UV light are all calculated using the sensors. These factors are used to describe the current location's climate. The data is collected and sent to the cloud for processing. The analyzed information helps to collect crucial weather data from outlying meteorological stations and is made publicly available at a central location. Farmers may gain prior exposure to environmental and soil data so that they can manage beforehand for the season change. Upon receiving alarming data from weather sensors, the system might transmit an alert about impending frost or rain. Figure 12.4 pictorially gives an idea about the working of an IoT-based weather station used to get information about the climatic condition.

Benefits of using IoT weather stations by farmers:
Crop dangers can be reduced by keeping an eye on severe weather conditions. It can also assist farmers in making the best use of their resources

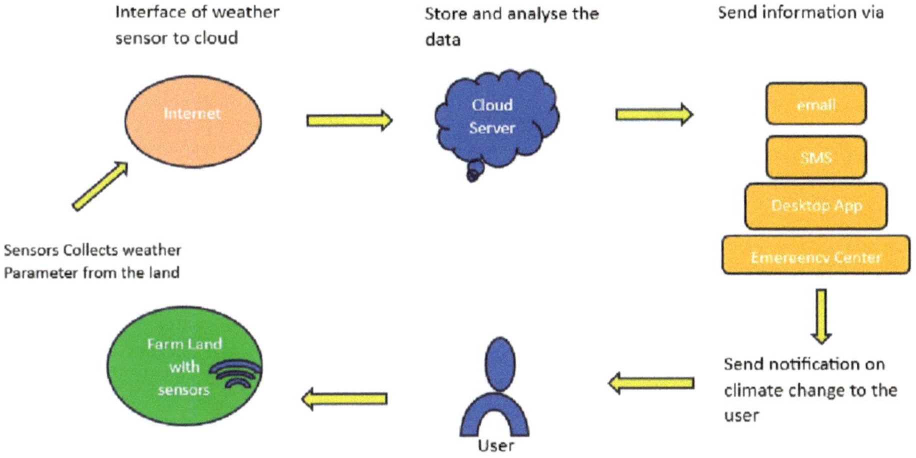

Figure 12.4 Working of IoT-based weather station.

and protecting their crops. IoT sensors can enhance production quality by recommending the optimal harvesting time. It can also send real-time reminders to numerous platforms and operating systems. Information collected using a weather station helps to get accurate data in the field that is appropriate to the location of a farm and the upcoming season. It also helps to access community data by collaborating with third-party sources and services.

Spatial data and hardware stations used for weather forecasting technology in agriculture:
Satellites aid in the gathering of climate prediction data for farmers in the agriculture industry. Farmers can use the data to forecast climate change and learn about the current climatic conditions in their area. Decision support systems assist farmers or users in taking suitable actions based on the information presented.

Farmers might use satellites to transmit information for apps they have on their phones or computers, or even to online support services for farmers. It also serves as a broadcaster, transmitting data from agricultural weather stations back to the ground. Figure 12.5 shows the use of satellites for weather forecasting in agriculture. Despite the fact that governments control satellites, the information they gather is very helpful in predicting agricultural yields based on the weather at the time and warnings of climate change. Additionally, these data can be utilized to forecast a crop's future yield and plan ahead for natural calamities. It also aids in smart irrigation management depending on weather fluctuations.

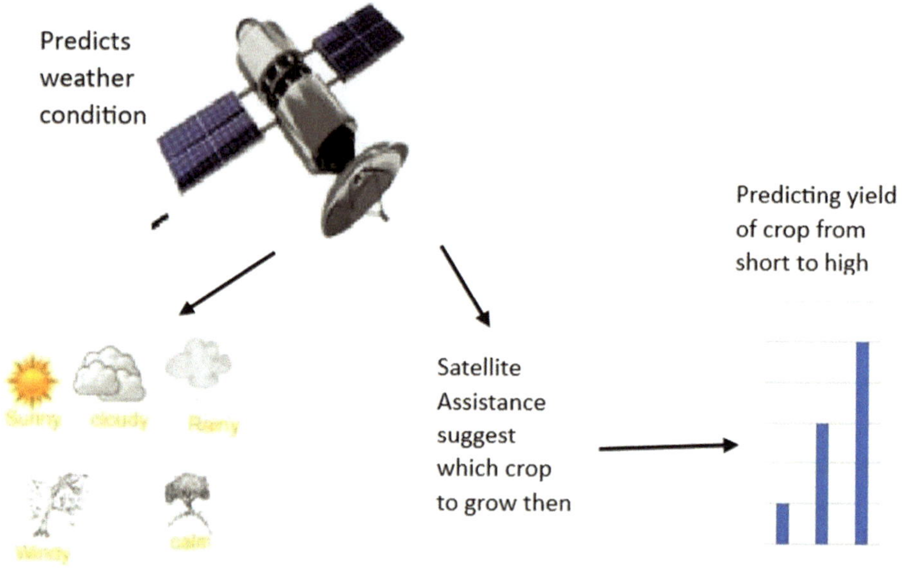

Figure 12.5 Satellite used for weather forecasting in agriculture.

Machine learning used in prediction of weather:
With more accurate weather predictions, farmers may make wise crop management decisions. To develop reliable localized weather predictions, a large amount of weather-related data is collected from networked sensors, satellites, and local hardware weather stations. For precise forecasting, machine learning algorithms focus heavily on the quality of the dataset; therefore, quality of the data and categorization were essential.

Sorting data and identifying weather patterns should aid in extracting precise insights into determining weather conditions. Machine learning models should be trained to identify weather patterns. Increased access to reliable data sources is critical for accurate weather forecasting. Figure 12.6 gives an idea of the working of the machine learning model in the prediction of weather changes.

12.6 Initiatives Measured by the Cultivators

12.6.1 Adaptation to climate change in agriculture

Changes in the weather pattern in the past had an adverse influence on cultivators. However, today's farmers may lessen the effects of climate change by using intervention programs. Farmers can improve their

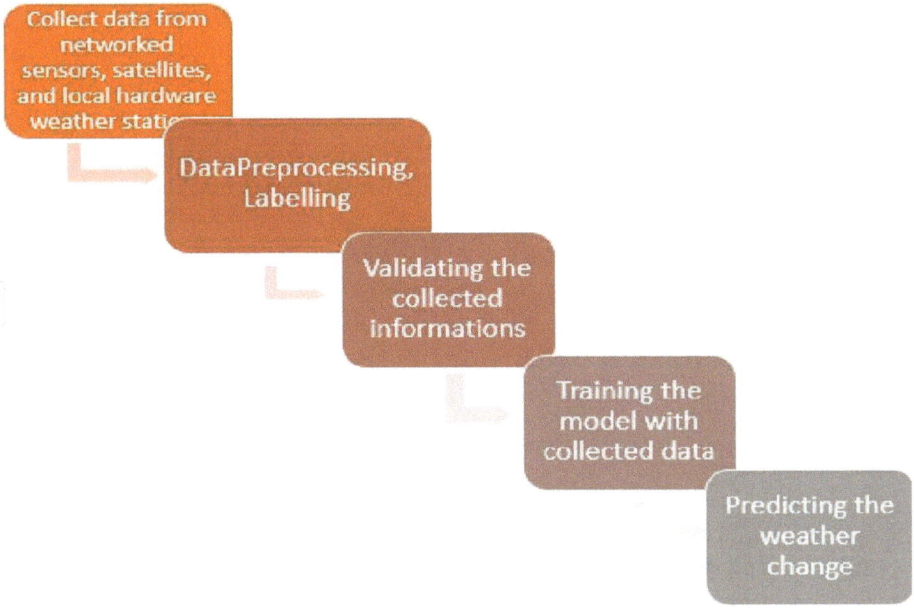

Figure 12.6 Working of ML model in the prediction of weather.

technical understanding of their crops by modifying crop management practices, improving water management, and embracing innovative farm techniques including resource conserving technologies (RCTs). Intercropping or crop diversification, enhanced pest control, updated weather forecasts, agricultural subsidies, and exploitation of indigenous seeds and crops are all part of RCT. The next section discusses a few of these approaches.

Farmers' primary objectives are to maintain product stability and create novel crop varieties with increased yields and pressure tolerance (famine, flood, and salinity) in order to adapt to the changing climate. One of the goals of captive breeding should be to improve the germplasm of major crops for thermal dissipation. Similarly, developing tolerance to a variety of biogenic stimuli as they exist is critical. Gene stacking is significantly aided by genetic engineering. It creates the "ideal plant type" by fusing all appealing traits in a plant. The gene might be a genotype that can adapt to harsh environments.

Utilizing the earth's resources, such as water, is crucial for coping with climate change. As the temperature rises and the pattern of the precipitation changes, water will become even more scarce. It is important to promote agricultural water-saving practices, micro-irrigation technologies, and crop-based irrigation decisions. New technology must be employed if water management is to become more effective. Rainwater harvesting can assist meet water demand in areas where water is scarce. Advanced irrigation systems, such as the utilization of laser-assisted ground flattening, drip watering, and spray irrigation, can also benefit water conservation. Laser-assisted leveling creates a smooth and even field, allowing for optimal water distribution with minimal water loss. Adopting these strategies enables producers to achieve uniformity in the planting of seed/seedlings and fertilizers, resulting in improved plant position, nutrient efficiency, and yield.

Modifying the harvesting strategy to make use of the rainy season and avoid extreme weather (e.g., typhoons and storms) during the vegetative stage are examples of adaptation techniques to mitigate the detrimental effects of high climate variability as witnessed in the desert and semi-arid regions

12.6.2 Farmer's predictions and adaptation to technology

In response, farmers who are aware of weather trends can cultivate more compatible crops or varieties. Farmers may choose the finest sustainable farming practices with the help of weather predictions. To reduce crop loss, harsh climatic occurrences should be predicted well in advance. Improved knowledge from worldwide monitoring and forecasting will be a breakthrough in response to climate variability. Enhanced microclimate modeling plays a major role in understanding the mechanisms of weather catastrophes more precisely. The climate change on agro-information can be distributed to farmers via audio–visual technologies, as well as through mobile telephony. Weather prediction and early warning will be extremely important in reducing the threats posed by climate change. Researchers and administrators could benefit substantially from utilizing digitalization technologies in developing mitigation strategies. For centuries, impoverished and marginal farmers in south Asia have experimented with climatic unpredictability. There is a lot of information on a variety of metrics that can aid in the development of solutions to address climate risks. It is necessary to harness this information and fine-tune it to meet present needs. People's traditional and indigenous awareness has withstood consistently for many years and may offer context and plausible options for improving behavior. Research in the study of human biological and physiological characteristics has emphasized the significance of community-based environmental management and social pedagogy in improving communities' ability to adapt to future climate change consequences. Potential indigenous strategies for rainwater retention and preservation, mineral and pest management, crop production, and plant breeding are all part of tribal and hill knowledge systems. For farmers, these technologies are tremendously helpful in weather forecasting and crop risk management.

12.7 Conclusion

Agriculture throughout the world is largely dependent on weather conditions. The expansion of the agricultural sector is directly affected by changes in weather patterns. Because of the unpredictable weather patterns, global food security has deteriorated due to insufficient food production and access to food. The effects of dramatic climatic transitions on agricultural production and yield will be stronger in rain-fed agriculture due to inter-decadal weather variation throughout the planting season. Plants are extremely vulnerable to floods and droughts, both of which will suffocate their growth. Extreme temperatures boost bug populations, necessitating more pesticide treatments and more water use.

Climate change has a wide range of consequences that influence the environment, human health, and biological diversity. Adapting to changes in weather patterns can assist farmers in producing higher-quality crops even while improving quantity. Botanists are also attempting to provide farmers with species that can withstand adverse weather conditions. Environmentalists are also promoting soil management practices to limit greenhouse gas emissions from farmland. To overcome the issues of climate change, adaptation to contemporary technologies and procedures should be practiced. Farmers can use online sites that provide weather pattern information to prepare themselves, their land, and for climate change.

Smart farming is one of the most significant achievements in climate change and farming solutions. This helps farmers save resources while also increasing agricultural yields. Precision farming also helps to reduce pollutants in the environment. Precision farming, which employs an increasing number of modern technologies, helps to minimize labor costs, enhance crop prediction, and solve problems quickly and efficiently. Agricultural technologies that use IoT devices allow researchers to focus on a specific issue area rather than the entire region. Drones and other contemporary IoT devices are utilized to water crops, spray herbicides, and even examine damaged crops. In addition, satellites are employed to help farmers process and interpret their data online and to get information from aerial images. EOS agricultural management approaches assist farmers in precisely calculating required inputs and short-term cost reduction for long-term environmental protection. By utilizing current IoT methods, agriculture can experience less of the harmful effects of climate change.

The primary goal of the government's agricultural policy is to raise agricultural yield and productivity at a fair price. India's agriculture strategy prioritizes modernizing the country's agricultural industry. In this case, policy assistance involves the implementation of contemporary technology in agricultural operations as well as the use of enhanced commodities for agribusiness, like HYV seeds, fertilizers, and drip and sprinkler irrigation systems to ensure more cost-effective and efficient use of water. Farmers must utilize government initiatives to increase productivity and address climate change issues.

Acknowledgement

"Good things come to those who believe,
Better things come to those who are patient,
Best things come to those who don't give up"

We are grateful to the All-Powerful for his countless blessings in all facets of our lives. Our profound gratitude goes out to the Vels Institute of Science, Technology, and Advanced Studies president for supporting us and giving us the chance to learn about and implement contemporary scientific technologies.

We would like to extend our profound gratitude to the Vels Institute of Science, Technology, and Advanced Studies' administration and personnel for their unwavering support and inspiration throughout the writing of this chapter. It gives us great pleasure to express our sincere appreciation to everyone who has supported our efforts to finish the task.

References

[1] M. A. Hossain, M. N. Amin , J. Sultana and M. N. A. Siddique "Climate Change Impact on Agriculture and Related Sustainable Land Management Practices in Bangladesh – A Review", International Journal of Environment and Climate Change, vol. 10, no. 2, pp. 53–69, 2020, DOI: 10.9734/IJECC/2020/v10i230181.

[2] Gurdeep Singh Malhi, Manpreet Kaur, and Prashant Kaushik, "Impact of Climate Change on Agriculture and Its Mitigation Strategies: A Review," *Sustainability. 2021, Vol. 13, Page 1318*, vol. 13, no. 3, p. 1318, Jan. 2021, doi: 10.3390/SU13031318.

[3] C. Carter, X. Cui, D. Ghanem, and P. Mérel, "Identifying the Economic Impacts of Climate Change on Agriculture," Annual Review of Resource

Economics, vol. 10, no. 1, pp. 361–380, Oct. 2018, doi: 10.1146/ANNUREV-RESOURCE-100517-022938.

[4] Raza et al., "Impact of Climate Change on Crops Adaptation and Strategies to Tackle Its Outcome: A Review," Plants 2019, Vol. 8, Page 34, vol. 8, no. 2, p. 34, Jan. 2019, DOI: 10.3390/PLANTS8020034.

[5] M.D.M.Kadiyala et al., "Modeling the potential impacts of climate change and adaptation strategies on groundnut production in India," Science of the Total Environment., vol. 776, p. 145996, Jul. 2021, doi: 10.1016/J.SCITOTENV.2021.145996.

[6] Pushan Kumar Dutta, Susanta Mitra, "Application of Agricultural Drones and IoT to Understand Food Supply Chain During Post COVID-19," Agric. Informatics, pp. 67–87, Mar. 2021, doi: 10.1002/9781119769231.CH4

[7] Kanderp Narayan Mishra, Shishir Kumar, and Nileshkumar R. Patel, "Survey on Internet of Things and its Application in Agriculture," Journal of Physics: Conference Series , vol. 1714, no. 1, p. 012025, Jan. 2021, doi: 10.1088/1742-6596/1714/1/012025.

[8] Chrysanthos Maraveas and Thomas Bartzanas, "Application of Internet of Things (IoT) for Optimized Greenhouse Environments," AgriEngineering 2021, Vol. 3, no. 4, pp. 954–970, Nov. 2021, doi: 10.3390/AGRIENGINEERING3040060

[9] Siddhartha Vadlamudi, Harish Paruchuri, Alim Al Ayub Ahmed, Md. Shakawat Hossain, Praveen Kumar Donepudi "Rethinking Food Sufficiency with Smart Agriculture using Internet of Things," Turkish Journal of Computer and Mathematics Education, vol. 12, no. 9, pp. 2541– 2551, May 2021, doi: 10.17762/TURCOMAT.V12I9.3738

[10] Pradeep Kumar Singh and Amit Sharma, "An intelligent WSN-UAV-based IoT framework for precision agriculture application," Computers and Electrical Engineering, vol. 100, p. 107912, May 2022, doi:10.1016/J. COMPELECENG. 2022.107912.

[11] Chander Prakash, Lakhwinder Pal Singh, Ajay Gupta & Amandeep Singh, "Smart Farming: Application of Internet of Things (IoT) Systems," Lect. Notes Networks Syst., vol. 221 LNNS, pp. 233–240, 2021, doi: 10.1007/978-3-030-74608-7_30/COVER/.

[12] Abhishek Khanna, Sangeet Kaur, "Evolution of Internet of Things (IoT) and its significant impact in the field of Precision Agriculture," Computers and Electronics in Agriculture, vol. 157, pp. 218–231, Feb. 2019, doi: 10.1016/J.COMPAG.2018.12.039.

[13] Antonis Tzounis, Nikolaos Katsoulas, Thomas Bartzanas, Constantinos Kittas "Internet of Things in agriculture, recent advances, and future

challenges", Biosystems Engineering, vol 164, pp 31-48, Dec. 2017, doi: 10.1016/J.BIOSYSTEMSENG.2017.09.007.

[14] Vincent Blok, Bart Gremmen," Agricultural Technologies as Living Machines: Toward a Biomimetic Conceptualization of Smart Farming Technologies", Ethics, Policy & Environment, vol. 21, no. 2, pp. 246–263, 2018, doi: 10.1080/21550085.2018.1509491.

[15] E. Navarro, N. Costa, and A. Pereira, "A Systematic Review of IoT Solutions for Smart Farming," Sensors, vol. 20, no. 15, p. 4231, Jul. 2020, doi: 10.3390/s20154231.

13

Cropping Pattern in Farming

K. Kalaiselvi[1], Pulla Sujarani[2], and V. Sakthivel[3]

[1]Department of Computer Applications, Saveetha College of Liberal Arts and Sciences, SIMATS Deemed to be University, India
[2]Department of Computer Science, Vels Institute of Science, Technology and Advanced Studies (Formerly Vels University), India
[3]Konkuk Aerospace Design-Airworthiness Institute, Konkuk University, South Korea
Email: kalaiselvik.sclas@saveetha.com; sujiraji873@gmail.com; sakthivel@konkuk.ac.kr

Abstract

Agriculture has a crucial part in the entire life of a given economy. Cropping pattern is the main important process in farming. Cropping pattern aids in preventing soil compaction, hence enhancing the soil's physical state. Crop rotation enhances both the soil texture and structure. This makes for favorable circumstances for root growth and seed germination. Cropping pattern, often known as the list of crops being grown in a region and their order in time, is the spatial representation of crop rotations. Because several crop varieties are used, one receives a general bountiful harvest each season in addition to a variety of crops. Farmers made their crop selections based on physical, social, and economic considerations. Due to unfavorable weather, improper seed sowing, unpredictable irrigation, and unanticipated locust infestations, crop output is a significant problem for farmers. Soil erosion, biodiversity declines, and growing consumer demand for higher-quality food are the bigger problems facing farmers today. These problems typically leave farmers feeling pessimistic about the future and on the edge of settling for less. Finally, this chapter discusses the types of cropping patterns, factors affecting cropping patterns, and increasing the crop yield. As a result, farmers now have a pressing need to choose technology like IoT to monitor their farms in real time and prioritize the production of high-quality food. Finally, this chapter gives a brief explanation of cropping patterns, and increasing the crop yield with modern agriculture techniques using IoT.

13.1 Introduction

Cropping pattern is a technique that aids in the production of more cereals each year. Because it alters throughout time and space, cropping pattern is a potent idea. It can be described as the percentage of land that is currently planted with different crops. The seasonal progression and spatial arrangement of sowing and fallow in a particular area are referred to as the cropping pattern. Rainfall, temperature, climate, the kind of soil, and technology all have an impact on the cropping pattern in India.

The main factors in the planning such as yield price, climate conditions, type of lands, crop demand, agricultural inputs, and cost of production are measurable and evaluated. It is difficult to predict the natural disasters like heavy rainfall, cyclone, flood, bad weather conditions, etc. [3]. It is crucial to map regional and worldwide cropland distribution quickly, yet due to their spectral similarities, differentiating between different types of crops and diverse cropping patterns is difficult. During the past three decades, agricultural production in Brazil has expanded dramatically due to the result of technical advancements, favorable commodity costs, and increasing worldwide demand [7]. Growing productivity is based on cropping patterns (CPs). The crop that is farmed the most frequently worldwide is rice. It takes up roughly 75% of the total farmed area, with the remaining 25% being split amongst all the other crops. There are several environmental variables in the country, which affects land use models. Single, double, triple, and quadruple crops can all be grown on a piece of land. In Bangladesh, the utilization of agricultural land is quite prevalent [5].

Improved soil fertility is necessary for higher agricultural yields. Soil organic matter is crucial for maintaining long-term soil fertility because it serves as a source of metabolic energy for the biological activities occurring in the soil that affect nutrient availability. Fertilizer use and food production are tightly related. In order to maintain the current per capita cereal production, the yearly global use of fertilizers will need to quadruple from around 130 million tons in the 1990s by the year 2030. The usage of fertilizers overall has expanded significantly over the past few years, but urea use has climbed significantly more than that of other fertilizers [4]. Each year, farmers must designate fields for various crops and decide on crop management practices. These choices, which are far from obvious, are crucial because they affect both short- and long-term farm productivity and profitability. Decision support models are created to assist farmers and effectively distribute limited resources [1].

The cotton-based cropping systems are very popular in the states of Andhra Pradesh, Punjab, Maharashtra, and Gujarat. 90.1% of the land in Punjab is used for growing wheat and rice, which generated 76.9% of the output in the year between 2014 and 2015 [6]. The effects of cropping pattern, sowing date, nitrogen on buckwheat seed yield are to be investigated. The outputs demonstrate that the pattern of planting had an impact on both the number of bunches and the amount of seeds in each bunch. The CGR is influenced by the weight of each bundle of thousands of seeds, the harvest index, and the sowing date for flounced seed yield. The nitrogen content of the CGR, the weight of 1000 seeds, the number of bunches in a plant, the number of seeds in each bunch, the planting pattern, the nitrogen effect on seed production, and the CGR were all considered [2].

13.2 Overview of Cropping Patterns

Farmers choose crops for cultivation based on a variety of criteria including physical, social, and economic considerations. They may plant a variety of crops on their farms and rotate a specific crop combination throughout time. However, it is worth noting that the best farming techniques always include certain cropping patterns and cropping systems to increase productivity and maintain soil fertility.

A cropping pattern is the percentage of land that is being cultivated for different crops at various times. This shows when and how the crops were planted on a specific acreage. The amount of land planted with various crops and the order in which they are planted in space and time would alter as a result of changing cropping patterns.

India is a sizable nation. Climate factors like temperature, humidity, and rainfall fluctuate from one place to region and influence the types of crops that are grown there. To get the highest yield, many cropping methods are practiced. Rainfall, temperature, climate, soil type, and technology all have an impact on India's crop patterns.

Depending on the seasons they grow in, the crops are divided into various types:

Kharif crops:
The monsoon season, which generally varies by crop and farming region, is when these crops are sown. In India, kharif crops are sown between the

months of June and July at the start of the rainy season. Between the months of September and October, at the end of the monsoon season, these crops are harvested. Paddy, rice, maize, bajra, jawar, cotton, jute, ground nuts, and other crops are examples of kharif crops.

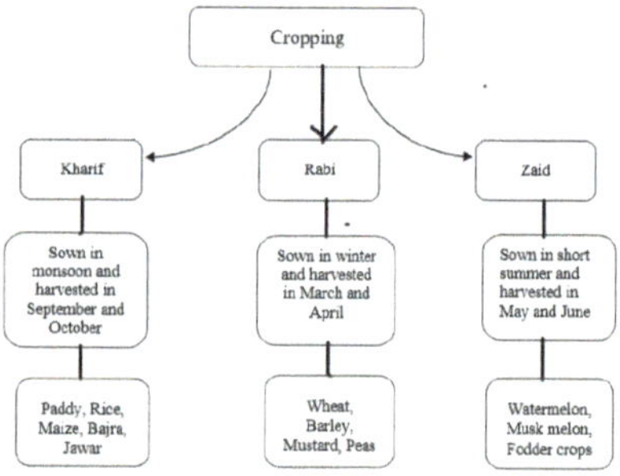

Figure 13.1 Types of cropping.

Rabi crops:

These crops are sown during the months of October and November, in the winter and following the monsoon. Crops for the Rabi season are harvested in India between the months of March and April. Wheat, barley, mustard, peas, and other types of crops are examples of rabi crops.

Zaid crops:

These crops are planted during the months of March and April, during the summer's brief sowing season. Zaid crops are harvested in India from May through June. Watermelons, musk melons, vegetable and fodder crops, etc., are some of Zaid's examples.

13.3 Types of Cropping Pattern

Any region's crop pattern is influenced by geographical factors such as the soil, climate, rainfall, etc. In addition, the type and accessibility of irrigation facilities must be considered. There are four different cropping patterns:

- Monocropping
- Mixed cropping
- Intercropping
- Crop rotation

13.3.1 Monocropping

Monocropping decreases soil fertility and ruins the soil's structure. The use of chemical fertilizers is necessary to

Figure 13.2 Monocropping.

Figure 13.3 Mixed cropping.

Figure 13.4 Intercropping.

increase production. The spread of illnesses and pests is made possible by this approach.

13.3.2 Mixed cropping

The practice of planting two or more crops simultaneously on the same plot of land is known as mixed cropping. Mixed cropping, for instance, is the simultaneous cultivation of wheat and gram on the same piece of land. This method reduces the likelihood that one of the crops will fail and provides protection against crop failure brought on by unusual weather.

The crops that will be cultivated together should demand varying amounts of water and maturity times. It is best to plant one tall crop and one dwarf crop together. The amount of nutrients needed by one crop should be lower than the amount needed by another. One crop needs deep roots, while another needs shallow ones. A successful mixed cropping pattern is the result of all these factors.

13.3.3 Intercropping

Intercropping is the practice of growing many crops at once in a predetermined row pattern on the same plot of land. Three rows of intercrops can be cultivated after one row of the main crop. As a result, productivity per area increases.

The different types of intercropping are as follows:

Row intercropping:
Row intercropping is the practice of planting the component crops in different rows. It aids in the most effective use of land area and the control of weeds during the initial stages of the primary crop.

Strip intercropping:

Strip cropping is the practice of growing two or more crops in broad strips so that they can be maintained independently. The crops are nevertheless close enough for interaction.

Relay intercropping:

When the existing crop has flowered but has not been harvested, this method of intercropping involves planting a second crop. For instance, rice, cauliflower, onions, and summer squash.

Advantages of intercropping:

- The soil's fertility is preserved.
- Pests and diseases are kept from spreading.
- Optimal use of the available resources.
- Growing more than one crop saves time and space.
- Maximum use of the soil's nutrient resources.

Some of the crops produced as intercrops are maize, soybeans, bajra, and lobea.

Crop rotation:

In this pattern, various crops are produced on the same plot of land in a predetermined order. Depending on how long they grow, the crops are divided into three categories: one-year rotation, two-year rotation, and three-year rotation.

The crop rotation program includes legumes to improve soil fertility. After the legumes, the crops that require a high level of fertility (such as wheat) can be cultivated. After the crops that demand large inputs are established, the low-input crops can be planted.

The following standards should be used when choosing the crops to be rotated:

- There ought to be enough moisture on the land.
- Existence of fertilizers, labor, and machinery.
- Facilities for processing and marketing.

Figure 13.5 Crop rotation.

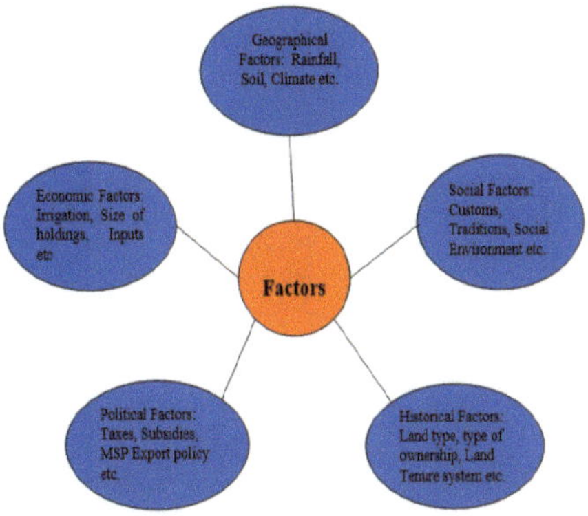

Figure 13.6 Cropping pattern factors.

- Nutrient availability in the soil.
- The length of the crop is short or lengthy.

The benefits of rotating crops are as follows:

- A long time is spent maintaining the soil's fertility.
- It stops the spread of weeds and pests.
- There is no need for several chemical fertilizers.
- The soil's chemistry and physical makeup are not changed.

13.4 Factors Affecting Cropping Patterns

The following elements influence the various kinds of cropping patterns:

- The cropping patterns are one factor in estimating agricultural output. This is a reflection of the local agriculture economy.
- Due to changes in agrarian policy, the availability of agricultural inputs, and advancements in technology, cropping patterns are suffering.
- As a result, by increasing crop output, cropping patterns play a key role in enhancing soil fertility. It guarantees both crop protection and nutrient accessibility for the crops.

The important factors in cropping pattern are as follows:

- Geographical factors

- Economic factors
- Political factors
- Historical factors
- Social factors

Geographical factors:
The different natural elements influencing an area's cropping pattern include the following:

Rainfall: Rainfall is a key factor in determining a region's cropping pattern. Depending on the area of rainfall, such as low, medium, and high rainfalls, different cropping patterns are produced.

Climate: Due to the dependence on jowar and bajra in some dry places, there may be insufficient rainfall and a significant degree of monsoon uncertainty.

Soil: A given type of crop should be cultivated in a specific type of soil, which is defined by the soil's constituents and the region's climatic effects.

Economic factors:
These play the largest role in defining the nation's cropping pattern.
The many economic factors include:

Irrigation: The north Indian plains are highly irrigated; two to three crops are produced there each year.

Size of land holdings: Small farmers are initially focused on producing food grains to meet their own needs. Smaller landowners give up less of their land to cash crops than larger landowners do.

Availability of inputs: Seeds, water storage, marketing, and transport have an impact on a region's cropping patterns.

Political factors:
The cropping pattern may also be impacted by the government's legislative and governmental policies. Acts such as those governing food crops, land usage, and intensive programs for cotton, oilseeds, and paddy have an impact on the cropping pattern. Farmers favor rice and wheat over food crops because they are offered higher minimum support prices (MSPs). The government's decision to promote or discourage a certain crop owing to different factors like a flood, drought, etc. The various political factors affecting cropping pattern are taxes, subsidies, MSP, export policy, etc.

Historical factors:
The historical aspects, which include the cultivation of various crops carried out in the area from long time due to distinct historical regions, also characterize the cropping patterns of the region. Although sugarcane is more widely farmed in north India, the circumstances are the best in south India. The type of land, the type of ownership, the system of land tenure, and other historical circumstances all affect the cropping pattern.

Social factors:
Food habit is different for various places. Rice is preferred in east and south India while wheat is preferred in north India. The various factors are customs, traditions, social environment, etc.

13.5 Crop Production and Management

The term "crop production and management" refers to a variety of techniques used to cultivate and harvest crops efficiently. It is protecting crops while they are either growing or have just been harvested. Crop productivity can be raised by managing nutrients, using irrigation, and changing cropping practices. Crop cultivation requires farmers to engage in a number of actions throughout time. The following list includes some of the actions or activities that are referred to be agricultural practices:

(i) Soil preparation

(ii) Sowing

(iii) Incorporating manure and fertilizers

(iv) Irrigation

(v) Weed protection

(vi) Harvesting

(vii) Storage

13.5.1 Soil preparation

The first step in cultivating a crop is soil preparation. The act of turning and loosening the soil is one of the most crucial chores in agriculture. This makes it possible for the roots to go deeply into the soil. Even when the roots are buried deeply, the loose soil makes it possible for them to breathe readily. The roots may breathe easily even when they are buried deeply in the soil due to the loose soil. Earthworms and other soil-dwelling bacteria benefit from the soil's loosening by growing.

Air, water, minerals, and some living things are all present in soil. Soil organisms break down dead plants and animals. The dead organisms' various nutrients are released back into the soil. Plants once more absorb these nutrients. Only a few centimeters of the soil's top layer supports plant growth, but churning and loosening the soil helps it rise to the surface where plants may access its nutrients. For the cultivation of crops, soil turning and loosening are crucial.

Tilling and ploughing are two terms for the loosening and turning operation. Plough is used for this and iron or wood is used to make ploughs. Before ploughing, the soil might need to be watered if it is too dry. There may be large soil clumps called "crumbs" in the ploughed field. These crumbs need to be broken, obviously. Field leveling is advantageous for irrigation as well as for sowing. With the use of a leveler, soil can be leveled.

To have a greater yield, it is vital to break up soil clumps prior to spreading the seeds and a variety of tools are used. The plough, hoe, and cultivator are the primary implements used for this purpose.

Plough: The plough has been in use since ancient times to cultivate the land, turn the soil, and fertilize crops while also eradicating weeds. A couple of bulls or other animals pull this wooden object (horses and camels). It has a ploughshare, a sturdy triangular iron strip. A ploughshaft, which is a long piece of wood, serves as the main component of the plough. The shaft has a handle on one end. The other end of the rope is fastened to a beam that is resting on the bulls' necks. The plough can be easily driven by a man and a single pair of bulls.

Hoe: A hoe is a straightforward instrument used for weeding and soil aeration. Strong, large, and bent iron plates are uniform to all ends, and they work like blades. It passes through animals. Animals drag it through them.

Cultivator: Cultivators powered by tractors are used to plough. Utilizing cultivators reduces labor and time costs.

13.5.2 Sowing

An essential component of crop production is sowing. A good range of clean, healthy, and high-quality seeds are chosen before sowing. Farmers favor using seeds that produce a high yield. Therefore, damaged seeds are lighter and float on water. This procedure works well for sorting healthy, excellent seeds from those that have been harmed. Before sowing, knowing the equipment needed to sow the seeds is one of the key duties.

Traditional tool: This funnel-shaped implement is used to traditionally plant seeds. The funnel is loaded with seeds, which are then fed via two or three pipes with pointed ends. To plant seeds, their ends puncture the ground.

Seed drill: Tractors are utilized to sow the seeds using the seed drill. This evenly distributes the seeds' sowing depth and distance. It guarantees that after sowing, dirt will cover the seeds. This prevents birds from eating the seeds. Using a seed drill to sow reduces labor and preparation time.

13.5.3 Incorporating manure and fertilizers

Manure and fertilizers are the materials that are put to the soil as nutrients to support the healthy growth of plants. The agricultural plants receive their mineral nutrients from the soil. These components are essential for the growth of plants. Crop after crop is planted on the same land by farmers in some areas. The field is never left fallow or untended.

The nutrients in the soil become depleted due to the ongoing agricultural cultivation. To replace the soil's nutrients, farmers must apply manure to the fields. Manuring is the name of this procedure. Plants that receive improper or insufficient manuring are feeble. Manure is an organic material created by the breakdown of animal and plant wastes. Farmers throw animal and plant waste in open pits and let it decompose. Some microorganisms are the cause of the breakdown. Organic manure is made from the decomposing material.

Chemicals called fertilizers are high in a certain nutrient. Fertilizer use has aided farmers in increasing the output of crops including rice, wheat, and corn. However, overuse of fertilizers has reduced soil fertility. Additionally, fertilizers are becoming a cause of water pollution.

Manure's benefits include: Manure made from organic materials is preferred over fertilizers. This is due to the following:

- It increases the soil's ability to water holding.
- It makes the earth permeable, facilitating the free gases exchange.
- It enhances the population of beneficial microbes.
- It enhances the soil's texture.

13.5.4 Irrigation

Irrigation is the process of regularly supplying crops with water. Depending on the crop, the soil, and the season, irrigation should be done at different times and more frequently. Wells, tube wells, ponds, lakes, rivers, dams, and canals are some of the places where irrigation water can be obtained.

Modern irrigation techniques: We can use water more efficiently, thanks to modern irrigation techniques. The following are the main techniques used:

Sprinkler system: When there is insufficient water, this technique works better on uneven terrain. The revolving nozzle-topped perpendicular pipes are connected to the primary pipeline at regular intervals. Water escapes from the revolving nozzles when it is allowed to move through the main pipe while being pushed through by a pump. The crop is sprinkled with it to simulate rain. For lawns, coffee plantations, and a variety of other crops, sprinklers are quite helpful.

Drip system: In this mechanism, the water trickles down toward the roots one drop at a time. So, the drip system is named. It is the ideal method for watering trees, gardens, and fruit plants. Water is not at all wasted.

13.5.5 Weeds protection

An unwanted plant could inevitably grow alongside the crop. Weeds are the name given to these unwelcome plants. Weeding refers to the removal of weeds. Even when being harvested, some weeds can interact and pose a risk of poisoning to both humans and animals. Farmers use a variety of techniques to get rid of weeds and limit their growth. Before planting crops, tilling aids in removing and eliminating weeds, which may later dry up and combine with the soil. It is best to remove weeds before they blossom and produce seeds.

Other methods of controlling weeds include the use of chemicals known as weedicides, such as 2,4-D. To kill the weeds, they are sprayed on the fields. The crops are not harmed by them. The necessary amount of water is used to dilute the weedicides before being sprayed with a sprayer into the fields. Farmers' health may be impacted by the use of weed killers. Farmers should thus use these chemicals with caution. When these chemicals are sprayed on them, they should use a piece of cloth to protect their mouth and nose.

13.5.6 Harvesting

Crop harvesting is a crucial duty. Harvesting is the act of cutting a crop after it has reached maturity. Crops are hauled out or cut close to the ground during harvest. A cereal crop typically takes 3–4 months to reach maturity. The grain seeds in the harvested crop must be distinguished from the chaff. Threshing is the name of this procedure. This is accomplished with the use of a device known as a "combine," which functions as both a harvester and a thresher. Winnowing is used by farmers with modest landholdings to separate the grain from the chaff.

Figure 13.7 Semi-automatic robot.

Figure 13.8 Drone.

13.5.7 Storage

Produce storage is a crucial responsibility. The stored grains must be protected from moisture, pests, rodents, and microbes if they are to be kept for an extended period of time. Grain harvests contain more moisture. Freshly harvested grains and seeds should be dried before storing them to prevent spoilage and microbial attack that renders them useless for use or germination.

For a better result, the grains are properly dried in the sun to remove the moisture before storing them. This stops germs, fungus, and insect pests from attacking. Jute bags or metal bins are used by farmers to store crops. To keep them safe from vermin like rats and insects, grains are stored in granaries and silos on a large scale. For preserving grains at home, neem leaves that have been dried are utilized. To keep them safe from pests and germs while being stored in vast warehouses, certain chemical treatments are necessary.

13.6 Modern Agriculture Technologies

13.6.1 Semi-automatic robots

Robots with arms that can detect weeds and spray pesticides on the afflicted plants can prevent major harm and lower the cost of pesticides overall. These machines can also be employed for lifting and harvesting. Heavy farming equipment may be operated from the comfort of homes using phone displays, and GPS can constantly track its whereabouts.

13.6.2 Drones

To image, map, and scan the farms, drones with sensors and cameras are used. They can be remotely controlled or they can fly autonomously using flight

Figure 13.9 IoT-based remote sensor.

Figure 13.10 Computer vision.

plans that are controlled by agriculture software and GPS in their embedded systems. Insights about crop health, irrigation, spraying, planting, soil and field, plant counts, and yield prediction, among other topics, can be gleaned from the drone data.

13.6.3 IoT-based remote sensing

IoT-based remote sensing collects data using sensors, such as weather stations installed on farms, and then sends it to analytical tools for study. They keep an eye out for variations in temperature, light, humidity, form, and size of the crops. The information gathered by sensors about humidity, temperature, moisture precipitation, and dew detection aids in forecasting the weather on farms so that appropriate crops can be cultivated. The analysis of soil quality aids in identifying the nutrient value and drier farmlands, soil drainage capacity, or soil acidity, allowing for adjustment of irrigation water requirements and selection of the most advantageous form of cultivation.

13.6.4 Computer imaging

The usage of sensor cameras placed at various locations around the farm or camera-equipped drones is necessary for computer imaging. They digitally process the photographs they take in order to extract valuable insights from them. They are used for produce sorting and grading after harvest as well as quality control, disease detection, irrigation monitoring, and crop grading. To control the quality of image processing using machine learning, the size, shape, color, and growth of standing crops are identified by comparing photos from a database with images of those crops.

13.7 Benefits of Implementing the Smart Solution in Farms

Remote control:
Farmers are looking for a better solution to their management issues because their farms are located in remote areas and distant lands. IoT technology offers a smart farming solution, allowing farmers to remotely manage their fields using smart gadgets. Furthermore, farm managers require proper visibility of their fields despite poor health, poor weather, travel restrictions, or labor shortages. Remote crop management enabled by smart technology provides transparency and real-time crop monitoring, resulting in higher yield.

Crop monitoring in real time:
Many places use indigenous farming methods, owing to a lack of proper knowledge among farmers. The agricultural state has recently become more critical, involving complex operations such as weather forecasting and soil quality checks. Such a situation necessitates a newer vision in agriculture, one that includes real-time monitoring through the use of sensors. Such devices play an important role in providing accurate results, which directly affect crop yield.

Quality of soil testing:
The IoT-based sensors are installed beneath the ground to monitor soil quality and test its suitability for various crops. This improves how the farming sector operates by providing the best solutions for achieving error-free results. The advancement of technology provides a complete package for analyzing soil quality and recommending farming options to farmers. Precision agriculture encompasses such processes, enabling smart measures to improve farming efficiency in all aspects.

Analysis of soil demand in real time:
Precision technology is made available via the Internet of Things for more efficient and productive farming practices. One of the most important tools for farmers to grow high-quality crops is accurate soil data. Additionally, IoT provides farmers with cutting-edge methods for sowing seeds as well as educational data on the weather forecast, necessary soil moisture content, suitable temperature, and humidity. The analysis of all of this is done in real time to assure agricultural productivity.

Protection of crops:
Modern farm managers can thoroughly check on their crops using the Internet of Things technology, which uses a data-driven approach. It enables farmers to combat unwanted pests and safeguard their crops from various diseases by

taking the proper measures. Additionally, the smart farming solution monitors each step of crop production, sending out prompt notifications regarding crop health, condition, and temperature requirements, and displaying all the information on a network of connected smart devices.

Smart greenhouses:
IoT technology and linked devices are used in smart greenhouses to autonomously adjust the environment for agricultural production. An atmosphere like this lessens the likelihood of predators invading the fields and the effects of the weather, safeguarding the crops at all costs. The autonomous function of smart greenhouses helps to achieve maximum efficiency by giving farmers real-time insights. It also has sophisticated sensors, gateway connectivity, and an intuitive interface for real-time updates on various crop operations as well as regular insights on irrigation, lighting, humidity, and temperature.

13.8 Conclusion

Cropping pattern is the classification of how crops are arranged throughout both time and area. Different plants require different climatic conditions, temperatures for growth, complete lifecycles, and improved plant diversity, which can be achieved through the selection process. To produce more grains in a year, different cropping patterns are used in the production process. With its ability to provide accurate crop data, IoT technology is leading the way in the farming sector. It is also predicted that nearly 12 million agricultural sensors will be installed globally by 2023, implying that a typical field could generate half a million data points per day. This is the cropping pattern of farming, with IoT increasing crop yields and profitability. Furthermore, the incorporation of IoT technology allows farmers to reduce their workload through automated processing. IoT primarily aids in the analysis and optimization of big data in agriculture, maximizing operational efficiency and lowering labor costs.

References

[1] Jerome Dury, Noemie schaller, Frederick Garcia, Arnaud Reynaud, Jacques Eric bergez , Models to support cropping plan and crop rotation decisions.Areview, Agronomy for Sustainable Development 2012, volume 32, pages 567–580.

[2] Mohammad Reza Sobhani, Gulahmad Rahmikhdoev, Dariush Mazaheri, Majid Majidian, Effects of sowing date, cropping pattern and nitrogen on

CGR, yield and yield component summer sowing buckwheat, , Journal of Applied Environmental and Biological Sciences 2012, 2(1)35–46.

[3] Mohammad Mansourifar, Morteza Almassi, Ali-Mohammad Borghaee, Reza Moghadassi, Optimization Crops Pattern in Variable Field Ownership, World Applied Sciences Journal 2013, 21 (4): 492–497,.

[4] M. H. Rahman, M. R. Islam, M. Jahiruddin, M. Y. Rafii, M. M. Hanafi and M. A. Malek, Integrated nutrient management in maize-legume-rice cropping pattern and its impact on soil fertility Journal of Food, Agriculture & Environment 2013, Vol.11 (1): 648–652..

[5] M Nasim, S M Shahidullah, A Saha, M A Muttaleb, T L Aditya2, M A Ali and M S Kabir ,Distribution of crops and cropping patterns in Bangladesh, Bangladesh Rice 2017, J. 21 (2): 1–55.

[6] R s Mann ,Cropping Pattern in Punjab (1966–67 to 2014–15), Punjab Exploring products 2017, Volume no 3.

[7] Yaoliang, Dengsheng Lu Chen, Emilio Moran, Mapping croplands, cropping patterns, and crop types using MODIS time-series data, International Journal of Applied Earth Observation and Geoinformation 2018, Volume 69, pages 133–147.

[8] Sylvain Ferrant, A. Selles, Mael Le Page, Al Bitar Ahmad, S. Mermoz, Simon Gascoin, Alexandre Bouvet,_Sentinel-1&2 for near real time cropping pattern monitoring on drought prone areas. Application to irrigation water needs in telengana, south-India, International Archives of the Photogrammetry, Remote Sensing and Spatial Information Sciences 2019, XLII-3/W6, pp. 285–292.

[9] Peter L. O'Brien, Jerry L. Hatfield, Christian Dold, Erica J. Kistner-Thomas, Kenneth M. Wacha, Cropping pattern changes diminish agroecosystem services in North and South Dakota, USA, Agronomy Journal 2019, Volume112, Issue1.

[10] Maria Kernecker, Andrea Knierim, Angelika Wurbs, Teresa Kraus, Friederike Borges, Experience versus expectation: farmers' perceptions of smart farming technologies for cropping systems across Europe, Precision Agriculture 2019, 21, 34–50.

[11] Muhammad Shoaib Farooq, Shamyla riaz, Adnan Abid, Kamran Abid, Muhammad azhar Naeem, A Survey on the Role of IoT in Agriculture for the Implementation of Smart Farming, IEEE Access, 2019, volume 7.

[12] K. A. Chavan, P. S. Bodake, C. B. Pande, A. A. Atre, S. D. Gorantiwar and A. D. Raut, Identification of Cropping Pattern in Khadambe bk. using Sentinel 2 Images and Arc GIS Software, International Journal of Current Microbiology and Applied Sciences 2020, Volume 9(9): 1139–1145.

[13] Sangeetha K, Narmada C, Karishma, Kishore karthi V, Smart Farming Using IoT, International Research Journal on Advanced Science Hub 2021, Volume 3, Pages 63–67.
[14] Benjamin Richard, Aiming Qi,Bruce D. L. Fitt, Control of crop diseases through Integrated Crop Management to deliver climate-smart farming systems for low- and high-input crop production, An International Journal Edited the British Society for Plant Pathology 2021,Volume 71, Issue 1.
[15] Dorna Jahangirpour, Mansour Zibaei, Cropping Pattern Optimization in the Context of Climate-Smart Agriculture: A Case Study for Doroodzan Irrigation Network-Iran, Department of Agricultural Economics 2022, Vol. 35, No. 4, p. 407–422.

14

Crop Welfare and Security to Farmers

Panem Charanarur[1], Srinivasa Rao Gundu[2], J. Vijaylaxmi[3], and Debabrata Samanta[4]

[1]School of Cyber Security and Digital Forensics, National Forensic Sciences University (NFSU), India
[2]Department of Computer Science, Government Degree College Sitaphalmandi, India
[3]PVKK Degree & PG College, India
[4]Department of Computational Information Technology, RIT Kosovo (A.U.K.), Rochester Institute of Technology – RIT Global, Dr. Shpetim Rrobaj, Kosovo
Email: panem.charan@gmail.com; srinivasarao.gundu@gmail.com; luckyvj2012@gmail.com; debabrata.samanta369@gmail.com

Abstract

Every ninth person in poor nations and rural regions suffers from malnutrition, which impacts 795 million people. All four areas of food security may benefit from technological advancements. It is possible to boost food supply via genetic engineering as well as soil fertility augmentation approaches and irrigation technology. It is possible to address food accessibility and nutritional value using science, technology, and innovation (STI) based climate solutions using precision agriculture and early warning systems, as well as post-harvest technology. Bio fortification and climate-smart solutions may also help alleviate food poverty. Artificial intelligence, synthetic biology, and tissue engineering may threaten the future of agriculture. Research and development, human resources, infrastructure, and information flows are needed to take use of these technologies' promise to improve food security. Gender-sensitive techniques of technology creation and dissemination as well as regional and worldwide cooperation and agricultural innovation tech forecasts and evaluation all contribute to a favorable environment for agricultural innovation. All aspects of food security may be improved with the aid of science and technology. Some of the new and growing agricultural technology, with an emphasis on smallholder farmers, was illustrated via the use of real-world instances. It is only those who have the ability to modify and share these tools and processes to meet local food security concerns that can use these tools and processes.

14.1 Introduction

They are more vulnerable to hunger in rural regions because they have a smaller food and finance supply. The majority of these people are small-scale farmers who are vulnerable to weather extremes, such as drought and floods, since their crops are grown on marginal ground. Landless farmers make up 20% of the population, while those who live in pastoral, fisherman, and forager communities make up 10%. About 20% of people live outside major cities in developing nations. More than 70% of the worlds impoverished are smallholder farmers, according to the United Nations Food and Agriculture Organization. Rural communities are plagued by hunger and malnutrition [1].

Over 90% of the world's 570 million farms are run by a single person or a family, and they depend heavily on family labor, says the Food and Agriculture Organization (FAO) of the United Nations. As of 2013, more than 80% of the world's food supply comes from farms in Asia and Sub-Saharan Africa; 84% of family farms are less than 2 hectares in size, and just 12% is managed by family farmers. The function of smallholder farms in developing nations may grow more problematic since large-scale farming has been the standard in wealthy countries and the number of people engaged in agriculture has decreased in recent decades. However, globalization and market liberalization are expected to encourage the development of increasingly specialized and large-scale industrialized production processes. Smallholder farming, based on knowledge-intensive agro-ecological production systems and eco-functional intensification, may have to play a more significant role in the face of environmental and socioeconomic concerns and fast population expansion. This means that while the importance of smallholder farms in ensuring food security may fluctuate over time due to structural changes [2], their contribution remains critical.

Reduced hunger can only be achieved by inclusive economic growth that gives impoverished people a chance to improve their situation. FAO et al. believe that the success of the global food system depends on smallholder family farmers, investment, and social protection (2015). Smallholder farmers all across the world are feeling the effects of globalization, market liberalization, technical advancement, and climate change. It is becoming more difficult to maintain political, social, economic, and environmental stability. Food systems have seen a considerable shift in recent years as a result of globalization, rising food commerce, technological advancements, longer food supply and processing chains, and unpredictable food commodity prices. Bio fuel production raises the possibility of increased deforestation and the displacement of food crops. In order to meet domestic demand for agricultural goods, countries must engage in international commerce. Trade may also provide benefits in terms of both absolute and relative costs. As a result, farmers' earnings may rise. Climate change will increase local production insecurity, which trade may assist to alleviate. Because some producers' conditions are worsened by the trade process, it might be a double-edged sword (e.g., in case products from other producers reach the markets with lower costs).

WTO and its supporting bilateral, regional or plurilateral liberalization agreements are critical to international law on international trade. Agricultural output, exports, and consumption are all affected by them. In poor countries, agriculture contributes less than 10% of GDP. Included under non-market and subsistence production, this sector generates 50%–80% of gross domestic product in many developing nations and employs half or more. With mining, agriculture is the primary economic sector in many nations and has a significant impact on society (for employment, income generation, nutrition, rural development, and the social fabric). This means that in many nations in the near future, the only possible drivers of economic and social progress will be primary production, selected service sectors, and mainly agriculture [3].

Global population growth and the resulting rise in food consumption are major contributors to global climate change. Using data from the Food and Agriculture Organization (FAO), food production has increased roughly 2.2 times faster than population growth over the last 50 years. Despite the fact that grain yields have increased by 2.2 times, the amount of fertilizer used has increased by five times. In industrialized nations, overconsumption and eutrophication of ecosystems are directly linked to biomass production (food, feed, fiber, and energy), whereas biomass production in supply regions has led to environmental challenges, land competition, and nutrient depletion. Global food production is both a cause and an influence on climate change. Due to climate change, the most common causes of food insecurity are natural catastrophes such as floods, tropical storms, prolonged droughts, and new bug species and illnesses. Due to global droughts, food shortages are a widespread problem. Crop failures and large animal deaths were common occurrences in 2011 in Eastern Africa as a consequence of recurrent droughts. In 2012, the Sahel area of Western Africa saw something similar. Australia, Central Europe, Russia, and the United States have all had similar-sized droughts. Soil loss due to wind and water erosion, insect and disease resistance to a growing variety of agrochemicals, and the loss of biodiversity are only some of the issues confronting agricultural production. Climate change may have a negative impact on future output growth, according to forecasts [4].

Climate change may have an impact on the FAO's food demand predictions in the future, according to future projections from the agency. For the first time, agriculture is the leading driver of environmental change, but it is also the leading victim of these changes. Using a new agricultural growth paradigm that emphasizes sustainable intensification within planetary boundaries, the authors argue for a worldwide food revolution based on sustainable intensification. We will not be able to feed the world's population while still maintaining a healthy and resilient Earth system unless we undergo a major shift. We must use science, technology, and innovation in every part of the food system if we are to eliminate world hunger by 2030. Also included in this research is an emphasis on the need of government investment in the development of talent, especially in poorer nations, in order to encourage innovation. Use of agribusiness and the broader food system's ingenuity may promote long-term economic growth and sustainability [5].

Crop calories available in 2006 are almost 70% lower than predicted 2050 calorie requirement, according to FAO (2006). Improve or maintain soil fertility and pastureland productivity and rehabilitate damaged land to narrow this gap in food production, as well as to decrease food loss and waste, and to alter diets (Ranganathan et al., 2016). According to the most recent data available (Ranganathan et al.), we must take into account environmental, ecological, and agronomic limits while dealing with food availability in this circumstance. It is up to the food supply to meet this shortage. Estimated pollution or erosion has caused around a third of the world's arable land to disappear during the last 40 years. Science, technology, and innovation may considerably enhance food production by discovering new plant kinds and optimizing the inputs needed to maximize agricultural output [6].

For example, plant genetic modification may be used to boost nutritional fortification and resistance to drought and herbicides, disease, and pests. Agricultural genetic modification has always relied on conventional cross-breeding methods. By crossing a main cultivar with a "related crop" with desired characteristics in the mid-1800s, Gregor Mendel was able to create new types of crops. Only within the same crop family may improvements be made (Buluswar et al., 2014); however, this technique is still beneficial for smallholder farmers throughout the world, particularly in poor countries. Nutritious Maize for Ethiopia is a current effort that uses conventional cross-breeding, helps farmers improve their abilities, and incorporates North–South collaboration into the equation. The first seeks to promote the broad adoption of high-quality protein maize (QPM) varieties in order to enhance household food security and nutrition for about 3.98 million Ethiopians. Farmer-focused seminars were held at 1233 locations in rural areas around the country for researchers, extension workers, local and regional government officials, and media representatives [7].

Traditional cross-breeding and technological transfer is two methods used by other nations to improve the productivity of basic crops in tough climatic and environmental circumstances. Since Peru's government began executing its genetic improvement program in 1968, Peruvian grains have been exposed to it. Peasants at high altitude regions where the climate and soil conditions prevent the growth of many food species often produce cereals (such as barley, wheat, and oats) and native grains (such as quinoa and amaranth) as staple meals. Farmers in Peru's highlands worked together with academics, government officials, business leaders, and members of the international community and civil society to use traditional plant genetic modification methods and technological assistance to develop rustic varieties that are better suited to sierra conditions. A participatory assessment of improved seed varieties aided in the dissemination of ground-breaking seed technologies to farming communities [8].

14.2 Making of Soil Management to Increase Yields

Genetically modified cultivars may not be able to boost yields if inadequate soil fertility is not addressed. Using rich soils is essential to preserving agricultural output and food security. Advances in technology are mostly aimed at combating pests and illnesses that affect crops. In addition, there is a lack of focus on soil management that is environmentally friendly. Healthy plants thrive in soils that are free of pests and illnesses; on the other hand, 1614 agribusiness have relied on synthetic fertilizers for decades to boost crop yields, but their high cost, reliance on natural gas, and negative effect on the environment make them unsustainable. Small-scale farmers that abuse fertilizer and water risk financial ruin as well as damage to the environment. According to the Intergovernmental Technical Panel on Soils, farmers are mining the soil, making it a non-renewable resource.

With the help of a variety of modern methods and tools, it is now possible to utilize fertilizer more sustainably. It is possible that new nitrogen fixation and other fertilizer component techniques that bypass the existing capital and energy-intensive approaches may make nutrient replenishment more sustainable nitrogen-fixing plants, according to a new research, and may improve yields by increasing soil water retention and water penetration (Folberth, 2014; United Nations, 2015b). Africa's legume crop producers may profit from "N2Africa," a multi-national effort to improve nitrogen fixation in their crops via scientific advancement and research [9].

Bio fertilizers (compost, manure, or dung) may be phased out in favor of newer, more efficient biological fertilizers. An organic and natural fertilizer derived from *Moringa oleifera* was produced by the Nigerian National Research Institute for Chemical Technology (NRICT). Sanitation may be required for the production of biodegradable fertilizers, especially those made from human waste. Improved yields and reduced environmental impact are both possible outcomes of precision agriculture, which may be used to apply inputs precisely according to crop type and soil conditions.

Crop yields are highly dependent on water availability and soil quality. Over 70% of the world's freshwater supply is used by agriculture. Because of physical or economic water shortage and a variety of other factors, farmers are frequently unable to get agricultural water. Low-cost drills, renewable energy-powered pumps, desalination, and water-efficiency technologies may be useful in solving these problems and increasing water availability for agricultural cultivation. Groundwater irrigation may become easier with the development of lightweight drills for shallow groundwater and groundwater detection devices.

The use of solar-powered irrigation pumps may be an option if manual pumps are insufficient or costly motorized irrigation pumps with ongoing fuel expenditures are out of financial reach (Buluswar et al., 2014). Another invention that might help with irrigation is a low-cost rainwater storage system. To irrigate fields in places where diesel or solar pumps are not possible, hydro-powered pumps may be used instead. It is possible to extend the growth season throughout the year by limiting the amount of water accessible in greenhouses due to unpredictable rainfall [10]. Using data rather than just physical devices and agricultural inputs may improve water availability and efficiency. Peruvian weather and climatic information is scarce and costly, making it difficult to get. According to weather, soil, and temperature data, the Institute for University Cooperation Onlus offers the appropriate watering strategy.

Because of a lack of trustworthy crop condition information, farmers in countries like Mozambique may be reluctant to invest in expensive inputs (such as high-quality seeds, fertilizer, and irrigation). Two weeks before crop stress is visible, Future Water's Flying Sensors use near-infrared sensors to identify crop stress early. In the first year of operation, some homes used 39% less water than they did the year before. It is also important to consider how women's lack of access to water and other inputs affects agricultural output as a last consideration when thinking about water for food [11]. In order to maintain a steady supply of food, it is essential to reduce production losses and retail and consumer food waste. As a consequence of poor storage conditions, African smallholder farmers who lack market access sometimes end up with grain that has been destroyed. Refrigeration is typically insufficient for meats, fruits, and vegetables. It is necessary to import much of the area's valuable crops, since there are no food processing facilities in the region.

As a result of a lack of agricultural job opportunities, farmers are forced to pay a higher price for the agricultural products they need. Due to a lack of readily available, low-cost refrigeration and energy, it is difficult to produce, store, and sell high-value perishables including fruits, vegetables, dairy, and meat. Because of its delicate nature, food must be transported through rugged and uneven terrain in order for it to retain its freshness (Buluswar et al., 2014; African Cashew Alliance, 2010). Agriculture losses may happen to any crop, but perishables are more likely to suffer. Harvesting equipment is out of reach for many smallholder farmers in tropical regions. Imported threshers may be a financial and operational hardship for smallholder farmers because of their high cost, large size, high power consumption, and high maintenance requirements. Maintenance and production of small- to medium-sized threshers may be managed by investing in the development of local expertise [12].

14.3 Image Processing in Farming

Advances in agricultural technology have been made in image processing. Enhancing the quality of a digital image or extracting useful information from it is accomplished via the use of photo processing. It has been shown that using image processing in agriculture is an efficient way to increase agricultural production and sustain global agricultural demand. Crop illnesses, weeds, and land mapping may be identified using infrared and hyper spectral X-ray imaging (Feng et al., 2018), which might save farmers money and effort. In order to combat agricultural diseases, farmers must recognize and prevent the spread of crop disorders. In order to meet market demand and offer farmers with timely information via different automated agriculture applications, image-processing technology will be used. Agricultural image processing has a wide range of uses, which will be explored in the next section. Image processing is often used to detect plant diseases in agriculture. Farmers face several difficulties as a result of pests and illnesses that affect their crops and crops in general. The most common sources of infection are fungi, bacteria, viruses, and nematodes [13].

Crop diseases have traditionally been difficult for farmers to recognize or suspect because they lack knowledge about them and the need for advice and assistance from experts. However, if illness is discovered early, crop loss may be minimized. Plants can only be diagnosed and detected by visual input; hence, image processing plays a crucial role in the process. Diagnostic and categorization of disease using neural networks may be advantageous (Jhuria et al., 2013). An intelligent system's diagnostic technique is provided for the identification of illness. Plants like apples and grapes are being used in the research. Two different databases were used by the system. The authors used color, texture, and morphology to classify and map illness. Students were scored based on how well the writers did in the course. Farmers may use the grade to calculate how much pesticide to use [14]. It was developed by Anuradha and Chahal (2015) to classify and identify plant and leaf diseases. They also presented a wide-ranging paradigm for image processing in their paper. Other efficient categorization methods, such as neural networks (NN), *K*-means, and principal component analysis (PCA), were also examined. In order to create a variable-rate chemical sprayer, they employed image processing (Tewari et al., 2020).

The primary goal of this investigation was to discover illnesses of the rice crop. Picture segmentation based on chromatic aberration was used to locate unhealthy areas in paddy plants. The author used a diagnostic technique and variable rate application to create a prototype that is both effective and environmentally friendly. Environmentally and economically, it is a win–win situation. Apps that assist in the identification of plant species may be beneficial to botanists, researchers, and even the general public. Image retrieval based on content is used to identify species photographs in a collection [15]. Morphological traits including texture, size, form, and color of leaves and flowers are used to identify plants.

To identify a species, an expert botanist is needed. Information technology, such as real-time picture capture equipment, may be used to overcome this problem. In order to identify plant species from a collection of species photos, feature extraction and image analysis are needed. Farmer and Jain (2005) say that leaf border shape analysis can be done. Both region-based and boundary-based techniques to leaf analysis are available. The border portions of the leaf may be used to identify plant species by employing boundary signatures (Femat-Diaz et al., 2011). Plants are classified based on more than just the color and texture of their leaves. The accuracy of the findings might be improved by combining color and form features [16].

Data from agricultural research and information technologies may now be used to get more accurate results. Farmers may be able to use this information to make better judgments about how to maximize crop yields. An in-depth knowledge of the area's natural resources and optimal usage is required. This approach maximizes profits and production while requiring the least amount of work and making efficient use of available resources. When it comes to farming, farmers must have a basic understanding of technology and how it functions. Anyone interested in learning more about precision agriculture will have to go through a rigorous educational process first. GPS and GIS technologies are widely available and used in precision farming equipment [17].

Using satellite signals to identify the data they contain, GIS and GPS work together to locate goods all over the globe. Photographs taken by satellites may be downloaded and analyzed digitally. As image processing technology has progressed, both remote sensing and GIS have advanced independently. There is an integrated system that uses remote sensing and image processing software for better precision farming. Automatic quality grading systems have been necessitated by growing customer awareness of and demand for high-quality products. The quality of the fruit must be evaluated in order to put it on the market in huge quantities. Color, shape, flavor, texture, and size all play a role in determining fruit quality (Freixenct et al., 2002). This computer vision job includes the capture, processing, and interpretation of images for analysis. Separating the fruit from its surroundings and studying its most important characteristics has become a need in determining the fruit's quality.

Depending on their quality, the fruits are divided into several categories. Grading is based on a variety of patterns and classifiers. A product's appearance is heavily influenced by its color, size, texture, and form (Mendoza & Aguilera, 2004). Because they are time-consuming, biased, and prone to mistakes, manual inspections are a waste of time. As a solution to these problems, scientists created a non-destructive way of evaluating fruit quality that, in contrast to other destructive procedures, is more accurate and faster. Many sectors are already using these computer-vision-based automated quality assessment systems to replace human inspections (Gao et al., 2010).

Fruit quality is assessed using classifiers including neural networks, SVMs, Bayesian decision theory, *k*-nearest neighbors (KNN), and PCA (Mans et al., 2010). Many food businesses have found that the automated system's findings on quality are accurate and valuable. Color, form, and size are the most critical criteria to consider when assessing the quality of a product (Moreda et al., 2012; Prabha & Kumar, 2013). Tofu-free soybeans may be made from peas and other legumes, according to current studies (DePalma et al., 2019) [18]. GIS relies heavily on remote sensing data from satellites. Remote sensing relies heavily on the amount of light energy reflected back from an external source. The passive system relies on the sun for its external supply of energy. Image sensors have made it easier to acquire data from satellites. Photographs taken by satellites may be downloaded and analyzed digitally.

GIS is no longer necessary for remote sensing because of advances in image processing technology, such as image enhancement and restoration. It is the primary goal of remote sensing to keep tabs on the planet's surface and collect data on its physicochemical geographic characteristics. Scientists say CaffeNet's maize yield prediction has a lower root mean square error (RMSE) than support vector regression (SVR) (Kuwata & Shibasaki, 2015). Using spectroradiometer-enhanced images of maize yields from 2001 to 2010, the climatic research team was able to gather the dataset. According to Kamilaris and Prenafeta-Boldu, CNNs have certain benefits, but they also have some drawbacks (2018). Costa Rican sugarcane fields are chosen for this study, but the future potential of DL for other sectors is examined as well. Smart farming relies on an integrated farm management system for processing, planning, and decision-making (Gardasevic et al., 2017). An investigation of the use of IoT and data analytics to enable smart agriculture was conducted by Elijah et al. (2018). The advantages and disadvantages of the Internet of Things (IoT) in agriculture have been discussed [19].

Focused on LPWA communications research, the main goal is to raise awareness of its importance. For small- and medium-sized farms that use IoT systems, the cost of the equipment, storage, processing, and transportation will keep decreasing [19]. UAVs, or unmanned aerial vehicles, have the potential to improve agricultural methods and operations. The agricultural process includes the evaluation of crop health, nitrogen measurement, spraying, and the monitoring of soil conditions. Using an IoT and GIS approach, drones might be used to map and photograph crop health, according to a research (Bodake et al., 2018). Bacterial and fungal spores may be detected using this method on farms [20].

14.4 IoT and AI Usage in Farming

Many of these applications are still in their infancy in wealthy nations. Whether used alone or in concert, these technologies have the potential to have a significant impact on food production in the future. CRISPR/Cas9 has revolutionized genome editing as a result of recent scientific developments. Genes from closely related wild plant species may be put into contemporary plants via this kind of genetic manipulation. Smallholder farmers in Africa benefit from the development of synthetic biology-based nitrogen fixation technologies. Synthetic fertilizer consumption may be reduced by using strategies like this one. Synthetic biology may be used to make food flavorings (like vanilla) while keeping the natural flavor.

There is the potential for new technology to significantly alter or perhaps completely replace conventional methods of raising cattle in certain situations. Decoding the genetic code of African naked-neck chickens is being investigated by the University of Delaware to see whether they can be developed into chicks that can endure climate change. Heat-resilient turkeys are also being researched at Michigan State University. Biotechnology, rather than the traditional factory farm method, may be used to create animal products in the laboratory. To reduce water and land use while maintaining the nutritional value of hen-bred egg whites, start-up firms are developing animal-free egg whites. It is not only corporations that employ plants to make meat and cheese; scientists are now 3D printing meat using tissue engineering results. It is conceivable that the production of beef in the laboratory might reduce greenhouse gas emissions while simultaneously conserving water and land. Though it may seem counterintuitive to think that the findings of this study will change the practice of raising cows in impoverished nations, it is possible. Due to advances in big data, the Internet of Things, drones, and artificial intelligence, modern agricultural operations may need less pesticide inputs, thanks to precision farming (AI). A number of businesses are combining cutting-edge genetic sequencing with artificial intelligence to increase agricultural productivity [21].

Machine learning is being used to produce weather predictions that farmers may use to maximize their crop yields using drone and satellite images. Plant genomic and phenotypic data may also be utilized to predict the performance of novel plant hybrids. Weeding row crops is being more automated by robots in an effort to be both environmentally friendly and financially feasible. Urban, indoor, and vertical farming may be able to boost agricultural output and water efficiency without the use of pesticides, herbicides, and fertilizers in certain cases, thanks to big data and the Internet of Things. For fully automated precision farming, sensors, AI, pictures, and robots can all work together. It is imperative that sound methods for analyzing the effect of converging technologies be put in place. For food security in

2030, risk and uncertainty management is essential. Assessment of projects' long-term influence on society is one of the most challenging aspects to consider when making decisions about their viability.

When faced with new challenges and technical failures, scientists and institutions must be able to react quickly with current knowledge. This has resulted in a change in the view of sexually transmitted diseases (STI). In the past, new technologies have been praised for creating new possibilities, but they have also been blamed for disturbing the status quo [22]. To address the most pressing challenges to global food security, UN bodies such as the Commission on Science and Technology for Development may play a greater role in collaborating with Member States to assess the potential benefits and risks of emerging technologies, focusing on both immediate and long-term implications. The planning and implementation of food-security-related STI must take into consideration a wider range of issues. From the international to the local level, these issues should be addressed in both developed and developing nations alike. If we want to abolish world hunger and malnutrition by 2030, each problem must be treated simultaneously since no issue is more important than the others [23].

The importance of preserving soil fertility and preventing erosion: Soils must never be lost, and soil fertility must be preserved or improved at all times. Monitoring plans for STI projects may be adjusted to include just a few easily measurable soil fertility and soil protection indicators and practical management changes that can be implemented if soil protection indications signal deteriorating conditions. The indication for climate change mitigation includes changes in water availability, and temperature variations. These catastrophic occurrences are to be considered in all STI planning and performance evaluations. There should be an evaluation of other potential sources of income as soon as possible. Encourage the use of agro-ecological, low-input, and large-scale agricultural techniques. Implementing low- or no-input sustainable agricultural practices is a critical strategy for ensuring food security. Smallholder farmers and agricultural workers might expect a more stable quality of living as a result of these systems' ability to foster agricultural variety and resilience.

As a result of the loss of biodiversity, our food system faces a severe threat to its future sustainability and resilience. STI approaches should include agro-ecological production and eco-functional intensification strategies because of the importance of functional biodiversity. Efforts to cultivate orphan crops each nation and region's tastes and demands must be taken into consideration while developing orphan crop breeding programs. A farmer's participation in the development of future breeding and seed programs is essential for these programs' success. A large amount of money and coordination are needed to ensure that these programs are run by the best-suited organizations [24].

From the local to the global level, policies should be put in place to aid the transition to more sustainable food production and agriculture systems. Since food security is linked to a wide range of environmental variables, agricultural practices, market actors, and consumer behaviors, policymakers should encourage adaptive system thinking and management. Everything that goes into making a product is a source of waste. More food produced and accessible to eat without affecting the environment is the same as increasing production and making it available for consumption. Examples include policies that promote healthy and sustainable diets, urban–rural connectivity, and local food processing and value creation. First, agriculture and environmental change are intertwined, as discussed in the first chapter, and a comprehensive strategy may be necessary to solve both issues. Interministerial and intersectoral coordination and cooperation will be possible if food security is seen as an integral part of a bigger innovation-driven development strategy. To show how ICTs and agriculture ministry's may work together, the FAO and ITU developed an e-agricultural strategy guide and toolbox [25].

Future agricultural productivity and sustainable consumption may be enhanced by including and cooperating with stakeholders in the development of locally appropriate R&D initiatives. Smallholder farmers' requirements need immediate investment in research and consultancy services linked with small-scale farming practices. An ever-expanding range of research goals should be pursued, including not just the most recent but also the oldest (e.g., climate change, renewable energy sources, efficiency in the use of energy, biodiversity, and the management of natural resources). Breaking the loop of "poor research and extension for impoverished farmers" is the most important lesson (CFS and HLPE, 2013). Government financing for extension programs is a crucial issue for public policymaking. Agricultural research stations and universities, as well as national policies, must better align with farmers' demands (e.g., women and young farmers) in order to construct long-term institutions that deal with technological development in a sustainable and trustworthy way.

Additional considerations should be made for marginalized or disadvantaged users of common land and pasture, water, and fisheries in food security and sustainable agricultural governance. All indigenous people and those whose rights are protected by customary agreements fall under this group here. Participating fully and effectively in important decision-making and planning is essential. If we are to end the "poor research for poor farmers" cycle, we must immediately expand agricultural extension services as well as educational opportunities for farmers and the public at large. Participatory development may benefit from the incorporation of contemporary technology, such as ICTs, big data, and other relevant innovations (for example, drones, three-dimensional printers, and remote sensing). If you are looking for a more effective use of resources and inputs and crop planning for eco-functional intensification techniques, you may utilize remote sensing and big data. Mobile extension services have been investigated by a number of programs, but there has to be some coordination in the way these difficulties are dealt with.

An increasing number of the essential websites that serve as entry points to these services should be developed and maintained by organizations that have a long-term commitment to hosting such sites, keeping them current with the changing environment of more software and hardware developments. Coordinating roles might be played by the FAO and CGIAR. It is also important to address concerns about data ownership and access [26]. Through UN-sponsored organizations, we must keep encouraging smallholder farmers to share important agricultural discoveries through the UN Technology Transfer Mechanism and the UN Technology Bank. Platforms like this may help facilitate and disseminate new technology. The focus should be on helping developing and least developed nations acquire access to innovative technology that can improve yields, minimize losses both on and off the farm, and promote more sustainable agriculture in general via these kinds of initiatives. Non-profit and civil society organizations like the African Agricultural Technology Foundation should keep working to make agricultural patent technology more widely available as long as food security is a problem. Technology may be transferred north–north, north–south, or south–south through a variety of different routes. Today, the Bangladeshi treadle pump, invented in the 1980s and still extensively utilized in Africa for irrigation, is still widely employed [27].

Along with the transmission of technological knowledge, organizations inside and outside the UN system should investigate how agricultural meteorology, the Internet of Things (IoT), satellites, and other data might be made more readily accessible to increase yields and improve rural life. While more data is being gathered passively and actively to improve agricultural practices, civil society organizations like GODAN are urged to continue their work. New educational and research programs, as well as the creation of new institutions, may all help to increase agriculture's innovative capacity. As an example, the Cuban Institute for Tropical Agriculture Study performs scientific research, but it also trains professionals from other nations, especially poor ones, in their fields. Applied and research institutions, as well as new university institutes, departments, and curriculum, may all provide specialized master's degree programs. This will take a long time and a lot of money. It is possible that the CGIAR and FAO centers, in conjunction with national agricultural research organizations, will assist and organize these initiatives.

It is possible to use knowledge aid to provide STI help as part of formal development aid. Donors may aid agricultural research in a number of ways, particularly in developing nations. Knowledge-based assistance may be used to improve industry and physical infrastructure in official development aid. For example, promoting international engineering societies and nonprofits and facilitating south–south interaction might be among these strategies. Research objectives in a certain area may be addressed via regional collaboration by organizations like the Forum for African Agricultural Research (FAAR), the LatAm Rice Fund, and FONTAGRO (the Regional Fund for Agricultural Technology for Latin America and the Caribbean). Tiny OS's open-source features are described in depth by Jao et al. (2013). Hardware and software of WSNs were fully discussed. The performance of the MDA300CA driver was tested in an experiment. A single node and a large number of nodes were studied in two separate experiments.

A mix of damp, saturated, and dry soil was detected by the sensors. Soil monitoring devices have been requested by the authors (Joshi et al., 2017). A web server powered by Raspberry Pi and the Internet of Things (IoT) is used to monitor and regulate soil parameters in the WSN system suggested here [28]. WSN paradigm for agricultural resource optimization and monitoring was created by researchers (Rawidean et al., 2014). The method outlined here is for obtaining real-time information about crops. By lowering the amount of water and fertilizer used, WSNs will increase agricultural productivity. WSN deployment was examined in another research (Ojha et al., 2015). Using case studies from the literature, the authors looked at how WSNs are being used in agriculture. In order to map pest and disease threats, one research suggests that "accumulated degree-days" may be constructed using land surface data from meteorological satellites (Marques da Silva et al., 2015). In addition, missing data and monthly degree-days data were dealt with using linear regression and logistic regression, respectively. Apple orchard parasites such as the apple moth, aphid, and apple blossom weevil, as well as the clearwing and the leaf sucker may be identified using a neural classifier based on MLP neural network design. There are a total of 23 parameters in the suggested classifier, which comprises 16 color characteristics, 7 shape factors, and 7 forms.

Many environmental variables, including the food price index, agricultural area, and annual precipitation, may be linked to each other using an approach and analysis presented by researchers (Sellam & Poovammal, 2016). Rice fields in India were used in the research. AdaSVM and AdaNAIVE were employed in one research (Balakrishnan and Muthukumarasamy, 2016) to construct a time series analysis-based prediction model that was both accurate and precise. There are numerous sensors, including invariant temperature and humidity sensors, that may be used for smart farming and intelligent irrigation (Varman et al., 2017), according to a research (Varman et al., 2017). A study of the optimum crop sequence and a recommendation for the next crop to produce in a specific area of interest have been offered by a model (Rajeswari and colleagues, 2018). Classification and association rules in the C5.0 algorithm is used to obtain maximum information gain from the incoming data. Soil samples are tracked in real time using the new approach. In the next decades, agriculture will be more important than ever.

Agriculture may be on the cusp of a revolution because of the advancements in precision farming and smart agriculture. The goal of smart farming is to bridge the gap between farmers in developing countries and those in industrialized nations. IoT and mobile deployment have had a huge impact on the usage of technology in farming. Most traditionally performed agricultural duties have undergone significant transformations, which is understandable. There may be a connection between the use of computers, software, sensors, and other forms of information technology in farming and the advancement of technology in general. In addition to better yields, improved quality, and lower costs, farmers may reap the benefits of smart agriculture, provided technology is correctly implemented.

Many resources, skills, and technological know-how are needed to achieve this level of innovation. We need more than a passion for farming to analyze agricultural data, maintain track of progress, and forecast changes in demand and pricing. Farmers and wise farmers might use this chapter as a starting point in their own study. It has been determined that peer-reviewed publications published in the last five years are of the highest quality. Each study relies on a unique collection of datasets, measurements, pre-processing procedures and models that cannot be generalized or compared. A more effective farming system can only be achieved by farmers who do study on the best practices that match their own agricultural goals and demands.

Innovation in farming practices is made possible by infrastructure including roads, power, cold storage, and agricultural processing facilities, as well as ICT, sanitation, and other amenities. In order to create novel food systems, more public money should be allocated to high-quality research and extension services related to agro-ecological production methods tailored to the needs of smallholder farmers. Innovation in agriculture may be encouraged by governments. The Pakistani government, for example, helped to establish a domestic tractor industry, which now supplies 95% of the country's needs. Since government and private initiatives, local manufacturing has been bolstered. UK-NI Food Innovation Network strives to overcome obstacles hindering UK agri-food and drink companies from innovating or improving productivity or development. Facilitation of the spread of genetic technology via the use of non-proprietary genetic material and research to build locally adapted genes that can produce under tough circumstances; changing manufacturing methods boosting agribusinesses and supporting the development of activities that improve value added at the smallholder level.

14.5 Conclusions

A new paradigm for WSNs in agriculture has been proposed for optimizing and monitoring resources. The MLP neural network design was used in another study to develop a neural classifier for detecting apple crop parasites. There are several environmental indicators, like the food price index, area under cultivation, and yearly precipitation that may be linked to one another (AP). Rice farming in India was the subject of the study. To help farmers in both developing and developed countries, smart farming was created. Resources, expertise, and technical skills are needed to achieve this level of innovation. Even while smart farming is crucial, it is also critical to do research on best practices that correspond with your agricultural goals. The goal of this chapter is to provide agricultural and smart farming researchers with a clear overview.

References

[1] J. Schöning and M. L. Richter, "AI-Based Crop Rotation for Sustainable Agriculture Worldwide," 2021 IEEE Global Humanitarian Technology Conference (GHTC), 2021, pp. 142–146, doi: 10.1109/GHTC53159.2021.9612460.

[2] S. Jain and D. Ramesh, "AI based hybrid CNN-LSTM model for crop disease prediction: An ML advent for rice crop," 2021 12th International Conference on Computing Communication and Networking Technologies (ICCCNT), 2021, pp. 1–7, doi: 10.1109/ICCCNT51525.2021.9579587.

[3] S. N and J. S, "AI Based Automatic Crop Disease Detection System," 2021 IEEE International Conference on Electronics, Computing

and Communication Technologies (CONECCT), 2021, pp. 1–6, doi: 10.1109/CONECCT52877.2021.9622700.

[4] S. Pérez-Carabaza, V. Syrris, P. Kempeneers and P. Soille, "Crop Classification from Sentinel-2 Time Series with Temporal Convolutional Neural Networks," 2021 IEEE International Geoscience and Remote Sensing Symposium IGARSS, 2021, pp. 6500–6503, doi: 10.1109/IGARSS47720.2021.9554358.

[5] A. Nowakowski et al., "AI Opportunities and Challenges for Crop Type Mapping Using Sentinel-2 and Drone Data," 2021 IEEE International Geoscience and Remote Sensing Symposium IGARSS, 2021, pp. 258–261, doi: 10.1109/IGARSS47720.2021.9553609.

[6] S. Qazi, B. A. Khawaja and Q. U. Farooq, "IoT-Equipped and AI-Enabled Next Generation Smart Agriculture: A Critical Review, Current Challenges and Future Trends," in IEEE Access, vol. 10, pp. 21219–21235, 2022, doi: 10.1109/ACCESS.2022.3152544.

[7] P. Savvidis and G. A. Papakostas, "Remote Crop Sensing with IoT and AI on the Edge," 2021 IEEE World AI IoT Congress (AIIoT), 2021, pp. 0048–0054, doi: 10.1109/AIIoT52608.2021.9454237.

[8] M. Ahmed, T. Mahajan, B. D. Sharma, M. Kumar and S. K. Singh, "AI-based Detection of Pest Infected Crop and Leaf," 2021 3rd International Conference on Signal Processing and Communication (ICPSC), 2021, pp. 402–406, doi: 10.1109/ICSPC51351.2021.9451698.

[9] R. Koppaka and T. Moh, "Machine Learning in Indian Crop Classification of Temporal Multi-Spectral Satellite Image," 2020 14th International Conference on Ubiquitous Information Management and Communication (IMCOM), 2020, pp. 1–8, doi: 10.1109/IMCOM48794.2020.9001718.

[10] P. Akiva, B. Planche, A. Roy, K. Dana, P. Oudemans and M. Mars, "AI on the Bog: Monitoring and Evaluating Cranberry Crop Risk," 2021 IEEE Winter Conference on Applications of Computer Vision (WACV), 2021, pp. 2492–2501, doi: 10.1109/WACV48630.2021.00254.

[11] A. Lytos, T. Lagks, P. Sarigiannidis, M. Zervakis and G. Livanos, "Towards smart farming: Systems frameworks and exploitation of multiple sources", Comput. Netw., vol. 172, May 2020.

[12] S. Misra and A. Chakrborty, "QoS-aware dispersed dynamic mapping of virtual sensors in sensor-cloud", IEEE Trans. Services Comput., May 2019.

[13] L. Pu, L. Jiao, X. Chin, L. Wang, Q. Xie and J. Xu, "Online resource allocation content placement and request routing for cost-efficient edge caching in cloud radio access networks", IEEE J. Sel. Areas Commun., vol. 36, no. 8, pp. 1751–1767, Aug. 2018.

[14] T. N. Gia, L. Qinqing, J. P. Queralta, Z. Zou, H. Tenhunen and T. Westerlund, "Edge AI in smart farming IoT: CNNs at the edge and fog computing with LoRa", Proc. IEEE AFRICON, pp. 1–7, Sep. 2019.
[15] C. Sarkar, V. Rao, R. Venkatesha Prasad, S. N. Das, S. Misra and A. Vasilakos, "VSF: An energy-efficient sensing framework using virtual sensors", IEEE Sensors J., vol. 16, no. 12, pp. 5046–5059, Jun. 2016.
[16] A. Gupta and N. Mukerji, "A cloudlet platform with virtual sensors for smart edge computing", IEEE Internet Things J., vol. 6, no. 5, pp. 8455–8462, Oct. 2019.
[17] T. Dinh and Y. Kim, "Information centric sensor-cloud integration: An efficient model to improve wireless sensor networks' lifetime", Proc. IEEE Int. Conf. Commun. (ICC), pp. 1–6, May 2017.
[18] L. Guijarro, V. Pla, J.-R. Vidael and M. Naldi, "Game theoretical analysis of service provision for the Internet of Things based on sensor virtualization", IEEE J. Sel. Areas Commun., vol. 35, no. 3, pp. 691–706, Mar. 2017.
[19] T. Wang, G. Zhang, A. Liu, M. Z. Bhuiyan and Q. Jin, "A secure IoT service architecture with an efficient balance dynamics based on cloud and edge computing", IEEE Internet Things J., vol. 6, no. 3, pp. 4831–4843, Jun. 2019.
[20] F. de Vita, G. Nocera, D. Bruneo, V.A. Tomaselli, D. Giacalone and S. K. Das, "Quantitative analysis of deep leaf: A plant disease detector on the smart edge", Proc. IEEE Int. Conf. Smart Comput. (SMARTCOMP), pp. 49–56, Sep. 2020.
[21] T. Wang et al., "When sensor-cloud meets mobile edge computing", Sensors, vol. 19, no. 23, pp. 5124, Dec. 2019.
[22] W. R. Heinzelman, A. Chandrakasan and H. Balakrishnan, "Energy-efficient communication protocol for wireless microsensor networks", Proc. 33rd Annu. Hawaii Int. Conf. Syst. Sci., pp. 10, Jan. 2000.
[23] T. D. Buird and R. W. Brodersen, "Processor design for portable systems", J. VLSI Signal Process. Syst., vol. 13, no. 2, pp. 203–221, Aug./Sep. 1996.
[24] A. Triantafylou, P. Sarigiannidis, S. Bibie, F. Vakouftsi and P. Vassilis, "Modelling deployment costs of precision agriculture monitoring systems", Proc. 16th Int. Conf. Distrib. Comput. Sensor Syst. (DCOSS), pp. 252–259, May 2020.
[25] T. Ojha, S. Misrai and N. S. Raghuwashi, "Sensing-cloud: Leveraging the benefits for agricultural applications", Comput. Electron. Agricult., vol. 135, pp. 96–107, Apr. 2017.

[26] S. Tyagi, M. S. Obadat, S. Tanwar, N. Kumar and M. Lal, "Sensor cloud based measurement to management system for precise irrigation", Proc. GLOBECOM IEEE Global Commun. Conf., pp. 1–6, Dec. 2017.

[27] A. Zervopioules et al., "Wireless sensor network synchronization for precision agriculture applications", Agriculture, vol. 10, no. 3, pp. 89, Mar. 2020.

[28] T. M. Banddara, W. Mudiyanselage and M. Raza, "Smart farm and monitoring system for measuring the Environmental condition using wireless sensor network - IOT Technology in farming," 2020 5th International Conference on Innovative Technologies in Intelligent Systems and Industrial Applications (CITISIA), 2020, pp. 1–7, doi: 10.1109/CITISIA50690.2020.9371830.

15

Urban Farming: Case Study

C. Augustine[1], K. Balaji[1], S.V. Dharanikumar[1], and A. Jose Anand[2]

[1]GRT Institute of Engineering & Technology, India
[2]KCG College of Technology, India
Email: austynineeleven@gmail.com; balaji.k@grt.edu.in; dharanikumar.sv@grt.edu.in; joseanandme@yahoo.co.in

Abstract

The creation, distribution, and marketing of food and other goods within the boundaries of a metropolitan area are broadly referred to as urban agriculture. Urban smart farms use aeroponics to grow fresh and healthy food in a cleaner, easier, and more effective manner. With 90% less land and 98% less water needed, our cutting-edge technology is the ideal option for farming at home or in an urban environment. The importance of greenhouse farming is also discussed in the chapter. A system is developed for agricultural field monitoring in IoT-based smart farming with the aid of sensors like light, humidity, temperature, soil moisture, etc. Farmers can keep an eye on the state of their fields from anywhere. Smart farming that is IoT-based is significantly extra effectual than traditional farming. MQTT and HTTP are used for the implementation of these models. Finally, supervising and scheduling the devices are discussed.

15.1 Introduction

Smart and resourceful farming construction or smart farming using Internet of Things (IoT) sensors is a well-grown technology in agriculture nowadays. With the support of big-data technology and cloud services, the agriculture space can be extended large [1]. The result will be based on weather conditions, humidity, temperature, water, society, and nature of the country.

In Indonesia, to do farming on the rooftop or even indoors, we use urban farming where urban people may not have enough specialized farming experiences. Whatever may be the technology like cloud computing, big data IoT can reduce the interference of human beings and produce accuracy and profit in agriculture [2]. Here a smart urban agriculture model is applied, which changes the TTA smart farming greenhouse methodology where the cloud computing technology is hybrid with IoT-based sensors. This control method includes both hardware and software. This methodology can be used by the government for the advanced agriculture method. Exactness agriculture, a farming controlling method with monitoring, calculating and replying to inter-domain and inter-domain-field variation in agriculture with a decision-making system, has been applied in past 20 years. This precision method of IoT in smart urban agriculture, which supports collecting the data, observing, monitoring, and changing the method according to the present situation, will accurately produce the result. This automation method in a smart urban area will produce better products for the farmers in crop production [3].

The smart farming method in agriculture based on IoT will reduce the interference of human beings and produce more profit for the farmers. In the processing of smart urban farming with IoT, the collection of data is done by cloud computing, monitoring process by sensors, and also completing the automation processes with actuators [4]. Hence, all American farmers were instructed to fix active radio frequency ID (RFID) tags on their farming and to state the activities on the farms to the online national database as mandatory. The process and involvements in applying IoT for a smart farming method with the various agricultural farming situations in the past 5 years have been noticed.

Wireless communication technology supports the imperative role of smart urban farming. After the global sensor network (GSN) is applied, sum-ups of sensor information are converted to destination-supporting form, with the semantic sensor network (SSN) ontology [5]. Further in wireless sensor network, ZigBee technology and low power wide area networks (LPWAN) are used for observing and regulating processes under the Message Queuing Telemetry Transport (MQTT), which is an IoT-based protocol. Overall, smart farming is applied in East Asia and South Africa regions. In Japan, smart agriculture farming grows the top five national agricultural methods included by the Ministry for Food Department, Agriculture, Forestry and Fisheries (MFAFF). The MFAFF plans to provide smart agriculture farm services to 5000 capacity cultivation (50% of facility rejuvenation), 830 cattle farms (20% of full-time farmers), and 600 fruit trees (25% of orchard-scale farms) by 2022 and with the support of artificial intelligence (AI) based monitoring system that robotically studies and analyzes the development and situation information robotically taken from IoT sensors [6].

In Japan, they approached an alternate farming method named urban farming agriculture. Urban agriculture farming is the process of calming, processing, and dividing fruits, flowers, and vegetables into nearby urban areas. With a random high-aging society and youngster job crisis, if well planned as a method, urban farming could be a policy for survival regional improvement, national safety overall construction, and part-time training. It is a way to include specialized methods in agriculture in the way of agriculture education and also it is also offered manually for youngsters or senior citizens.

Smart urban farming will be required for the economical profit in the metropolitan area in the government visualization. Each coin has two sides. There are some drawbacks as well in this technology. MFAFF dispersed individual method smart agriculture apparatus to 740 urban farms in the year 2016 [7]; however, only 75 planters were achieved in the appraisal, with only a 10.1% success rate. This occurred due to the rural population of agriculture. Most agriculturists fear learning new technologies in agriculture. They have fear regarding their immediate output and casual profit in the routine agriculture process. Moreover, the output will not be the same all the time. It will differ according to the climate, water, soil, and environment. Just a sample; therefore, 95% of homesteads accept cut-glass greenhouses in Korea. From this, the bulk of agriculturists in Japan rests on plastic greenhouses [8]. Therefore, plastic greenhouses achieve less productivity when compared with the glass greenhouse. It also has less durability compared to a glass greenhouse. Whether a glass greenhouse or a plastic greenhouse, it is tough to implement, broadcast, and enlarge data technology in the field.

Due to the shortage of agricultural land in urban areas, people started to move to smart urban farming agriculture with the support of IoT technology. The proposed method formation of urban smart farm agriculture and it was integrated with cloud computing service with the support of IoT service platform, where it is varied from traditional combined control system type [9]. Due to this, both traditional and new technologies were integrated. The sensor node, control node, IoT, and cloud computing technologies were combined. The controller directly interferes with field farms. The proposed combined system simplifies the formation and construction of the whole system and unreliable connectivity.

15.2 Smart Urban Farming System Configuration

The smart urban framing system combines a sensor field and control field, integrated with IoT technologies that are used for the automation process

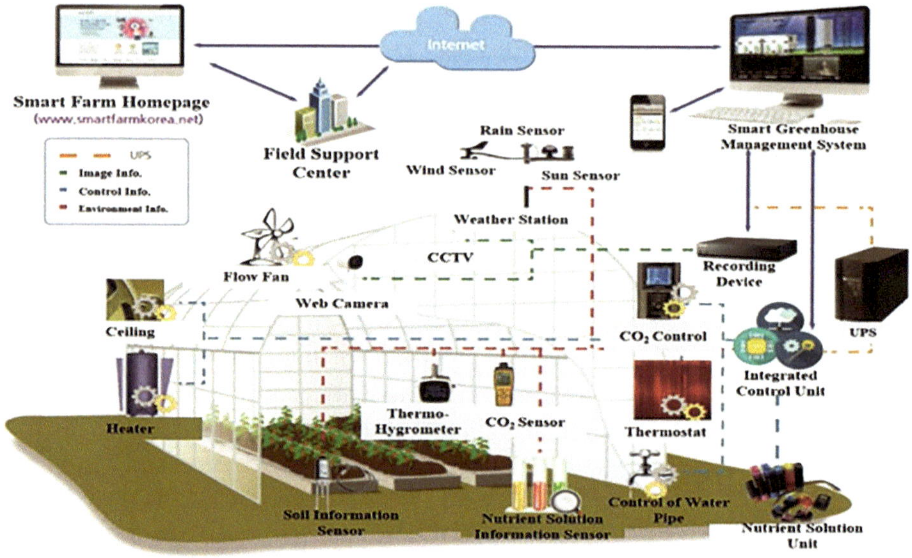

Figure 15.1 Smart farming model.

and cloud computing technology for integrating all the data information as shown in Figure 15.1 [10]. The main objective of this model is to make farming automation. With the support of cloud computing, the farm field can be extended. Moreover, this model needs to tackle organization schemes, information organization schemes, model organization schemes, and agriculture observing service and control techniques. The organization scheme service must be processed in multicasting the ratio of (1.*n*) forms and provide the virtualization between different component platforms. Moreover, it should support data admittance with outer general data and with external models and relevant services. Unexpectedly, the target level of this model is not achieved in the real-time application [11].

With this model, smart urban agriculture productivity and service chain were integrated into the urban areas. The integration of traditional and smart urban farming methods has different features in different aspects. More divergent distinctiveness of urban farming from the traditional method can be estimated, which is as follows:

- In comparing traditional agriculture farming with plastic house farming, smart urban farming does not have sufficient area, size, etc.
- In agriculture farming, the agriculturist gardening plants were more than 8000 m^2 and in smart house urban farming, the agriculturist gardening was less than 100 m^2.

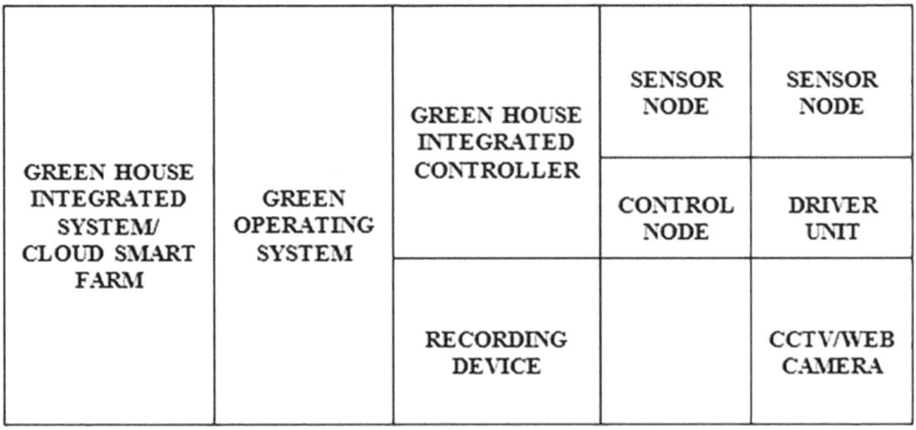

Figure 15.2 ICT Smart farming information.

- Moreover, agriculture farming lies in the family business and the land can be extended, whereas urban agriculture shares the common land.

Thus, urban farming should be simplified in configuration and it must provide more productivity in less area. In our proposed method, the gardening structure configuration can be classified into several types, such as indoor gardens, rooftop gardens, and box-type gardens. As per the different aspects of different environmental situations, the smart urban farming solution must take care of parameters like humidity, storms, carbon dioxide, water, etc. It is observed that one common single management solution will not be suitable for urban farming [12]. Due to this drawback, we proposed a smart urban farming method that is integrated with different technology and provides a reliable connectivity structure. In 2018, Telecommunication Technology Association (TTA) set a standard information and communication technology (ICT) model as shown in Figure 15.2.

It is compulsory to ensure that different components in smart urban agriculture have reliable communication between the components. The model must satisfy more than a few principles such as Greenhouse Control Data Standard (RUCFS-0009) and Smart Farm Greenhouse Control System Requirements Profile (TTAK.KO-06.0288) [13]. From the above standard, the proposed model is shown in Figure 15.2. It provides a reliable connection between the components and integration between the technologies. To extend the farming area, cloud technology has been integrated. This model combined the greenhouse controller, sensor device, and control device of the previous system into a mono shrewd farm combined supervisor under the typical requirement.

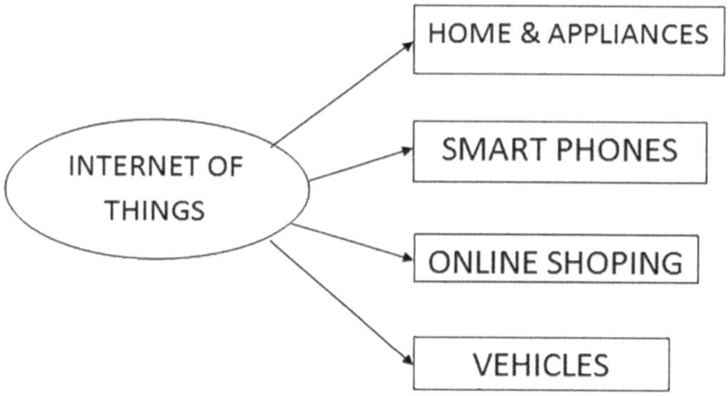

Figure 15.3 IoT-based applications.

15.2.1 An optimal solution to monitor smart farming conditions using IoT

Smart urban area farming will be fulfilled with the support of IoT. IoT is the most emerging technology nowadays, which fulfills whatever you want and wherever you are. It also supports all real-time applications. It makes all types of real-time applications in the automation method. IoT supports all the fields like business, agriculture, medical field, etc. IoT research and developed provides more advancements in the agriculture field. It is necessary to ensure that food safety through the global population is increasing rapidly. Farmers first started to apply ICT-based techniques to produce more profit with the limited area. As a result of ICT methods in agriculture with the support of IoT technology, the agriculturist can achieve more profit within a limited area [14]. Farming needs applications like soil moisture monitoring, climate condition censoring, temperature monitoring, moisture monitoring, supply chain organization, and infrastructure organization. Various smart farming applications are shown in Figure 15.3.

15.3 IoT in Smart Farming

In smart urban farming, with the support of the IoT, the agriculturist can achieve more productivity within a limited area. With the help of smart farming, agriculturists can utilize the use of fertilizer and other resources effectively. The agriculturist cannot be available to monitor the field all the time. Moreover, the agriculturist cannot use different tools to measure different parameters like temperature, humidity, etc. IoT can provide automation processes without the interference of humans. And it can monitor all the parameters simultaneously and control the process if it is necessary. Moreover, it can transmit the data information to the farmer even though the farmers are not available in the farm fields [15].

15.3.1 Benefits of smart farming

Agriculturists started to work on smart urban farming with the support of IoT. Smart urban farming with the support of IoT makes the automation process simpler. Productivity will be increased within the limited area. The data will be transmitted to the agriculturist even if he/she is not available in the farm field. The system continuously monitors the field and controls the process if it is necessary and transmits the data with the support of cloud computing.

15.3.2 Shortfalls of smart farming

- Farming is a natural process that depends on nature, and man predicts the parameters like humidity, rain on the need-based, pest control, etc. It is an evergreen method of implementing IoT systems in agriculture.
- Smart agriculture will be fulfilled when an internet connection is provided. Good internet coverage will not remain the same in all places.
- Faults in sensors and any other devices will lead to the wastage of water and some other resources.
- To process smart farming, the farmer needs basic education regarding IoT and smart urban farming. Extending large-scale agriculture in smart urban farming is a major challenge.

15.3.3 Components used in smart farming

The main function of smart farming using IoT is to monitor the system and take control action if necessary and transmit the data to the farmers even though the farmers are not available in the farm field. For monitoring, the sensors like temperature, humidity, soil, UV, moisture, etc. are used [16]. The farmers will receive different types of messages. Once the data information reached the farmer, quick action is to be taken by the farmer to avoid losses and to increase productivity. The product will be increased; the monitoring system has to be observed continuously in various conditions. The observation of different parameters can be seen on tablets and mobile phones.

15.4 Design Concept to Control and Monitor Greenhouse Temperature by an Intelligent IoT-based System

Farming in the country of Saudi Arabia (KSA) meets a lot of problems based on the parameters like high temperature, humidity, soil, lack of water, etc. [17]. To overcome this problem, a lot of alternate solutions were taken by the Saudi government. Self-resource is the major backbone of agriculture in that country. For example, in a country, agricultural self-sufficiency can be provided by alternate solutions and providing alternate resources like water, shelter, fertilizer, etc. Greenhouse agriculture farming is interesting because greenhouse farming is isolated from other environments by providing alternate climates and parameters like shelter, water, temperature, etc.

Due to the desert climate for more than half a year, the temperature is very high in Saudi Arabia. The consolidated temperature in June is around 47 °C, and the average temperature in February is about 15 °C. It is very difficult to produce the vegetables like tomatoes, cucumber, papaya, carrots, etc., in a degree less than 14–18 °C [18]. From that, it can be observed that it is compulsory to give a suitable monitored microenvironment, for different kinds of vegetables and fruits, which are needed for four important climate parameters, such as temperature, humidity, carbon dioxide level, and light density. The greenhouse environment is isolated from traditional agriculture. The greenhouse environment allows cultivating varieties of fruits and vegetables that require low-temperature for its growth. This meet the customer demand with high quality in a fast process.

15.4.1 Big data

In recent days, with random technology advanced in the information area, information data is available and accessible whenever you want and where ever you are though collected in huge quantities with various formats and from different resources in a semi-structured or unstructured way [19]. Big data is the technology that converts or accepts different types of data formats. It can transfer the data to the farmer even though the farmer is not available on the farm. The data can be visualized by a tablet or mobile phone. With the support of IoT, the data can be transferred, and a centralized huge amount of data was transferred by the sensor device. A huge amount of data is possible with the support of big data technologies [20]. The information data can be captured in different formats like texts, images, documents, sounds, videos, global positioning system (GPS) points, route information, and more. On the other hand, at the backend of the process, these different types of information data must be content kept, and linked, before being analyzed, processed, and retrieved. These operations and the other basic CRUD (create, read, update and delete) brought different challenges in the domain of big data. Rending to old relation-based information data models does not serve the needs of gathering and scheduling megabytes or petabytes of big data. Big data is a technology that supports huge data transmission from farms to agriculturists. The sensor observes the data and transmits the data in different formats. Big data accepts different formats of data and converts from one data format to another data format. The data can be transferred from the farm to the farmer even though the farmer is not available in the field. The agriculturist can visualize the message through laptops, tablets, and mobile phones [21].

The data can be observed in the format of the graph. According to the graph theory, the graph gives the connection to the data for the different environments and different situations. The graph theory connects the different parameters like productivity, gain, losses, temperature, humidity, etc. It is observed that smart urban farming system has integrated with big data, cloud computing, IoT, etc. Neo4j, TITAN, and Orient DB are some of the examples of the graph database. This system follows the distributed system [22]. The distributed system accepts different types of data formats and conversion of data formats. Every system is responsible for observing, monitoring, generating data, collecting data, processing data, controlling the process if it is necessary, and transferring the data to the agriculturist. Graph representation is a perfect and easy method to explain and model the distributed data due to its nature. In defense of old and proposed relation database systems, the efficient performance of the graph data model gives an idea about the performance of smart urban farming.

According to the data increasing, the different operators can join and maintain the common database. The graph format database gives the connectivity between different entities and parameters like temperature, soil, humidity, water level, etc. It gives the relationship between the data in different environments and different situations. The supporting graph model format Neo4j, TITAN, and Orient DB are some examples of the graph database. Traditional agriculture can extend the land size, whereas smart urban farming cannot extend the land size. The main function of smart urban farming over IoT is to monitor and observe the field farm by using sensors like temperature, soil, humidity, moisture, etc. After observing the data, controlling process takes place if it is necessary and transmits data to a centralized database. The database can be extended with the support of cloud computing and big data technologies [23].

15.4.2 Security

Security takes an important role in IoT. Smart base agriculture is automated with the help of IoT. So confidentiality and authentication are required in urban smart farming based on IoT. For security purposes, different algorithm methods are used; for example, advanced encryption method, data encryption method, Blowfish, RC5, etc. The data can be encrypted based on the key type. Two types of keys are used. One is a public key and the other one is a private key algorithm. In a public key, both the encryption and decryption are done by a single key, whereas in the private key algorithm, encryption and decryption are done by different keys. By increasing several blockchain-based frameworks, it is easy to store and audit access control policies [24].

15.5 Overview of Indian Smart Agriculture by IoT

The increase in the worldwide population demands better inventions to supply food in all sectors, predominantly in agriculture. Even though next to a certain period, the requirement and delivery will not match. Administrating and supporting the funds, manpower is unmoving in demand to improve agricultural production. So, smart agriculture will and should be the finest choice for mounting food invention, supply administration, and labor [25]. Considering a farmer's prior experiences, these works provide some general idea of predictive analysis, IoT cloud management, and safety units designed for multiple cultures in the farming sector [26]. While integrating the present technology into the traditional farming practice experience, expected challenges and complications will be highlighted. With the help of the arithmetical and quantitative approach, this work provides better innovative changes in the current cultivation system. Moreover, the usage of drone along with IoT technology makes the irrigation, plant shrubberies illness identification and harvesting makes easy in smart farming. The sensors that are triggered for different determinations in smart agriculture through IoT were deliberated. The main idea of this work is to develop modern agriculture with recent IoT techniques. The organized evaluation will provide existing and future trends in the agriculture sector. The drastic growth in the global population will be approximately 10 billion by the year 2060 according to an investigation. On the other hand, the requirement for food grain increases rapidly in recent years because of population growth. Regrettably, population growth is indirectly proportional to food grain. Hence, foodstuff production has to be improved in the future time internationally. Figure 15.4 illustrates the summary of IoT-based elegant farming factors.

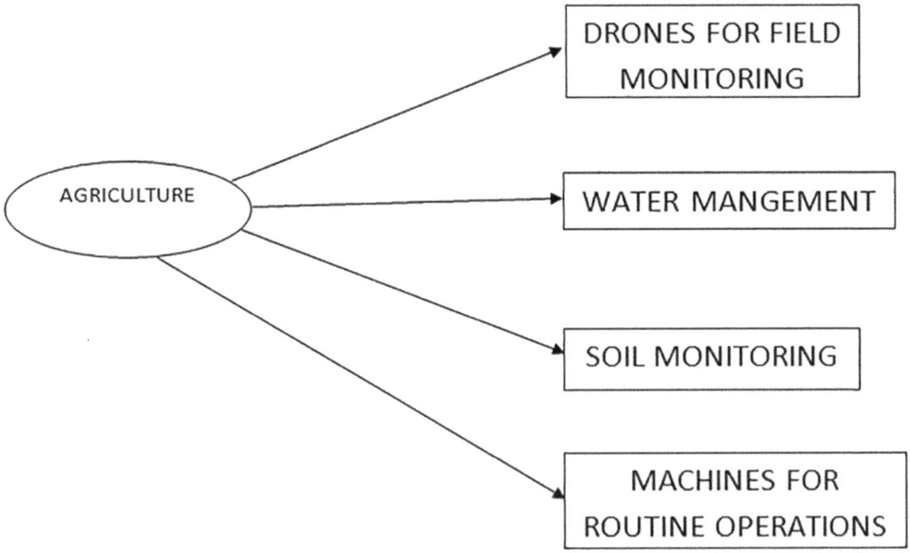

Figure 15.4 IoT in agriculture.

Utmost IoT recently gave a well-built thought of the farming division with an ample range of sensors used for a variety of smart cultivation targets. There was a rapid increase in IoT applications continuously year by year. Figure 15.4 shows the monitoring and controlling of IoT devices for neat farming. In the agriculture sector, different types of sensors play an important role in IoT technologies by connecting numerous interrelated procedures, seeing that numerous sensors, device drivers, and elegant substances to movable gadgets using internet usage. The information allocation with cleverly managing and supervisory amenitics cntails IoT facilities due to countless cloud-based inaccessible data acquisitions. Such capability knows how to supply effective products to the canny agriculture production. The expected tactic of farming is to increase rationalized crop growing with the examination of the IoT area of attention in the agricultural arena. IoT growth gave plenty of compensation in all segments concluded in the previous decade [27]. The IoT is a significant component where the ascendable software, hardware, lucrative process, the self-sustainable process, and canny choice for smart farming will be combined. Figure 15.4 indicates the diverse measuring workings for smart architecture. The development includes all the activities, for example, irrigation, vegetal progression, acknowledgement of disease by its leaf, and fabrication administration in the canny agriculture segment. In general, the implementation charge is exceptionally sensible for all farming resolutions with IoT canny agriculture. Investigators familiarized numerous integrated highly developed equipment to boost production throughput in the agricultural domain. Consequently, to reach the objective gradually, countless fresh inventions can be pooled with old-style farming. Using several sensors, discussed in green nature, the IoT can stylishly assemble agriculture [28].

15.5.1 Methodologies

Figure 15.4 illustrates the group drawing of future impression. This persistent mass diagram consists of the units of ethnicity examination, extrapolative investigation, IoT clouds, gadgets and sensor units, Agaric computerization robot, and defense administration for all integrated components [29]. Through IoT systems information's are easily gathered from the field using sensors and are classified using algorithms, and yield predictions are made and also the information's are stored in national dispensation servers and delivers to green research devices. To display the output of the system, all the IoT devices integrate with all other sensor substructures, the audio and video interface. The data in the sensor creates uncooked statistics from soil; otherwise, any suitable location is treated by IoT CPU with the finest planned time [30].

The cloud computing procedure is performed to create demand in extrapolative examination with big data dealing out from IoT on behalf of multiple beliefs investigation. The probabilistic procedures deliver the augmented fabrication in the following downpour called predictive examination [31]. Detailed block illustration of investigative examination for the projected system is the conventional agriculture that should have thoughts about grassland areas together with topsoil nutrients, temperature, rainfall details, and outlook climatic circumstances with an extremely knowledgeable farmer's communal. At this point, the extrapolative analysis skeleton is performed using numerous segment data found on the discovery ability for forecasting the possible conditions that happened. In conservative agriculture, previous data are used for predicting the insect issues and promotes minimum soil disturbances. An optimal forecast technique is used to foretell the circumstances before big data examination [32]. This configuration breakdown also predicts the usage of the vehicle for transporting all wrought commodities from the plant invention in which we have a good turnover and an optimistic impact on the sale of goods in the marketplace. This guess will enlighten the position of revenue or damage that occurs at the current and also in the forthcoming days. With the sustenance of this extrapolative scheme, the farmer will restrain numerous risk issues. For the unbeaten implementation of the new age of farming, this method expresses procedures [33].

15.5.2 Components and services

Based on TTA smart greenhouse standards, a smart urban farming integrated controller (SUFIC) supports the interconnection of actuator equipment, and sensor strategies are discussed. Here the hardware is designed in such a way as to please the standard terms of smart greenhouse beneath the driver interface (TTAK.KO-10.0845) and sensor boundary (TTAK.KO-10.0903). The foremost organizer embraces the STM32-32-bit microcontroller MCU segment and delivers 72-MHz performance intended for rapid interrupt comeback for real-time sensing and control. The hardware configuration of SUFIC is summarized [34]. The typical MQTT communication procedure of IoT is offered between SUFIC in addition to the service stage. We employ user and controller information as a topic for the MQTT statement to handle status/control in a row for all integrated controller devices. It further organizes the topic as per the function near to make out position message and control message and identify a set of rules according to communication standards between Smart Greenhouse Integrated Controller and Greenhouse Operation System. In this, we use MQTT and HTTP designed for supervising and scheming devices, and the proposed system uses MQTT protocol and HTTP [35].

In an urban smart farm platform server, an MQTT agent for communication swap based on the MQTT code of behavior and records for user and service data executive is put together to built on web server environment [36]. The overhaul provision server connected to the MQTT agent receives sensor and device data available from the incorporated controller, and makes available a particular service as per the API request of the application [37]. Control data (ID) of the controller is transmitted to the respective organizer through the MQTT agent by the appeal of a specific controller received from the integrated managing system [38]. As per the pattern of the topic, an unique ID is used for every incorporated controller as a topic, permitting stretchy monitoring and control services for each controller [39]. Such policy server operations are summarized in Figure 15.5. Service request to endow with smart agricultural overhaul is offered as a web application using HTML/CSS/JavaScript and provisions service in a variety of client environments such as PC and mobile [40]. This procedure routinely builds modules and boundaries on the controller data received by the provision boards, which wires the ability to add/change mechanisms and boundaries rendering to user locations.

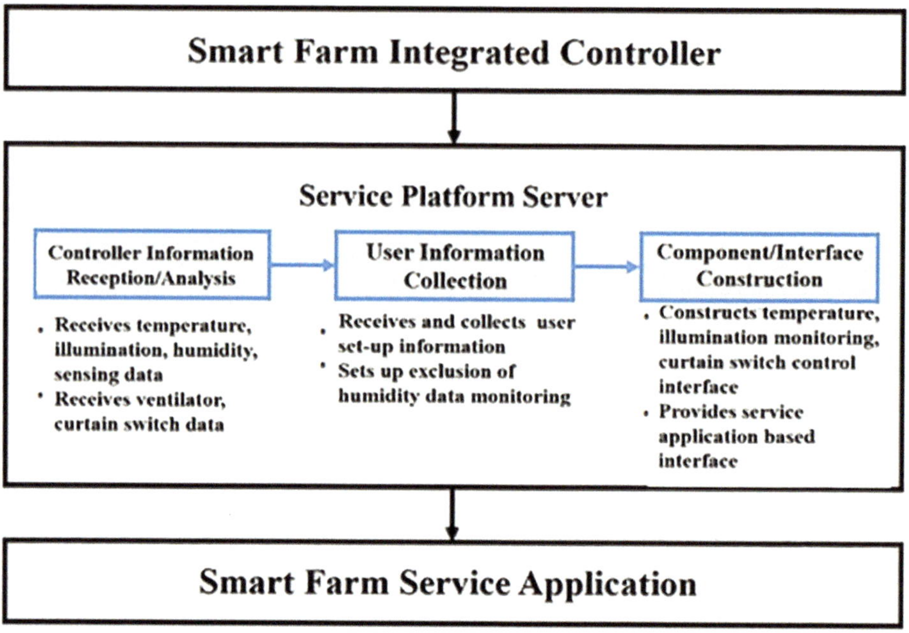

Figure 15.5 Example of service application supports.

15.6 Conclusion

IoT-based smart farming systems further became popular worldwide. Up to date, the progress in this field is owing to big data applications and cloud technology. On the other hand, in Korea, there exists an additional requirement for a skill-based cultivation system. Smart urban farming is unlike production/marketing slanted elegant farming, in which it makes use of tiny collective regions in the harvest and also the crop growing condition is unlike common outdoor farming within various aspects. Therefore, there is a demand in that to expand a new mold of smart farming in an urban environment as rooftops and indoor gardening. Here, in this chapter, we proposed a cloud scheme that supports the stylish metropolitan farming archetypal that abridges the typical ICT greenhouse mock-up recognized in 2016 by TTA. The hardware blueprint and layout of the stylish urban agricultural integrated controller are specified and a succeeding service request with message usage is clarified. Hence, the simple procedures for linking the sensors with IoT and other tools makes the smart farming process simpler and easy for the farmers. Smart urban farming is predictable to be very trendy, and, thus, this work contributes to the nationwide procedure to improve knowledge in the following five years.

References

[1] A. McBratney, B. Whelan, T. Ancev, and J. Bouma, "Future Directions of Precision Agriculture", Journal of Precision Agriculture, vol. 6, pp. 7–23, 2011.

[2] Gershenfeld, N., Krikorian, R. and Cohen, D., "The Internet of Things", Scientific American, vol. 291, pp. 76–81, 2014.

[3] A.Kamilaris, F. Gao, F. X. Prenafeta-Boldu, and M. I. Ali "Agri-IoT: A Semantic Framework for Internet of Things-enabled Smart Farming Applications, "2016 IEEE 3rd World Forum on Internet of Things, pp. 442–447, 2016.

[4] A. Ponmalar, Jose Anand A, P. Saravanan, S. Deeba and Jyothi BR, "IoT Enabled Inexhaustible E-vehicle using Transparent Solar Panel," 2022 International Conference on Communication, Computing and Internet of Things, pp. 1–5, 2022.

[5] F. Balducci, D. Impedovo, and G. Pirlo, "Machine Learning Applications on Agricultural Datasets for Smart Farm Enhancement," Machines, vol. 6, pp. 1–22, 2018.

[6] A. A. R.Madushanki, M. N. Halgamuge, W. A. H. Surangi Wirasagoda, and A. Syed, "Adoption of the Internet of Things in Agriculture and Smart Farming towards Urban Greening: A Review," International Journal of Advanced Computer Science and Applications, vol. 10, pp. 11–28, 2019.

[7] R. Venkatesan and A. Tamilvanan, "A Sustainable Agricultural System using IoT," International Conference on Communication and Signal Processing, pp. 763–767, 2017.

[8] K. Taylor, C. Griffith, L. Lefort, R. Gaire, M. Compton, T. Wark, and D. Lamb, "Farming the Web of Things," IEEE Intelligent Systems, vol. 28, pp. 12–19, 2013.

[9] J. Anand, Gowtham H., Lingeshwaran R., Ajin J., and Karthikeyan J., "IoT Based Smart Electrolytic Bottle Monitoring", Advances in Parallel Computing Technologies and Applications, IOS Press, Vol 40, pp. 391–399, November 2021.

[10] S. Vaishali, S. Suraj, G. Vignesh, S. Dhivya, and S. Udhayakumar, "Mobile Integrated Smart Irrigation Management and Monitoring System using IoT," International Conference on Communication and Signal Processing, pp. 2164–2167, 2017.

[11] M. N. Rajkumar, S. Abinaya, and V. V. Kumar, "Intelligent Irrigation System - An IoT based Approach," International Conference on Innovations in Green Energy and Healthcare Technologies, pp. 1–5, 2017.

[12] J. Anand, and J. R. P. Perinbam, "Automatic Irrigation System using Fuzzy Logic", AE International Journal of Multidisciplinary Research, Vol. 2, Issue 8, pp. 1–9, August 2014.
[13] A. J. Rau, J. Sankar, A. R. Mohan, D. D. Krishna, and J. Mathew, "IoT based Smart Irrigation System and Nutrient Detection with Disease Analysis," IEEE Region 10 Symposium, pp. 1–4, 2017.
[14] S. Salvi, S. A. F. Jain, H. A. Sanjay, T. K. Harshita, M. Farhana, N. Jain, and M. V. Suhas, "Cloud-based Data Analysis and Monitoring of Smart Multi-level Irrigation System using IoT," International Conference on IoT in Social, Mobile, Analytics, and Cloud, pp. 752–757, 2017.
[15] M. Jhuria, A. Kumar, and R. Borse, "Image Processing for Smart Farming: Detection of Disease and Fruit Grading", 2013 IEEE Second International Conference on Image Information Processing, pp. 521–526, 2013.
[16] Karl Aberer, Manfred Hauswirth, and Ali Salehi. "A Middleware for Fast and Flexible Sensor Network Deployment," Proceedings of the 32nd International Conference on Very Large Databases, pp. 1199–1202, 2006.
[17] Michael Compton, P. Barnaghi, L. Bermudez, R. G. Castro, O. Corcho, S. Cox, J. Graybeal, M. Hauswirt, C. Henson., et. al., "The SSN Ontology of the W3C Semantic Sensor Network Incubator Group," Science, Services, and Agents on the World Wide Web, vol. 17, pp. 25–32, 2012.
[18] C. Nwabueze, S. A. Akaneme, and I. R. Nwabueze, "Enhancing Agricultural Production Using Wireless Sensor Network," Iconic Research And Engineering Journal, vol. 2, pp. 274–284, 2019.
[19] Augustine C., J. Anand, Vinoth K. K., Kaleeswaran D., Muneeswari G., Santhoshkumar S. P., "6G-Based Mobile IPTV Using Adaptive Hybrid Transmission", Handbook of Research on Design, Deployment, Automation, and Testing Strategies for 6G Mobile Core Network, Edited by D. Satish Kumar, G. Prabhakar, R. Anand, IGI Global Publisher, pages 20, 2022.
[20] C. Yoon, M. Huh, S.G Kang, J. Park, and C. Lee, "Implement Smart Farm with IoT Technology," 2018 20th International Conference on Advanced Communication Technology, pp. 749–752, 2018.
[21] Y. Waché, et al., "Prospects for Food Fermentation in South-East Asia, Topics from the Tropical Fermentation and Biotechnology Network at the end of the AsiFood Erasmus+ Project," Frontiers in Microbiology, vol. 9, 2018.
[22] R. Vaishnavi, J. Anand, and R. Janarthanan, "Efficient Security for Desktop Data Grid using Cryptographic Protocol", Proceedings of the IEEE International Conference on Control, Automation, Communication

and Energy Conservation, Kongu Engineering College, Erode, vol. 1, pp. 305–311, 4–6 June 2009.

[23] M. V. Arokiamary, and J. Anand, "Analysis of Dynamic Interference Constraints in Cognitive Radio Cloud Networks", International Journal of Advanced Research in Science, Communication and Technology, vol. 6, Issue 1, pp. 815–823, June 2021.

[24] J. Anand, D. Srinath, R. Janarthanan, and C. Uthayakumar, "Efficient Security for Desktop Data Grid using Fault Resilient Content Distribution" International Journal of Engineering Research and Industrial Applications, vol. 2, No. VII, pp. 301–313, 2009.

[25] C. Dupont, M. Vecchio, C. Pham, B. Diop, C. Dupont, and S. Koffi, "An Open IoT Platform to Promote Eco-sustainable Innovation in Western Africa: Real Urban and Rural Testbeds," Wireless Communications and Mobile Computing, vol. 2018, pp. 1–17, 2018.

[26] B. Y. Kim, et al., "Top 10 Issues of Agricultural Policies, Focus in Agricultural Policy," Published by Korea Rural Economic Institute, vol. 142, pp 1–27, 2017.

[27] M. Bailey and J. Nasr, "From Brownfields to Greenfields: Producing Food in North American Cities," Community Food Security News, 2000.

[28] Aditya R. Rao, Ajay H., Balavanan M., Lalit R., and J. Anand, "A Novel Cardiac Arrest Alerting System using IoT", International Journal of Science Technology & Engineering, Vol. 3, Issue 10, pp. 78–83, April 2017.

[29] C. Jamroen, P. Kumkum, C. Fongkerd, and W. Krongpha, "An Intelligent Irrigation Scheduling System using Low-Cost Wireless Sensor Network Toward Sustainable and Precision Agriculture", IEEE Access, vol. 8, pp. 172756–172769, 2020.

[30] T. -Sai, B. Proeung, S. Tep, S. Chhorn, R. Pec, and V. Nail, "Prototyping of Smart Irrigation System using IoT Technology, 2021 7th International Conference on Electrical, Electronics and Information Engineering, pp. 1–5, 2021.

[31] J. Anand, K. Sivachandar, and M. M. Yaseen, "Contour-based Target Detection in Real-time Videos" International Journal of Computer Trends and Technology, vol. 4, Issue 8, pp. 2615–2618, August 2013.

[32] I. K. Oh, et al., "Perspectives on the Potential of Job Creation Based on Participation Motivation, Attitudes, and Barriers to Urban Farming: Focused on Young Farmers," Journal of The Korean Regional Development Association, vol. 3'1, pp. 55–70, 2019.

[33] C. Uthayakumar, K. Sarukesi, and J. Anand, "A Review on Knowledge Management and Knowledge Audit" International Journal on The

Technology World, Malaysia, Vol. V, Issue I, pp. 228–233, Oct–Nov 2009.
[34] U. Yeo, I.B.Lee, K. S. Kwon, et., "Analysis of Research Trend and Core Technologies Based on ICT to Materialize Smart-farm, "Protected Horticulture and Plant Factory, vol. 25, No. 1, pp. 30–41, 2016.
[35] S. Y. Lee, "Smart Farm Technology based on Cloud," Journal of The Korean Institute of Communication Sciences, vol. 34, no. 1, pp. 51–57, 2016.
[36] S. Wolfert, L. Ge, C. Verdouw, M.J Bogaadt, "Big Data in Smart Farming– AReview," Agricultural Systems, vol. 153, pp. 69–80, 2017.
[37] J. Ji, and Y. Chung, "Collaborative Filtering Algorithm Based on User-Item Attribute Preference," Journal of information and communication convergence engineering, vol. 17, no. 2, pp. 135–141, 2019.
[38] J. Anand, R. Janarthanan, Ponkoodaligam Kannan, and Amit Konar, "Efficient Data Storage in Desktop Data-Grid Computing using Real-Time parameters" International Journal of Computer Science and Technology, Vol. 2, Issue 3, pp. 392–397, September 2011
[39] J. Anand, B. Thyla, and B. Sharmila, "Coverage Area based Routing in MANETs" International Journal of IT, Engineering and Applied Sciences Research, Vol. 3, No. 2, pp. 41–44, Feb 2014.
[40] V. Gotarane, and S. Raskar, "Smart Irrigation Environment", Smart Irrigation System, June 2020.

16

IoT: Applications and Case Study in Smart Farming

V. Kanchana Devi[1], E. Umamaheswari[2], A. Karmel[1], Nebojsa Bacanin[3], and R. Sreenivas[1]

[1]Vellore Institute of Technology, India
[2]Center for Cyber Physical System, Vellore Institute of Technology, India
[3]Singidunum University, Belgrade, Serbia
Email: kanchanadevi@vit.ac.in; umamaheswari.e@vit.ac.in; karmel.a@vit.ac.in; nbacanin@singidunum.ac.rs: sreenivas.r2021@vitstudent.ac.in

Abstract

Generally, the agriculture sector contributes significantly to the Indian economy. The common problem faced by the farmers is that of decreased production because of dynamic infrastructure, feeble storage, lack of access to the market, and apt delivery. There is a demand for digital transformation in order to convert the traditional farming system toward a smart farming system based on IoT technology. There are several areas in agriculture where IoT can be implemented. Some of the focus areas are: predicting weather conditions, field monitoring, crop health and growth analysis, seedlings, fertilizer sprinkling, pesticide sprinkling, irrigation, soil condition monitoring, animal trespassing, etc. With the help of IoT sensors, a huge amount of data can be collected. Decisions can be made manually or automatically to monitor the fields with the aim of achieving equivalent growth and higher production. Smart farming based on IoT technology helps monitor real-time data, which improves the smartness of the agriculture system. The main objective of this chapter is to propose a case study in smart farming using IoT technology. Incorporation of such IoT technology leads to data collection and better control over internal processes, and, as a result, lowers production risks, better cost management, and waste reduction. Thanks to increased business efficiency through process automation, enhanced product quality and volumes are by-products.

16.1 Introduction

There are numerous IoT applications that use IoT technology for agricultural monitoring. Some of them are as follows:

- Automated climate condition monitoring and management.
- Monitoring of lightning, temperature, soil condition, and humidity for greenhouse automation.
- Crop health monitoring and management.
- Livestock monitoring is done for stock health, well-being, and location tracking.
- Efficiency and making exact data-driven decisions in precision farming.
- End-to-end farm management systems.

People require food to survive, which leads to their reliance on food. Industries are occupying agricultural lands, which affects the production of food. Crops will not adapt to the varying climatic conditions. Labor would also decrease in the field of agriculture, keeping urbanization a priority. To overcome these, smart farming has been brought into the spotlight [6]. This involves various technologies and techniques to scale up the productivity of farming. This method is primarily based on the concept known as the "Internet of Things" (IoT), which involves sensors, actuators, processors for data, etc. This would also require networking protocols and communication devices for the transfer of data. The ultimate aim is to increase the productivity of crops, the sustainability of crops, and profit. These depend on technologies such as the cloud, big data analytics, unmanned aerial vehicles (UAVs), etc.

Smart farming, along with the Internet of Things (IoT), is also paving the way for the so-called "Third Green Revolution." The revolution consists of all the topics that are to be discussed in this chapter, i.e., the role of UAVs, the role of data processing (big data), the farming equipment used, and many others [7]. According to the smart revolution of farming, otherwise known as "smart farming," the usage of pesticides and fertilizers will drastically come down, and the overall yield and efficiency will shoot up. The use of IoT along with farming techniques will ensure food safety. It would also benefit the environment in many ways; for example, the usage of water would be reduced. Smart farming has many advantages. Farmers with an unassailable amount of land can monitor crop growth, the health of their cattle, etc., using wireless technologies with the help of various networking protocols (discussed already in this book). As already discussed, unmanned aerial vehicles

(UAVs), otherwise known as drones, play an important role in monitoring the health of the crops, spraying disinfectants, and so on. Smart farming has an important role to play in the near future and has abundant potential to deliver sustainable farming and resource-efficient farming. It is also precise and productive when compared to the traditional form of agriculture.

16.2 Background Study

This chapter provides an in-depth comparison of key performance indicators as well as the challenges that smart farming will face during implementation [1]. It also discusses the various IoT application layer protocols. This chapter also discusses the various applications of smart farming. Application layer protocol plays an important role in smart farming. In the IoT, the various application protocols used are CoAP, MQTT, XMPP, DDS, and AMQP. These play an important role in communicating messages, and in applications that have limited hardware, etc.

This chapter discusses the use of unmanned aerial vehicles (UAVs) incorporated with the Internet of Things and their short-range connectivity, as well as various communication technologies used in smart farming to achieve sustainable farming [2]. Unmanned aerial vehicles play an important role in smart farming. It helps farmers monitor the health of crops, monitor livestock, and spray fertilizers to ensure that the crops stay healthy. They give some important insights to farmers and help them in various aspects. Applications of inter-connected devices and technologies are also discussed in this chapter.

This chapter primarily focuses on wireless networks and their technologies to improve the efficiency of smart agriculture and, therefore, reap its fruits [3]. This chapter also discusses the control system using nodes and sensors. These are also used to extract data, process the data, and finally send it to the control box. This chapter discusses the role of data mining, which involves the processes of data extraction, data cleaning, and data analysis, in smart farming.

This chapter discusses the role of digital twins in improving the efficacy of smart farming and in the advancement of smart farming [4]. Digital twin is basically used to decouple the physical flow and incorporate it with the virtual world. According to this chapter, digital twins help farmers remotely monitor various parameters of crops, such as the crop growth rate, the health of crops, and so on.

In this chapter, the role of biomimicry and its six principles in agriculture is discussed [5]. Biomimicry helps to alleviate the damage caused by climate change to crops, ensuring that the health of the crop is maintained. It also discusses how the concept of biomimicry is used to depict smart farming technologies as ecological innovations.

16.3 Applications and Use Cases

16.3.1 Role of drones in agricultural field

Unmanned aerial vehicles (UAVs)

UAVs are otherwise known as drones. When sustainability is considered, UAVs play an important role. UAVs capture images of fields and analyze those images, which would be of great help to the farmers. These images are then processed. It also serves as a monitoring tool for the same. Field monitoring, livestock monitoring, and compost management are some of the applications of UAVs. Wireless sensors and other wireless devices are used in UAVs. These are classified as perception layers. A UAV is also used to spray pesticides on the field. It is used for various things, such as weed detection, image processing of the field, etc. UAVs can be integrated with 3D cameras to monitor the growth parameters of crops. Several UAV applications are listed below.

A. Monitoring:

Monitoring helps in sensing different parameters of crops, such as the leaf area index, pH level of crops, salinity of crops, etc., which would be of extreme help to farmers. High-resolution data can be obtained.

B. Mapping:

Insightful 3D or 2D maps can be obtained. This helps in determining the precise profitability of crops. It also determines the status of crops, soil conditions, etc.

C. Weed detection:

Weed detection is the most important factor in farming. Weeds can damage crops. To minimize the damage to crops, weed detection is used with the help of multi-spectral cameras.

D. Seed planting:

When a large paddy field is taken into account, it would be tedious for farmers to plant seedlings. It would require a lot of labor. UAVs help reduce labor costs by evenly distributing seeds, fertilizer, etc. This is done using image recognition technology.

E. Forecasting:

With the help of different machine learning and artificial intelligence algorithms, detection of plant diseases, prediction of soil moisture, etc., have been made possible. Thus, forecasting is one of the important applications of smart farming.

16.3.2 Predictive analytics for smart farming, such as crop harvesting time, the risks of diseases and infestations, and yield volume

Generally, data analysis plays a huge role in smart farming. The input is in the form of data, and the output is also generated in the form of data. The processing of data is definitely a huge challenge as large amounts of data come into play. As a result, the concept of knowledge data discovery (KDD) is used here. This concept falls under data mining. KDD involves various steps:

I. Data pre-processing:

The data might be messy or incomplete, or there might also be metadata involved. To increase the accuracy, consistency, and integrity of data, it is pre-processed. This method is applicable when huge amounts of data come into play. The relevant data is extracted from the metadata and used for the process.

II. Data reduction:

Data reduction is the process of chopping off unnecessary data. There are various reduction techniques involved in this process. Smart farming uses the technique of numerosity reduction. Histograms can also be used for smart farming.

III. Data discovery:

This step involves the extraction of knowledge from the reduced data. Some of the examples are clustering, classification, association rules, etc. Out of the mentioned examples, association rules, comes in handy when smart farming is considered. They find the relationship between the parameters of the soil and those of the plants or the crops. The association rule has two basic criteria: (i) minimum support and (ii) confidence.

IV. Solution analysis:

This is the real-time implementation and analysis of smart farming.

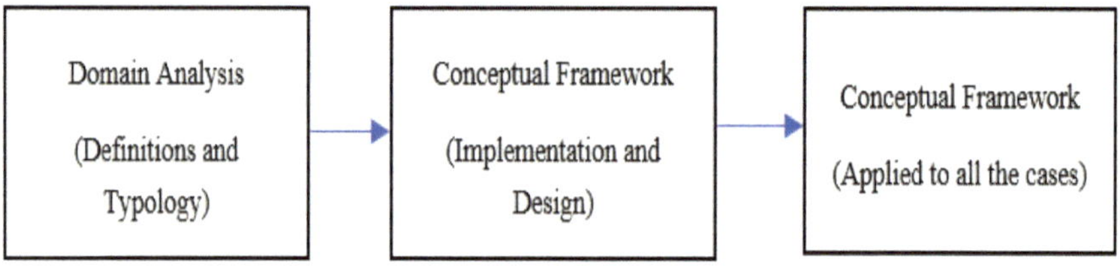

Figure 16.1 Basic block diagram of the digital twin.

16.4 Digital Twins

Digital twin is created by integrating the real-life behavior of objects with the virtual world. This comes in handy in smart farming. Digital twins can also be used to implement smart farming. Digital twins include (i) domain analysis (definitions and the typology) and (ii) the conceptual framework (implementation and the design). Figure 16.1 shows the basic block diagram of the digital twin.

16.4.1 Product life cycle phases

What is a product in IoT?

The product design in IoT is the design of products that are made viable for the Internet of Things, merging the products that contain sensors, actuators, and software for the functioning of the same. These products also require some networking protocols, which are discussed later in this chapter. Data must be stored in the cloud and must be processed. Data mining techniques are used for this purpose. Knowledge data discovery (KDD) is the most important aspect of this process (also discussed earlier in this book). For a product to function well, it must have a life cycle. This is known as product life cycle. The product life cycle phase of the digital twin includes the following:

1. Design: The design stage includes the model of the product.
2. Physical deployment: This includes the deployment of the model on the field.
3. Operational stage: This is used to track the behavior of the product.
4. Disposal stage: In this stage, the developed product is disposed of.

The digital twin consists of the product model, the implementation, the behavior of the product, and the disposal. The above life cycle is integrated with the digital twin and is then implemented in the physical world. The block diagram of the product life cycle is shown in Figure 16.2.

Digital twins as central hub for information from various sources:

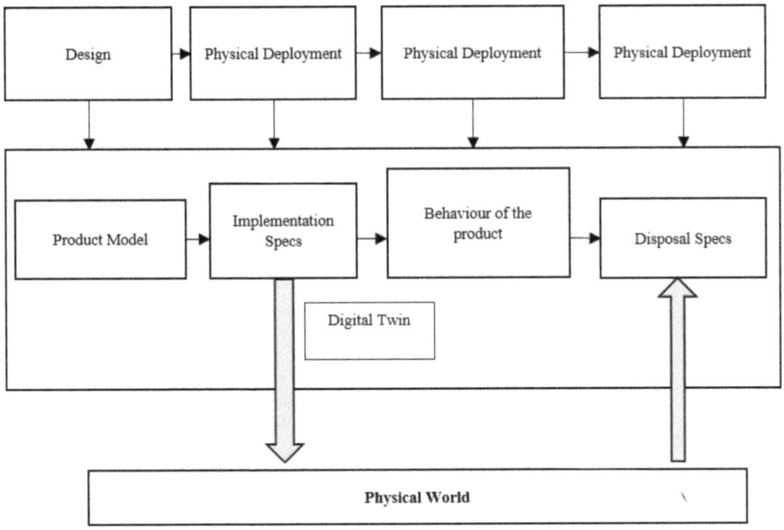

Figure 16.2 Product life cycle.

The digital twins act as a central hub for information from various sources. Farm management, which is discussed here, is an example. Other applications include business applications, etc. Figure 16.3 shows the central hub of information from various sources.

16.4.2 Virtual control of farming enabled by digital twins

There are two sources involved in smart farming: (i) supplier and (ii) consumer or the customer. There is a mediator for this. The mediator collects the data and passes it on to the sensor. Sensors are used in the monitoring of the temperature of the soil, the climatic conditions, etc. Sensors used in smart farming are known as agricultural sensors. These assist farmers in optimizing and monitoring the crops. Data from sensors are then segregated as the input data, the throughput data, and the output data. Input data are the measurements obtained from sensors. Throughput data is the amount of data transmitted successfully. The output from the sensor is also produced in the form of data. This helps the farmers in assessing the parameters of the plant productivity.

16.4.3 Basic control model

The basic control model of the system sensors, known as agricultural sensors, is used to help farmers monitor their crops. It is then subjected to discrimination. The effector activates the data, which is then passed on to the physical

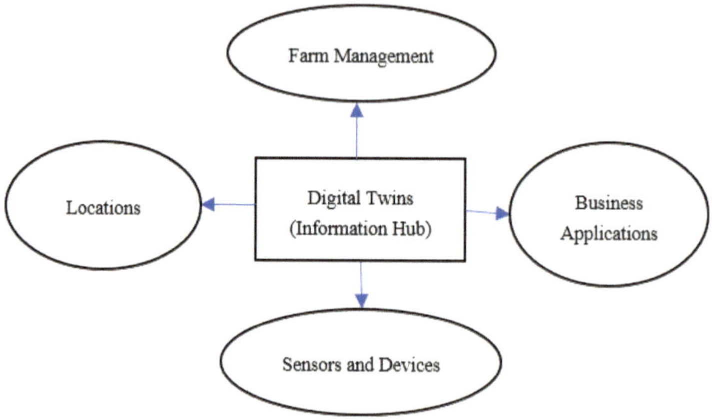

Figure 16.3 Central hub for information.

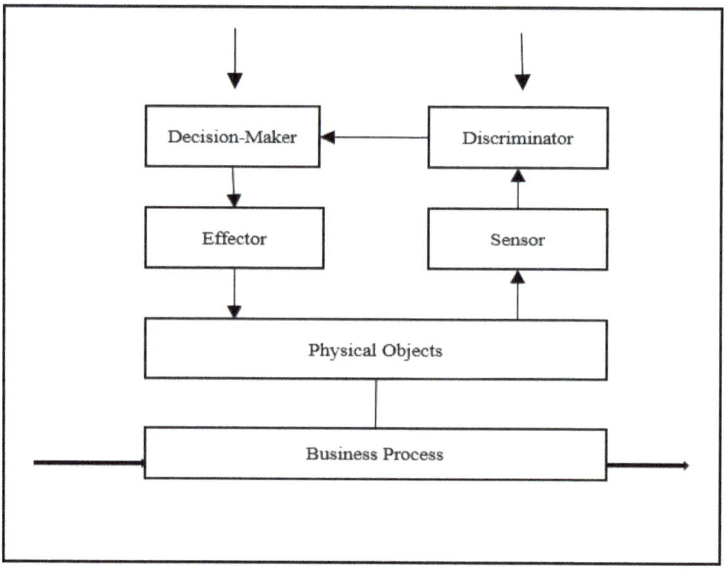

Figure 16.4 Basic control model.

objects. It is then used for business purposes. The diagrammatic representation of the basic control model is shown in Figure 16.4.

16.4.4 Conceptual model based on digital twins

The physical objects send the information to the sensor. The sensor senses the information and is passed on to the model, along with the data collected. It is then passed on to the digital twin, which decouples the physical flow. The

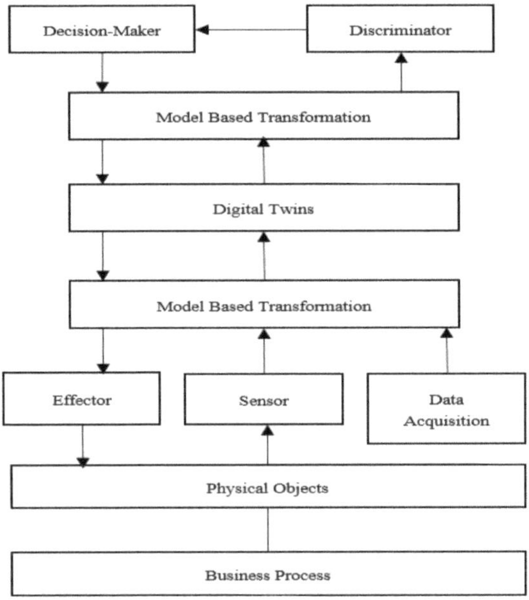

Figure 16.5 Block diagram of digital twins.

model then transforms the received data and sends it to the discriminator. A discriminator is a classifier that classifies real-time data from the data created by the generator. The data is then passed on to the decision maker (used to ensure consistency). The data from the decision maker is then transformed according to the model. Effector is a sensor-based compensation unit, which actively measures functions. The data is implemented in the real world and then used for business processes. Figure 16.5 represents the block diagram of a conceptual model based on digital twins [16].

16.4.5 Simplified control models of digital twins and its typology

16.4.6 Integrated control model of a digital twin

Data from the physical object is passed on to the reference object. The reference object would be present in the IoT reference architecture. A reference architecture consists of sensors, actuators, etc. The sensor senses the data, and additional data is transferred onto the model. The data is then passed on to the combination of digital twins (imaginary, present, future, and past). The selected data is again channeled to the model and then finally to the discriminator. Based on the facts and the data discriminated, a decision is made, providing insights on smart farming. Data are simulated, passed on

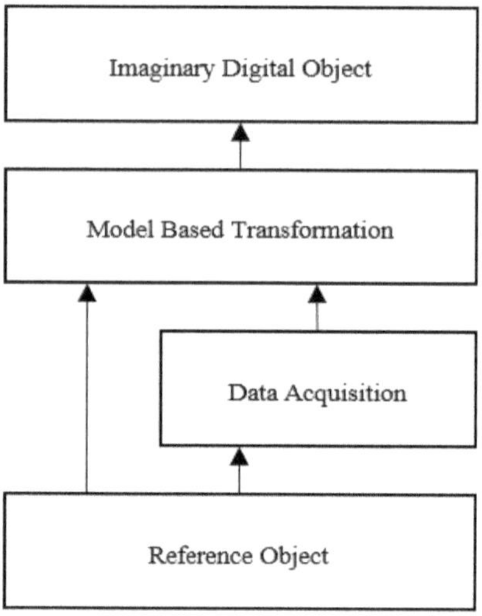

Figure 16.6 Imaginary twin.

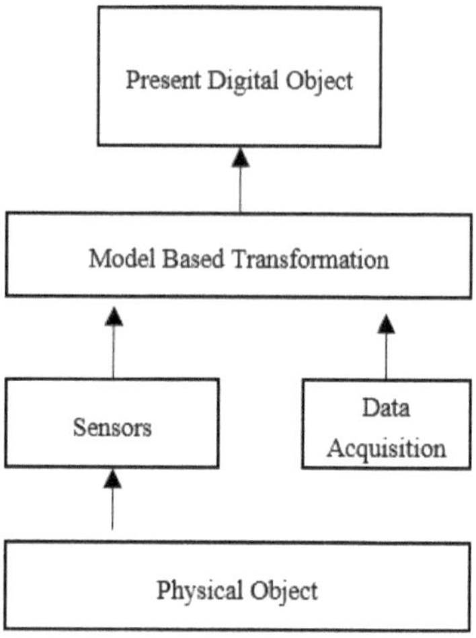

Figure 16.7 Monitoring twin.

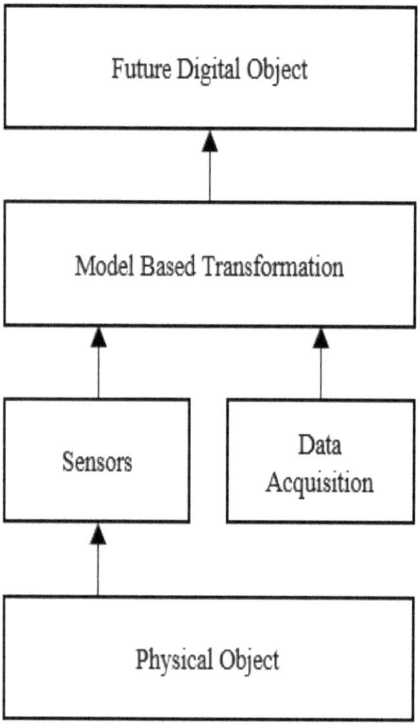

Figure 16.8 Predictive twin.

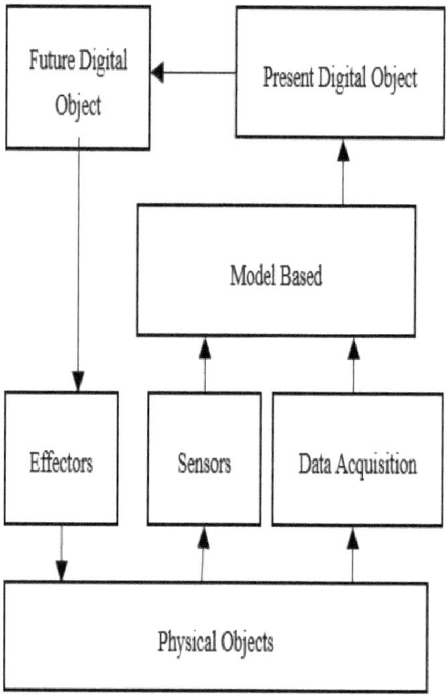

Figure 16.9 Prescriptive twin.

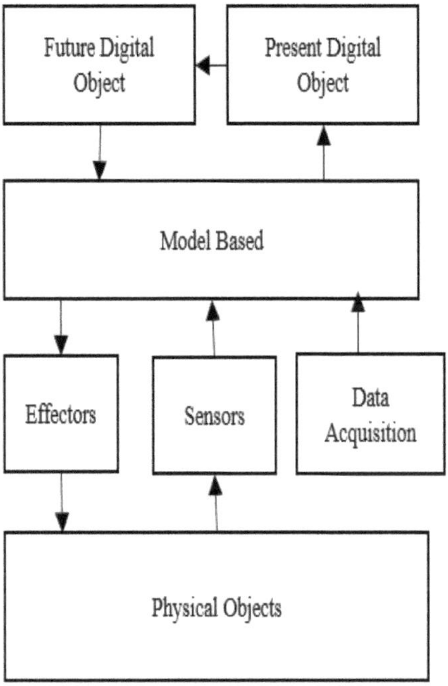

Figure 16.10 Autonomous digital twin.

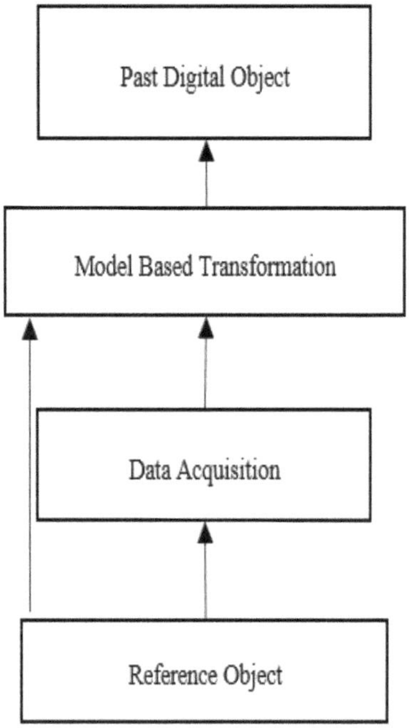

Figure 16.11 Recollected digital twin.

to the model, and passed on to the digital twin. The planned model changes after this process, passed on to the effector, which measures the data, and is passed on to the physical objects. This has finally been implemented for the business process. The application layer is the top-most layer, and this is the layer in which smart farming is implemented. The device layer is the lower-most layer. The security layers consist of a reference model and security management information. The service layer consists of database processing, and the network layer consists of the devices used for the medium of communication. This is also known as the functional system, which provides the system for identification, sensing, actuation, communication, and management [15]. Figures 16.6 to 16.11 shows the various types of digital twins such as imaging, monitoring, predictive digital twins, perspective digital twins, autonomous digital twins, and recollected digital twins.

Devices: Devices contain sensors, actuators, and various other functions. The chapter discusses the use of sensors and actuators.

Communication: The communication block handles the models and devices used for communication. There are four types of communication models, namely:

1. Request response

2. Push–pull
3. Publisher–subscribe
4. Exclusive pair

The communication devices are used to establish communication between the client and the server.

1. **Services:** Various types of services are used in IoT. The most apt example of this would be the monitoring of crops.
2. **Security:** Security consists of various protocols that govern the application. Some of the protocols that are used are LoRa, cellular networks, etc.
3. **Application:** The application block acts as an interface.
4. **Management:** The management block is used to govern various systems in the application.

16.4.7 Implementation model for digital twins (Figure 16.12)

16.4.8 Control model of the weeding use case (Figure 16.13)

The below use case is that of weed control. The decision maker consists of the prediction of yield and the harvest plan. The crop growth is monitored, and the weed pressure is also monitored. The digital twin holds the field, weeds, weeding machine, and lettuce. The effector controls the weeds and the harvesting. These are taken care of by the weeding machine and the harvesting machine, respectively. The sensor is used to sense the weed pressure and crop growth rate from the weeding machine and the field, respectively [14]. These data are collected by the sensor and sent to the digital twins. The physical objects consist of the weeding machine, the planting machine, the field, and the harvesting machine. The planting machine corresponds to young plants. The weeding machine corresponds to the weeds. The harvesting machine displays the amount of lettuce harvested. The field displays the amount of lettuce growth and the weeds. The data is collected from the customers and passed on to the digital twin, which processes the data in the corresponding areas and is then sent to the discriminator. Finally, the source produces the lettuce, which is harvested and finally delivered to the customer from the supplier.

The figure is that of the digital twin implementation model of the above use case. The explanation is going to be a bottom-up approach. The functional block diagram has already been discussed in this chapter. Now, some of the use cases and applications are discussed below. The management and

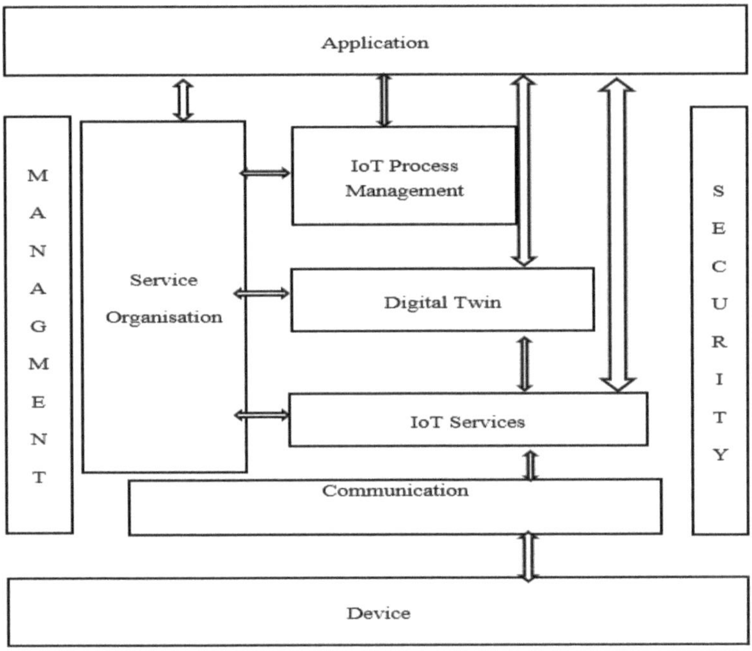

Figure 16.12 Implementation model of a digital twin.

security layers correspond to the suppliers and the customers. The suppliers manage the production. The source plant, harvest, and lettuce fall under the device block. These are the gadgets. The source plant, harvest, and lettuce are considered devices [13]. The communication blocks involve various communication activities, like planting machines, etc. The IoT services block consists of various types of data. These consist of the effector, sensors, and data acquisition. The effectors consist of weed control and harvest control. The sensors consist of data on pressure and the growth rate of the crops. The meta data such as those of weather fall under data acquisition. The digital twin block consists of the weeding machine, field, lettuce harvest, and weeds. The top-most layer is the application layer, which consists of the basic requisites for this use case. These include harvesting, weeding, crops, prediction of yield, and so on.

The most important component is the digital twin. Only if a digital twin is available can smart farming be effective. As discussed earlier in this chapter, a digital twin uses real-world data to create simulations, which are then used for prediction. Smart farming is all about prediction, e.g., crop yield prediction, weed prediction, etc. Only if the prediction is correct, can the application be further processed.

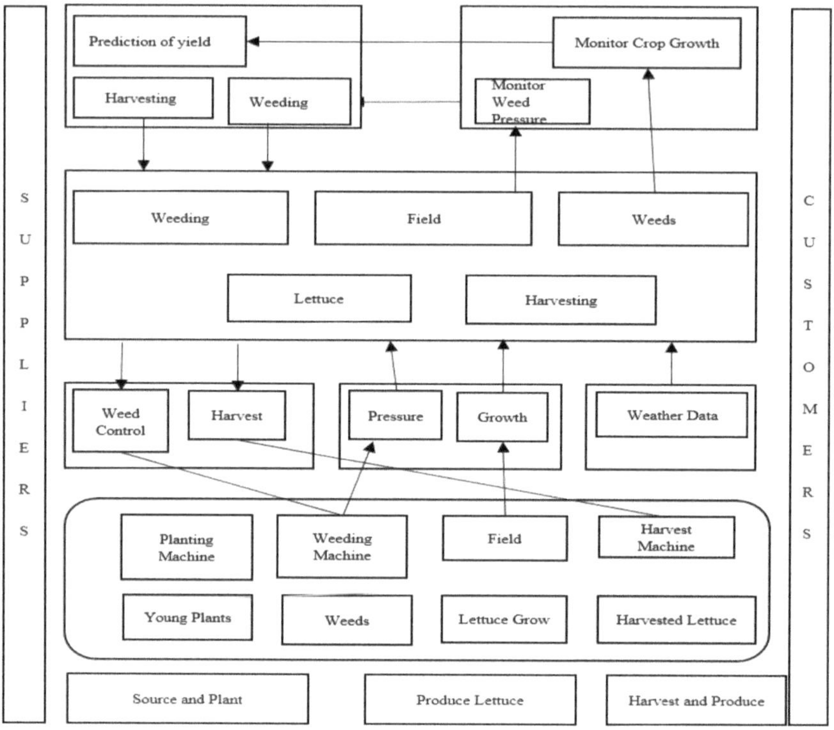

Figure 16.13 Weed use case.

The application layer is the top-most layer, and this is the layer in which the smart farming is implemented. The device layer is the lower-most layer. The security and management layers consist of reference models and security management information. The service layer consists of database processing, and the network layer consists of the devices used for the medium of communication.

16.4.9 Digital twin implementation model of the use case (Figure 16.14)

In general, IoT is combined with agriculture (smart farming) [12]. Smart farming is the future and has many benefits. Some of the benefits are listed below.

1. Monitoring of climate conditions is automated, as are monitoring and management.
2. Monitoring lighting, temperature, soil condition, and humidity for greenhouse automation. This is achieved with the help of actuators

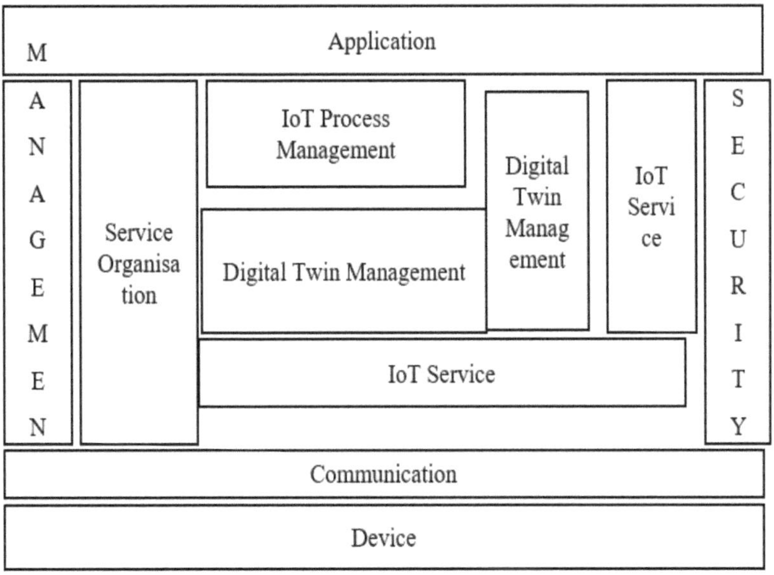

Figure 16.14 Digital twin implementation.

and UAVs. The electric signals are converted into physical signals. For instance, lighting is an electric signal that is converted into physical signals using an actuator. This information is fetched by the UAV, and the final data is delivered to the farmer using wireless technologies.

3. Crop health monitoring and management. For this purpose, sensors are integrated with wireless devices through microcontrollers, and data is transmitted.

4. Livestock monitoring is done for stock health, well-being, and location tracking. This can be accomplished by tagging an animal in order to monitor its health and track its location. Sensors, once again, come into play. Sensors are integrated with the network protocols (already discussed in this book). These protocols, which are used by wireless technology devices, send notifications about what is going on with farmers via their cellular devices.

5. Efficiency and making exact data-driven decisions in precision farming. This is achieved by KDD with the help of sensors. Data is saved in the cloud. This is, in turn, connected to a network. Sensors monitor this network, known as the Internet of Things. Finally, data is extracted and delivered to the farmers with the help of sensors. The

main advantage of using sensors is that data can be retrieved as and when farmers need it.

6. End-to-end farm management systems.

Table 16.1 represents the various protocols applicable with its performance parameters such as latency, bandwidth, energy requirements, reliability, computational requirements, and data security. Table 16.2 lists the wireless technology used for short-range applications and long-range applications and its protocol usage, networks type, data rate, frequency value, line of sight distance with an example. Table 16.3 shows the use cases for the various protocols used.

16.5 Biomimicry

Biomimicry is the concept of combining agricultural systems with natural ones while leaving the natural species alone. This was brought forward as a sustainable technology [10].

The six principles of biomimicry in agriculture are given below:

1. Biomimicry includes the self-perseverance of natural creatures' identities as self-constitution (autonomy and headstrongness).
2. Self-perseverance is included in biomimicry as self-regulation, self-healing/self-repairing, and adaptability of natural entities' identities to new or changing situations.
3. Biomimicry is flawed because it copies or uses the characteristics of natural things that come into being, change, and go away because of nature's complexity and variety.
4. Biomimicry recognizes the limitations of human monitoring and control in human-assisted eco-systems.
5. Biomimicry is flexible in how it looks at the identity of natural things; so it can work with new or changing arrangements.
6. Biomimicry addresses the eco-system in which they develop, unfold, and fade away, rather than just imitating or incorporating the esthetic shape or function of a natural organism.

16.6 Conclusion

Agriculture is the backbone of almost all the countries around the globe. People require food in order to survive. The use of pesticides and fertilizers

Table 16.1 Various protocols applicable.

Protocol	Latency	Bandwidth	Energy requirements	Reliability	Computational requirements	Data security
CoAP	Lowest	High	High	–	Simple	DLS protocol
MQTT	Low	High	High	Best fit	Simple	TLS protocol
AMQP	Significantly lower	Significantly lower	Significantly lower	–	–	SASL and TLS
Web Socket	–	–	–	–	–	TLS protocol
XMPP	Significantly lower	Significantly lower	Significantly lower	–	–	Built-in TLS
HTTP	–	–	–	Best fit	–	–

Table 16.2 Wireless technology used [8].

Distance	Protocol	Network	Data rate	Frequency	Line of sight	Examples
Short-range applications	IEEE 802.15.4	LPWAN	20–250 kbps	433 MHz, 868 MHz, 915 MHz, 2.4 GHz	100 m	Bluetooth, RFID
Long-range applications	IEEE 802.11ah	LoRa/ LoRaWAN	500 Mbps	900 MHz	1 Km	Cellular applications

Table 16.3 Use cases [11].

Soil trial	Use-case	Use-case challenge	Focal country	Chain role	Adopter type	Type of the soil
Arable	Field management	Developing field management by actuating devices with external data	NL	Farming and logistics	Early adopters and majority	Conventional/ organic
Dairy	Happy cow	Improving dairy farm activity through 3D cow activity sensing and cloud	NL	Farming	Early adopters	Conventional/ organic
Vegs	Chain-integrated greenhouse production	Integrating the value chain and quality innovation through full-sensor-actuator	SP	Farming, logistics, and compression	Majority	Conventional/ organic
Vegs	Added value, weeding data	Boosting value chain by harvesting of weeding data	NL, AT	Farming	Majority	Organic
Meat	Pig farm management	Optimization of pig production management	BE, NL	Farming, processing and consumption	Both	Organic/ conventional

would degrade the quality of crops and also affect their yield. This would also, in turn, affect humans when consumed. To overcome all these, IoT is combined with agriculture (smart farming). Smart farming is the future and has many benefits [9]. Some of the benefits are: monitoring of climate conditions is automated, as are monitoring and management, monitoring lighting, temperature, soil condition, and humidity for greenhouse automation, crop health monitoring and management, livestock monitoring is done for stock health, well-being, and location tracking, efficiency and making exact data-driven decisions in precision farming, and end-to-end farm management systems.

This chapter focuses on the technologies used in smart farming, such as the various protocols required, the various implementation models required, and so on. This chapter also explains the need for smart farming and the importance of smart farming in the near future. This chapter also discusses how data is processed and how the output is drawn. This chapter discusses the applications of various technologies and the need for them. It also discloses the various parameters taken into account during the monitoring of the health of crops, monitoring live stocks, and so on. It consists of an explanation of the digital twin implementation and the use case for the same. The figures of various levels of the digital twin are illustrated. This chapter has a brief explanation of the use cases and the various types of wireless technologies required for smart farming. This chapter is concluded by defining bio-mimicry and its six principles used in agriculture. Overall, this chapter discusses the various technologies required; the various implementations models, and briefly discusses a digital twin, a use-case, and bio mimicry and its six rules in agriculture. It also discusses the various applications of smart farming and the devices used.

References

[1] Glaroudis, D., Iossifides, A. and Chatzimisios, P., 2020. Survey, comparison and research challenges of IoT application protocols for smart farming. Computer Networks, 168, p.107037.

[2] Islam, N., Rashid, M.M., Pasandideh, F., Ray, B., Moore, S. and Kadel, R., 2021. A review of applications and communication technologies for internet of things (Iot) and unmanned aerial vehicle (uav) based sustainable smart farming. Sustainability, 13(4), p.1821.

[3] Muangprathub, J., Boonnam, N., Kajornkasirat, S., Lekbangpong, N., Wanichsombat, A. and Nillaor, P., 2019. IoT and agriculture data analysis for smart farm. Computers and electronics in agriculture, 156, pp. 467–474.

[4] Verdouw, C., Tekinerdogan, B., Beulens, A. and Wolfert, S., 2021. Digital twins in smart farming. Agricultural Systems, 189, p. 103046.

[5] Blok, V. and Gremmen, B., 2018. Agricultural technologies as living machines: toward a biomimetic conceptualization of smart farming technologies. Ethics, Policy & Environment, 21(2), pp. 246–263.
[6] Navarro, E., Costa, N. and Pereira, A., 2020. A systematic review of IoT solutions for smart farming. Sensors, 20(15), p. 4231
[7] Balafoutis, A.T., Beck, B., Fountas, S., Tsiropoulos, Z., Vangeyte, J., Wal, T.V.D., Soto-Embodas, I., Gómez-Barbero, M. and Pedersen, S.M., 2017. Smart farming technologies–description, taxonomy and economic impact. In Precision agriculture: Technology and economic perspectives (pp. 21–77). Springer, Cham.
[8] Idoje, G., Dagiuklas, T. and Iqbal, M., 2021. Survey for smart farming technologies: Challenges and issues. Computers & Electrical Engineering, 92, p.107104.
[9] Virk, A.L., Noor, M.A., Fiaz, S., Hussain, S., Hussain, H.A., Rehman, M., Ahsan, M. and Ma, W., 2020. Smart farming: an overview. Smart Village Technology, pp. 191–201.
[10] Farooq, M.S., Riaz, S., Abid, A., Abid, K. and Naeem, M.A., 2019. A Survey on the Role of IoT in Agriculture for the Implementation of Smart Farming. IEEE Access, 7, pp. 156237–156271.
[11] Walter, A., Finger, R., Huber, R. and Buchmann, N., 2017. Opinion: Smart farming is key to developing sustainable agriculture. Proceedings of the National Academy of Sciences, 114(24), pp. 6148–6150.
[12] Boursianis, A.D., Papadopoulou, M.S., Diamantoulakis, P., Liopa-Tsakalidi, A., Barouchas, P., Salahas, G., Karagiannidis, G., Wan, S. and Goudos, S.K., 2022. Internet of things (IoT) and agricultural unmanned aerial vehicles (UAVs) in smart farming: a comprehensive review. Internet of Things, 18, p.100187.
[13] Rajasekaran, T. and Anandamurugan, S., 2019. Challenges and applications of wireless sensor networks in smart farming—a survey. In Advances in big data and cloud computing (pp. 353–361). Springer, Singapore.
[14] Boursianis, A.D., Papadopoulou, M.S., Diamantoulakis, P., Liopa-Tsakalidi, A., Barouchas, P., Salahas, G., Karagiannidis, G., Wan, S. and Goudos, S.K., 2020. Internet of things (IoT) and agricultural unmanned aerial vehicles (UAVs) in smart farming: a comprehensive review. Internet of Things, p. 100187.
[15] Giua, C., Materia, V.C. and Camanzi, L., 2022. Smart farming technologies adoption: Which factors play a role in the digital transition? Technology in Society, p. 101869.
[16] Virk, A.L., Noor, M.A., Fiaz, S., Hussain, S., Hussain, H.A., Rehman, M., Ahsan, M. and Ma, W., 2020. Smart farming: an overview. Smart Village Technology, pp. 191–201.

17

Future of Farming

M. Gomathy[1], K. Kalaiselvi[2], and V. Sakthivel[3]

[1]School of Computing Sciences, Vels Institute of Science, Technology and Advanced Studies (VISTAS), India
[2]Department of Computer Applications, Saveetha College of Liberal Arts and Sciences, SIMATS Deemed to be University, India
[3]Konkuk Aerospace Design-Airworthiness Institute, Konkuk University, South Korea
Email: marimuthu.gomathy@gmail.com; kalaiselvik.sclas@saveetha.com; sakthivel@konkuk.ac.kr

Abstract

Duc to the rising global population, farmers must work harder to produce enough food. In this century, farmers must overcome numerous difficulties. In addition to dealing with climate change, they must also accept and understand new technology, increase farm productivity, and satisfy the demands of an expanding population. To fulfill the increased demand for food worldwide, farmers in rural areas should increase production despite the lack of workforce. The majority of food produced is not eaten. Food waste also drives up resource and labor demands, which are essential to agriculture. This chapter explores the factors driving the rise in global food demand as well as how farmers are embracing Agriculture 4.0, the agricultural industry's revolution. Upcoming agriculture will rely on smart farming, which employs cutting-edge IoT techniques to solve all the problems encountered by producers. Agro businesses should overcome the demographic transition, the effect of greenhouse gases from agricultural lands, and the impact of climate change. Future farming should address these challenges by adapting to modern techniques and technology at affordable prices. New techniques like urban farming, minimizes food miles by producing products nearer to the consumers and vertical farming, crops are grown vertically, requiring less water, soil, and space has shown impressive results in the adapted countries. Genetic engineering, and applying 3D printing technologies help up new possibilities, automated cooking, induces nutrition value, as well as these techniques, reduce food wastage. Following integrated cross-industry technologies like IoT, nanotechnologies, drone technology, and blockchain in agriculture helps to produce high-yield crops in lowering expenses. As precision agriculture expands and farmland becomes more networked, efficiency and production will rise in the future.

17.1 Introduction

By 2050, there will be 9.7 billion people on Earth. As a result, the growers must generate more than what is required to meet the food requirements. For farmers, climate change is a persistent challenge. Farmland output has decreased as a result of decreasing labor availability due to urbanization. Farmland deteriorates as a result of improperly managed fallow periods, crops, an imbalanced application of fertilizer, and excessive livestock grazing. The past 50 years have seen a nearly two-fold rise in greenhouse gas emissions due to agribusiness, forest management, and other land uses. Around 30%–50% of the food produced globally is never consumed. Landscapes have indeed been destroyed of trees, species have gone extinct, native communities have been relocated, and the soil has deteriorated to produce food that is ultimately squandered. All of these pose difficulties for farmers' crop production and yields.

To meet the need for food from the expanding population, modern agribusiness should focus on the global scenario. Agriculture 4.0 and smart farming provide various solutions to overcome the barricades of the farmers and increase the production level. Implementing modernized farming practices and embracing new technologies can help farmers solve issues and boost their ability to produce more.

Vertical farming utilizes a soil-free methodology and maximizes productivity while using the least amount of natural resources. Algae feedstock is a substitute for protein-rich foods, which helps to be delicious, nourishing, and fully sustainable. Some of their main benefits include their ability to grow in both freshwater and saltwater, their resistance to effluent, and their capacity to reuse water naturally. Desert farming and seawater farming have proved to utilize the arid land and saline water for agriculture in many countries like Saudi Arabia and Australia. Adapting to smart farming with modern technologies helps the farmers to have sustainable growth in the crops and productivity.

Sensors and drones will be used for precision farming, which increases the quality and environment cost-effective crops. Small and marginal farmers will also be using these technologies and will make agriculture more profitable, easy, and environmentally friendly. By minimizing nutrient losses during fertilization and reducing chemical waste, nanotechnology will be employed in agriculture to boost production through pest and nutrient control. The forthcoming agricultural business implementing smart farming practices has the potential to turn all of agriculture's difficulties into advantages.

17.2 Obstacles in the Farming Sector

Several problems, including food and nutrition security, poverty, and agro-ecology, are encountered in the agricultural industry. In addition to demographic pressure, resource depletion, climate change, and food security, global development have also put farming under pressure.

17.2.1 Demographic transition will increase demand for food

Population growth:
While the population of the planet is still increasing, there are already 7.3 billion people still living, and during the next 15 years, another billion people will be added. In 2050, there would be 9.7 billion people on Earth. Food demand will rise steadily in economically developed nations due to population expansion. Due to the demographic changes, the way they consume food has also altered. The majority of individuals enjoy eating foods that are high in protein. This is a result of urbanization and rising income.

Increase in urbanization:
As urban populations grow and urban residents' diets and preferences alter, there are significant changes in the demand for agricultural products. Due to this, there have been and still are significant changes in how expectations are addressed. Food security in rural and urban areas may potentially be severely hampered.

Metropolitan growth will eventually engulf some rural land as markets and property values move closer to urban areas. Many landowners decide to keep their properties vacant in the hopes of making money when they sell them or put them to use for purposes other than agriculture. Most metropolitan areas in low- and middle-income countries grow haphazardly due to the lack of a land-use plan. Where various homes, businesses, and government-related activities choose to legitimately or illegally locate and expand determines this expansion.

Greater demand, which results from a growth in the population, drives up output. According to the Food and Agriculture Organization (FAO), farmers

Figure 17.1 Agriculture and the impact of desertification.

must produce 70% more food by 2050. The entire agriculture business model must be considered to tailor this food to the demands of the expanding urban population.

Agriculture's productivity is rising, thanks to investments and innovation, but yield growth has slowed to levels that are insufficient for the general public's comfort. Despite a population decline, there is a rise in demand for food in rural areas. Rural populations are also aging quickly, which has a significant impact on the labor force, production patterns, land tenure, and economic growth.

17.2.2 Current uses of natural resources are highly stressed

Cropland around the world is getting less and less productive. According to certain measurements, a quarter of all agricultural land is already considered to have extremely deteriorated, and another 44% is severely or somewhat disrupted. Deterioration of farmland is mostly brought on by agribusiness, which also indirectly suffers from it. Inadequately planned fallow times, plantings, and overgrazing by animals are all factors that contribute to soil erosion. Nutritional imbalance is brought on by the unbalanced use of fertilizer to increase yield.

Nutrient imbalance is brought on by the unbalanced application of fertilizer to restore yield. Agribusiness concerns account for over 80% of worldwide deforestation. And while removing vegetation to find space for crops is essential for land clearance and does not directly result in soil deterioration, it does so indirectly by depleting water supplies. Irrigation technologies have increased consumption effectiveness, yet expanding populations raise legitimate concerns about water availability and security. For irrigation water management alone in developing nations, it is estimated that investments totaling $1 trillion will be required until 2050. Figure 17.1 shows the impact of deforestation and degraded land, which raises on the investment of irrigation.

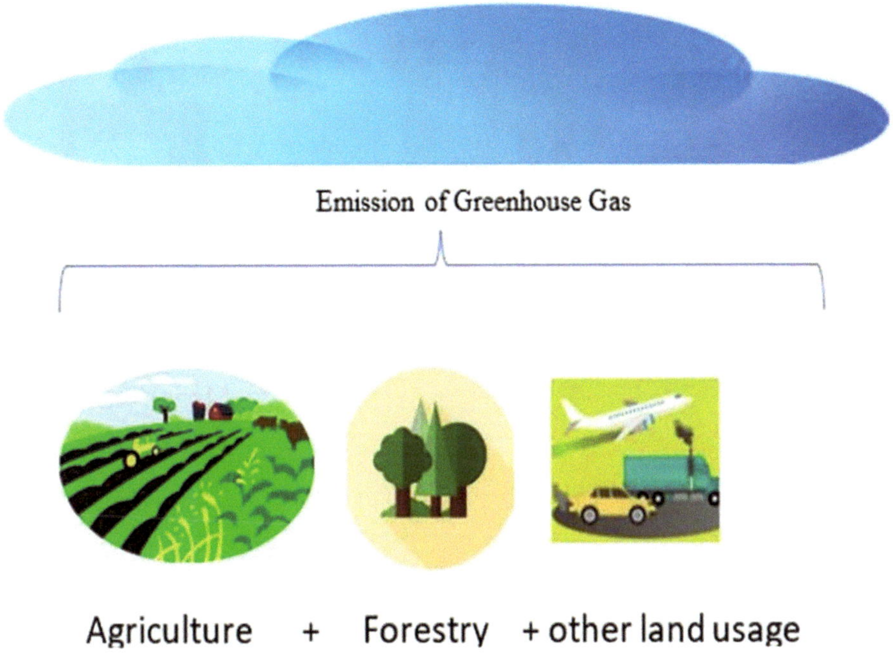

Figure 17.2 Rationale for climate change.

All of these issues are the product of poor foresight and planning. Land shortage and poverty yield unsustainable land management practices, the direct causes of degradation named above. Poor farmers are led to clear forests, cultivate steep slopes without conservation, overgraze rangelands, and make unbalanced fertilizer applications.

17.2.3 Impact of climate change on agricultural productivity

The effects of climate change are causing a significant evolution of the earth. Human-caused Green House Gas (GHG) emissions make the greenhouse effect stronger and contribute to climate change. One of the main sources of GHGs is the agricultural sector. Figure 17.2 clearly explains that agribusiness, forest management, and other land uses have contributed to a nearly two-fold increase in greenhouse gas emissions during the previous 50 years. Methane and nitrous oxide emissions from agriculture account for the majority of these emissions. Furthermore, predictions point to a rise by 2050.

Figure 17.3 Impact of food waste.

17.2.4 Food waste – a huge ecological hazard

Approximately, 33%–50% of the foodstuff produced worldwide is never consumed, and this waste disposal is worth more than $1 trillion. Moreover, biowaste is harmful to the ecosystem. To produce food that is ultimately wasted, a land area larger than China is needed. Additionally, 25% of the world's freshwater use is made up of food that is never consumed. In addition to being a complete waste of resources, leftover food goes to the landfill, where without oxygen, organic matter breaks down and releases methane, a gas that is 23 times deadlier than carbon dioxide. Figure 17.3 shows the wastage of precious water resources used for the wasted foodstuff.

Outcome: poverty and hunger:
The issue of food scarcity is a result of these macroeconomic changes. Global poverty and famine are the results. 800 million people experience chronic hunger, 700 million people live in extreme poverty, and 2 billion people have nutritional deficiencies. One-third of the 800 million people reside in rural areas in emerging nations.

Over the past 50 years, the food and farming industries have advanced significantly. Since 1960, the use of Green Revolution innovations has increased the world's agricultural output, as well as the sector, which has become much more globally diversified. By utilizing these tendencies and combating income disparity, we can abolish world poverty and starvation. It will be vital to implement growth strategies that take into account not only agriculture but also employment creation and revenue diversification.

Significant investments must be made to help small-scale and family farming in emerging nations to eliminate this hunger war, enabling people to learn about and access services for sustainable agriculture. In emerging regions, the focus of sustainable agriculture switches from environmental concerns to issues about agricultural yields, crop diversification, and farmer revenue. People can maintain themselves as a result of the adoption of these farming practices because there is more supply of food. Farmers are urged to produce an abundance of food to feed their families and earn from the market by implementing the new effective agricultural techniques. Among them are techniques for insect management, developing sources of potable water, and emphasizing effective cropping systems. Following agricultural improvement that can support the population and the country is the key goal.

17.3 Agriculture 4.0: Future Farming with New Technologies

The food corporation's conventional strategy is going through a major shift. Agriculture's first technological revolution achieved significant progress. Modern farming techniques, such as irrigation, the use of fertilizers and pesticides, as well as the creation of new and more productive crop types, allowed cereal yields in East Asia to increase by nearly 300% between 1961 and 2004 (World Bank 2008). The upcoming agricultural revolution, known as Agriculture 4.0, needs to be technologically advanced and environmentally friendly. Agriculture 4.0 must consider both the consumers and producers of the issue of food scarcity.

Instead of just being an innovation, reengineering the value chain and utilizing technology in addressing the genuine demands of consumers is the key to Agriculture 4.0. The advancement of technology, including sensors, devices, machinery, and information technology, will revolutionize the operation of modern farms and agricultural enterprises. Robots, humidity and temperature sensors, aerial photographs, and GPS technology will all be used in agriculture in the future. These innovations will make it possible for enterprises to operate more profitably, effectively, safely, and sustainably.

The need to saturate entire fields with water, manure, and pesticides will be eliminated by Agriculture 4.0. As an alternative, farmers will employ the absolute necessities or possibly cut them out of the distribution chain entirely. They will be able to cultivate crops in desert regions and produce food using plentiful and pristine resources like the sun and ocean. Governments have a significant role to play in resolving the grain shortage problem. In addition to their conventional regulatory and facilitation roles, they must assume a wider and more prominent role. By adopting such a program and contesting the conventional legacy approach, governments can:

- make sure there is food security and lessen reliance on imported goods;
- increase your exports of fresh ideas and intellectual property in addition to goods;
- boost productivity and encourage the transition to a knowledge- and innovation-based economy.

The technical and digital developments are reshaping the sector and improving every link in the food production chain. AgriFunder reports that since 2012, the number of startups developing agricultural technologies has increased by more than 80% annually. Startups in the agritech industry are thriving, and business owners and investors are fawning over it. Business moguls Jack and Suzy Welch, Richard Branson, Bill Gates, and the VC firm DFJ, as well as the food giant Cargill, have all made significant investments in Memphis Meats, a cutting-edge clean-meat startup. Additionally, a start-up involved in vertical indoor farming will receive $200 million from the SoftBank Vision Fund of Japanese billionaire Masayoshi Son. The Agriculture 4.0 technology and solutions have offered hope to the issue of food scarcity.

17.3.1 Produce differently using new techniques

Hydroponics: In hydroponics, crops are grown without the need for soil and instead in nutrient-rich water that serves as the plant's root system. It enables farmers to cultivate food year-round, wherever in the universe, and with less input while generating higher yields. In hydroponic systems, the water being used can be purified, replenished with nutrients, and then sent back to the plants once again. This way, water is continuously recycled rather than being lost. 98% less water is used in vertical roots than in conventional soil-based systems. Most commonly, soil-based farming is where most of the pests and plant diseases are found. However, since there is a lower chance of pest infestation while growing in a hydroponic system, producers avoid using pesticides. Additionally, plants receive the necessary nutrients precisely through the solution, allowing them to grow more quickly and disease-free.

For instance, the Australian company Sundrop has created a hydroponics seawater technique that combines agriculture, distillation, and solar energy to grow veggies anywhere. This system is landless, dependent on the sun for energy, and sustainable. Instead, its innovations combine hydroponics, freshwater production, power production, and solar energy. The amount of food produced is comparable to that produced using conventional techniques. By using hydroponics, Sundrop can put a seawater greenhouse – a combination of solar, desalination, and agriculture – to grow vegetables anywhere in the world.

Algae feedstock: Algae can be added to cuisine as an ingredient. Brown algae are gathered, ground, and dried to make algal flour (porridge), a feed supplement for animals. The abundance of minerals including algae enhances the soil and provides a source of vitamins and minerals, while the high fiber content helps to retain moisture. As a result, algae can also be utilized as a premium fertilizer. Because algae do not require fresh water, they are an excellent alternative, especially for the water. The current global situation dictates that 70% of this water is used for livestock raising and crop irrigation. Algae, in contrast, may thrive in a variety of environments, including trenches, ponds, and oceans. They also require very little water to grow and can even flourish in the desert. Some algae species are so rich in proteins that it makes up 40% of their weight. This indicates that these algae produce seven times as much protein as soybeans in the same region. Thus, the future is getting ready to be both scrumptious and nutritious, as well as completely sustainable. Their ability to be grown in fresh and saltwater, their ability to withstand effluent, and their ability to organically filter water are just a few of their key advantages. Aside from that, algae naturally perform photosynthesis, which transforms CO_2 from the atmosphere into oxygen and purifies the air by lowering the number of greenhouse gases in the atmosphere. Additionally, fuel naturally degrades; so if there is a spill, the environment is not negatively impacted. The supply of algae is not reliant on the capture of fish, making it a more dependable source of feedstock. Producers can estimate the potential investment and have more control over costs as a result. An algae-based feedstock is a good replacement that is also reasonably priced.

Desert agriculture and seawater farming:
Oceans make up the majority of the water that covers the surface of the earth. The Earth's remaining landmass makes up about 29% of its total surface. One-third of the remaining 29% is made up of deserts of all kinds. The world has to convert deserts and oceans into food-producing areas to address the food issue, which will require the combined brainpower of the best minds, academic institutions, and research centers.

Research in desert agriculture is being led by Saudi Arabia's King Abdullah University for Science and Technology (KAUST). At KAUST, the Desert Agriculture Initiative aims to address the whole range of challenging issues that agriculture in a desert setting brings. Both biotic and abiotic variables are being studied by KAUST. Important biotic research areas include genome engineering technologies to control the growth and development of plants and biological systems; regulators of plant growth that enhance resilience to adversity; and hormones in plants that influence stem and root morphology depending on the availability of nutrients.

Harvest yield is lost due to heat, salt, and drought to the tune of about 60%. Abiotic stress helps to increase tolerance, which is key for crop productivity. The interaction with particular bacteria influences a plant's capacity to respond to harsh stress situations. KAUST is seeking to: discover bacteria related to plants that thrive in famine, salinity, and high heat find the molecular processes by which plants can adjust to the adverse environmental conditions brought on by the microbial interaction; and utilize the proper rhizosphere allies to assist boost crop food production while improving plant stress resistance.

Saltwater agriculture and arid farming are two scenarios where farming would be impossible, but recent technological advancements have enabled some businesses to grow food in the desert using only sunshine and seawater. The seawater greenhouse utilizes cardboard buildings that have been dampened so that air can flow over them and chill the environment. Although this idea has been used for ages to keep people cool, employing it to keep plants cool is a more recent development.

Seawater is forced up to the top of the cardboard building by a solar-powered pump, allowing water to cascade down the sides. The salt evaporates outside the building, adding to its structural support and providing a resource that can be taken advantage of. This less expensive approach to farming in the desert might even be done on a smaller scale by a small farmer!

To grow fruits and vegetables, an Australian, UK-based corporation has started planting initiatives in arid regions of the world. These cases illustrate that it is feasible to grow food even in some of the driest climates, despite the specific issues and challenges with living and farming in a desert.

17.3.2 Use new technologies – to increase efficiency in the food chain

17.3.2.1 Vertical and urban farming

The demand for food is rising in tandem with the population. More creative and effective farming techniques keep emerging as farmers try to close the gap between crop demand and supply by utilizing conventional agricultural practices. Growing food vertically is one of these techniques. Vertical farming is the practice of growing crops in layers that are stacked vertically rather than on a flat area like a field or greenhouse.

Utilizing controlled environment agriculture (CEA) technology, this farming technique keeps track of the necessary indoor humidity, temperature, gases, and light levels. For instance, farmers simulate natural sunshine using artificial lights and metal reflectors. Four primary factors affecting vertical farming are as follows:

Layout: To increase food production despite utilizing fewer resources, farmers cultivate crops using tower-like constructs.

Light: To guarantee maximum lighting effectiveness, cultivators employ a combination of artificial and natural light as well as tools like rotating beds.

Sustainability: The goal of vertical farming is to reduce the use of natural assets. This agricultural method reuses irrigation water as a result, reducing consumption by up to 95%. Additionally, farming indoors uses hardly any chemicals and pesticides, which conserves more commodities.

Growing medium: Aquaponics, aeroponics, and hydroponics are instances of soil-free farming techniques used in vertical farming. Peat moss and coconut husks are two additional frequent planting mediums used by farmers.

Maximizing productivity while using the least amount of natural resources possible is the primary goal of vertical farming. Compared to conventional farming methods, it strongly supports sustainable agriculture and precision farming. The difficulties presented by traditional agricultural methods are intended to be overcome by the vertical way of cultivation. These are as follows:

Unpredictable climatic conditions: Crops are grown vertically inside in regulated climatic conditions. It helps farmers grow crops throughout the year-long without bothering about yield loss due to bad weather.

Plant diseases: Pest infestations or hereditary plant diseases are reduced to a minimum under controlled indoor growth settings, incorporating non-soil growing media. Thus, growers can use fewer sustainable agricultural products.

Long supply chain: Considering the nature of vertical farming, farms can be established in densely populated areas with good transportation infrastructure. It cuts down on the journey time, geography, and intermediaries needed to deliver finished goods to buyers.

The main advantages of vertical farming are as follows:

Best space utilization: Ample plain and fertile land is needed for traditional farming techniques. Nevertheless, using vertical cultivation methods does not require that. This farming technique grows crops on layers of terraces that are tilted vertically. Additionally, because they use non-soil mediums, these crops can grow in a variety of settings and environments. Additionally, they may thrive in extreme and unfavorable environmental circumstances.

Lowered commuting costs: Vertical farms' delivery outlets are conveniently located in urban areas. Contrary to traditional farming, crop producers do not have to transport finished goods over international borders and oceans. Instead, they may establish farms near the sites of their clients. The supply chain is shortened, which lowers transportation costs and lowers the carbon footprint.

Sustainable and increased crop production: Since these farming methods are independent of environmental elements like sunlight, seasons, water, and area, among others, the most attractive aspect of vertical farming is likely its consistent food output throughout the year. A higher yield is also produced overall as a result of such reliable crop production. CropIn SmartFarm and SmartRisk are effective digital technologies that assist farmers in precisely forecasting crop output.

Reduced water use: Vertical farming uses hydroponic growing techniques, which use less water for irrigation. Additionally, even after use, the used water is immaculate, making it recyclable for the following farming cycle. As a result, these agricultural techniques only use 10% of the water that is available.

Use of no agrochemicals: Crops grown in regulated environments do not get invaded by weeds or bugs that spread illness. As a result, a farmer may require fewer crop protection products. In turn, this makes the finished product safer to eat, further enhancing food safety.

Energy efficiency: Some vertical farms have built-in energy production systems that cut energy expenses and lessen their carbon emissions. Other farms can connect with enterprises using green sources to combine heat and power and generate a ton of sustainable energy. Additionally, farms can send any surplus power to their local power grid or use it to power other commercial machinery.

Reduced workplace risk: In addition to being environmentally friendly, vertical farming lowers occupational dangers related to using large agricultural machinery. They also do not have to cope with dangerous agricultural chemicals or pathogenic bacteria. Because it has no negative effects on nearby flora and fauna, this farming technique also supports diversification.

Reduced labor costs: The vertical cultivation method is mostly reliant on technology. As a result, a farm that adopts a fully automated method of operation will only require a small amount of manual work. Farmers benefit from a better yield for lower labor costs, thus maximizing returns.

17.3.2.2 Genetic modification and cultured meats

Meat production is a new business known as "cellular agriculture," which uses cell-based nanotechnology to substitute conventional animal-derived goods like dairy, leather, flesh, and fish. The prospect of cultured meat was made possible by the discovery of stem cells, which allowed for *in-vitro* cell creation. Mammalian cell lines, cell and gene therapy goods, as well as current cell culture and biomanufacturing techniques, can be used to create flesh or nourishing components for personal consumption.

Enhanced food quality, such as vitamin or mineral content, better crop structure and function for greater production or progress in hostile climates, herbicide tolerance to lead to improved weed management with broad-spectrum herbicides, and insect infestation resistance are all potential advantages of genetically modified (GM) crops. Investment in genetic engineering research is justified by the prospective advantages of GM crops, particularly for underprivileged farmers in emerging regions. Innovations in genetic engineering enable the crops to be more susceptible to certain pesticides and disease resistant.

17.3.2.3 Applying 3D printing technology to food

The use of 3D printing in the food industry opens up new possibilities, like individualized nutrition, automated cooking, less food waste, etc. The food business can use 3D printing to address unmet needs related to individualized nutrition, food waste, consumption, and food supply. It is a rapidly developing technology that offers a wide range of advantages, including time savings, extreme efficiency, ecology, and many more. To prevent food waste, food production organizations are now focusing on approaches or processes that can assist them to utilize foodstuffs correctly to create wholesome meals. The population is expanding quickly, increasing the demand for food. But food shortages are also a result of food waste. Since 3D printing can efficiently use food resources with little to no waste, it is being used as an innovative solution to address this issue. Various prototype printers for food manufacturing are accessible on a worldwide platform. The food business will benefit greatly from 3D printing as it continues to advance.

According to experts, printers utilizing hydrocolloids (materials that gel with water) might be used to print food employing renewable raw materials like algae, duckweed, and grass in place of the traditional foundation components. Microalgae are a natural source of high-quality protein, carbs, colors, and antioxidants. The Netherlands Organization for Applied Scientific Research has created a printing technology that utilizes phytoplankton to produce meals like carrots. In essence, technology is transforming "mush" into food. In one investigation, scientists modified a shortbread biscuit recipe to include milled mealworm.

Future grocery shops may store "food cartridges" that survive for decades instead of perishable complete items, freeing up shelf space and lowering the need for shipping and warehousing. Meat replacements may be the most innovative and technically challenging 3D food printing solution. Some scientists are testing algae as an alternative to livestock protein, whereas others are attempting to produce meat from cow cells cultured in a laboratory.

17.3.3 Integrate cross-industry technologies and applications

As "precision agriculture" expands and farmland becomes more networked, efficiency and production will rise in the upcoming years. The number of agro IoT devices in use is predicted to reach over 75 million by 2020. In 2050, the typical farm will produce 4.1 million data points each day, which is an increase from 190,000 in 2014. However, even though the rise in network devices creates a significant potential for farmers, it also creates complications. Smart technologies have a huge potential to help farmers to produce higher and superior yields and develop additional opportunities to encourage resource-rich, ecologically sound agriculture. More advanced technology exists than others. However, the advances show a lot of promise. Some major game-changers are listed below.

17.3.3.1 Internet of Things (IoT)

IoT solutions are geared toward assisting farmers in bridging the resource gap by guaranteeing good harvests, financial success, and environmental protection. Precision agriculture is the process of utilizing IoT technology to guarantee the best allocation of resources to produce high crop yields and lower operating expenses. Advanced gadgets, remote access, programming, and IT services make up IoT in agro innovations. IoT-based smart farming helps farmers and producers minimize waste while boosting productivity. Through the use of sensors, IoT smart farming technologies is a device developed for plant field monitoring (light, moisture, temperature, transition level, crop health, etc.) and the irrigation system is automated. The state of the fields is always accessible to the farmers. Based on this data, they can choose between conventional and automatic solutions for taking the appropriate steps.

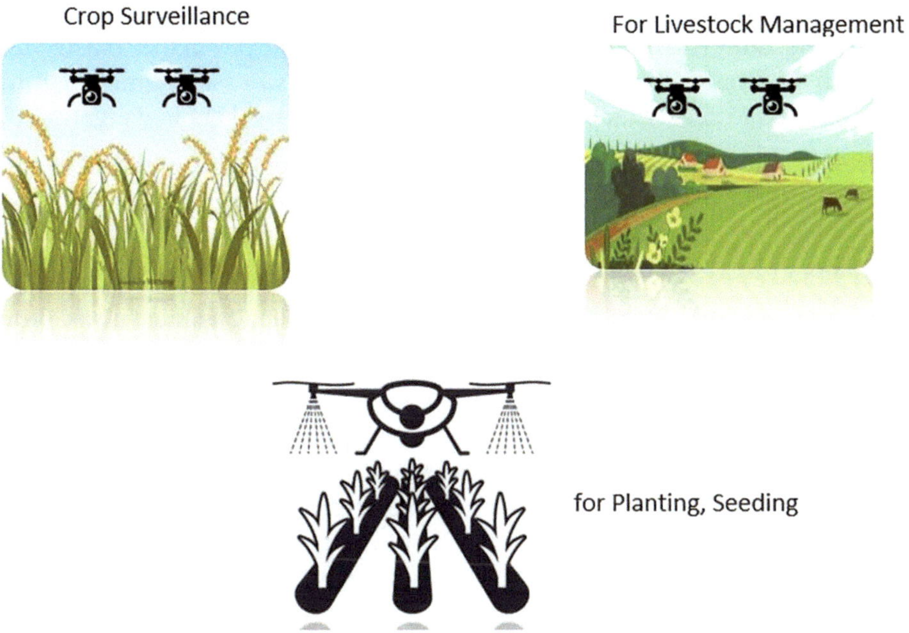

Figure 17.4 Significant usage of drones in smart farming.

17.3.3.2 Drone technology in agriculture

Drone technology in agriculture is rapidly expanding. The drone is an efficient method for achieving agro-ecological life. This streamlines operations and provides farmers, agricultural technologists, and agronomists with useful information on crops.

Crop surveillance: Drones are primarily utilized for monitoring crops. It involves monitoring the growth of a crop from the seed sown until harvest. The normalized difference vegetation index (NDVI) is a specialized imaging system that the drones are outfitted with. The NDVI provides precise color information to show the state of the plant. This enables farmers to keep an eye on crops as they develop and address any issues promptly to protect the crop.

Drone field monitoring: For healthy farming, the nature of the land needs to be closely monitored. However, it becomes challenging for farmers to keep an eye on such a huge expanse of land. The state of the soil and the state of the fields are monitored by the agricultural drone. The drone offers precise field data, including aerial imagery that aids in identifying field abnormalities. Figure 17.4 shows the significant usage of drone technology in smart farming.

Planting and seeding: Four million trees may be planted daily by ten drones. Currently, automated drone seeders are largely employed in the forestry sector, but this could change in the future as more industries adopt them. Drones can be used to access the locations without putting workers in danger.

Livestock management: Animals used as livestock are plentiful among farmers. Drones, sensors, and cameras can be used in this situation to effectively monitor and control the animals. Additionally, the drones may spot predators before an assault and can spot sick animals.

Spray treatment: To enhance effectiveness and reduce chemical costs, drone sprayers offer very thin spray treatments that can be focused on particular locations. It guards against farmers applying potentially harmful chemicals to their fields.

Security: With the aid of a drone, surveying the region may be done quickly instead of walking for a long time. To make sure everything is functioning properly, the drone camera offers a daily overview of agricultural operations. They can also be used to find objects and track wounded or endangered animals in distant viewing regions.

Drone use in agriculture is crucial for soil and field analysis, water management, and crop management. Drones fly and gather multispectral, thermal, and visual imagery. Farmers may use these data to learn more about a range of variables, including plant development and production forecast, plant height measurement, and more.

17.3.3.3 Blockchain in agriculture

Blockchain, coupled with IoT, is remodeling the food production industry. It is set to make farming a sustainable practice by using a simplified approach to optimize farming resources like water, labor, and fertilizers. Blockchain combined with IoT can facilitate farmers and other stakeholders in making optimum decisions.

Better farm inventory management: A lot of farming businesses lack the technology necessary to effectively manage their supplies. Waste of materials and commodities may result from this. Farmers lack the equipment needed to address the issue. Agro blockchain technology has the potential to permanently alter the situation. Inventory management is an excellent use case for blockchain technology.

Enhancing agricultural supply chain efficiency: The improved efficiency of the sector as a whole is another fantastic application of blockchain technology in agriculture. Small- and medium-sized farmers are more likely to lack access to expensive technologies that might improve the production process as a whole. It can swiftly solve all the factors, lower the cost of agricultural procedures, and improve the overall efficiency of the output by using a blockchain immutable ledger system.

Modernizing farm management software (FMS): Another benefit of blockchain in farming is the practice of updating agricultural software solutions. The client–server approach is the traditional one used by farm management software. Its effectiveness suffers as a result. Blockchain advances FMS to the next level. Furthermore, with the protection of blockchains, FMS software incorporates necessary security features. Consequently, farmers need not worry about cyber-attacks any longer.

Security for IoT optimization in agriculture: IoT devices are used by the agriculture sector to monitor yields, record weather, and land conditions, and take corrective measures. Perhaps more so, some gadgets can predict natural disasters in advance. Information was stored in cloud services by IoT devices. Cloud data are particularly susceptible to online attacks. Agricultural data can be well preserved with the aid of blockchain because it provides a better networking system for working and protecting the gathered data.

Providing fair pricing: Blockchain in the agricultural sector allows farming businesses to receive the remuneration they deserve for their output. They can sell their goods to honest customers using a blockchain-based marketplace that has already been established. They can even reach more customers than they could previously. This will enable them to negotiate the price. The farmers will then receive what they are legitimately owed.

Transparency in agricultural subsidies: Farmers receive unequal amounts of subsidies. Regarding the incentives, there is no openness. Larger groups, therefore, receive more than small-scale farmers. To increase transparency for this important issue, blockchain technology in the agriculture sector can be helpful. The public can monitor whether or not the money is going to where it should be by using a public blockchain to help the government transfer subsidies to the proper farms.

Therefore, blockchain technology applications in agriculture can help improve quality control and food safety, increase traceability in the supply chain, increase efficiency for farmers, and provide fairer payments for farmers.

Providing microloans for small- to mid-sized farmers: Obtaining microloans is a fantastic application of blockchain farming techniques. Blockchain technology can be used to provide essential peasants with microloans at a low interest rate. They can use blockchain to access microloans from creditors all across the world through the network. They can run their firm for a very long period by taking on the small mortgage rates associated with a little mortgage balance.

Bonus payment from consumers to farmers: Direct payments to farmers are a fantastic application for blockchain technology. Customers may simply connect to the blockchain network to reward the farmers if they are satisfied with the goods. Here, the technology will remain connected to the farmers' wallets so that they can collect tips from the patrons. Numerous blockchain applications for agriculture are already developing this kind of capability.

Incentivizing sustainable practices: Although many agricultural business practices are unsustainable, they nonetheless exist. In actuality, the problem is excessive pesticide use, which eventually contaminates water. If sustainable methods are not promoted, this undesirable behavior will continue. Therefore, the usage of blockchain in the agriculture sector can aid in providing incentives to farmers who are ready to utilize environmentally and consumer-safe practices rather than harmful pesticides. Typically, this incentive feature can persuade underdeveloped nations to use fewer chemicals and concentrate more on organic foods.

Therefore, blockchain technology applications in agriculture can help improve quality control and food safety, increase traceability in the supply chain, increase efficiency for farmers, and provide fairer payments for farmers.

17.3.3.4 Nanotechnology in agriculture

The agricultural industry faces many difficulties, including significant weather change, declining soil fertility, a lack of macronutrients and micronutrients, excessive use of agrochemicals, and the accumulation of heavy metals in the soil. But as the world's population has grown, so has the demand for food. By boosting crop yield and replenishing and upgrading soil health, nanotechnology has made a significant contribution to sustainable agriculture. Figure 17.5 explains the use of nanotechnology and its implementation in agriculture for the enhancement of growth and productivity. Nanotechnology is used in several facets of agriculture, including nanopesticide administration, the controlled and gradual release of biofertilizer-containing nanoparticles, potential relevance of nano biosensors for quick detection of phytopathogens and other biotic and abiotic stressors, and transfer of specific genes for crop development.

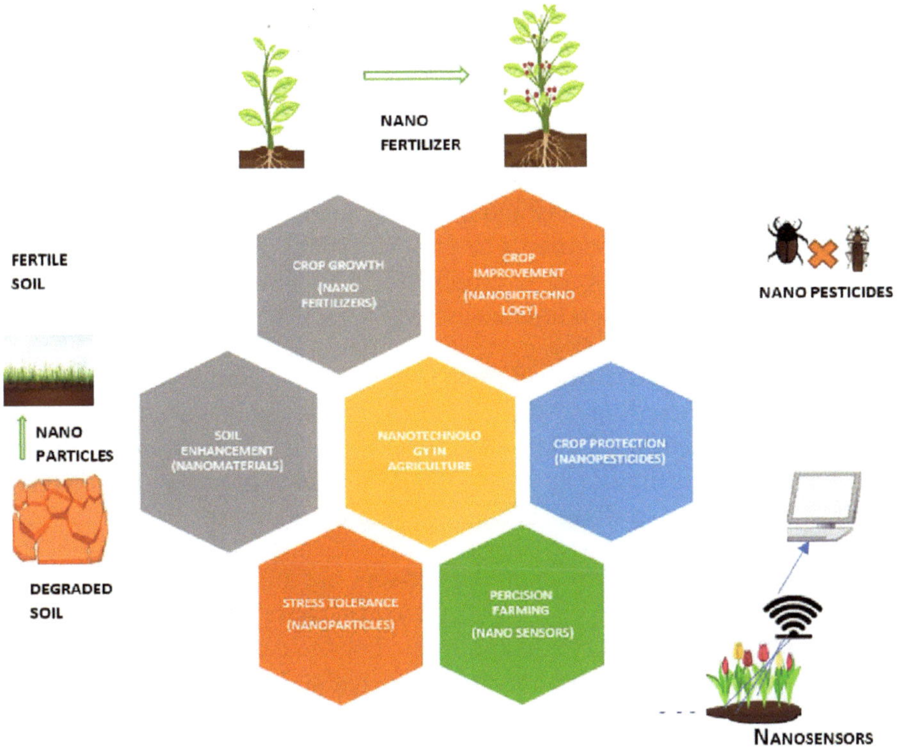

Figure 17.5 Nanotechnology in agriculture.

Implementation of agricultural nanotechnology to significantly raise crop productivity:

Nanopesticides and nanoherbicides: Crop yield has greatly risen as a result of the use of nanopesticides and nanoherbicides for weeds and pest management. The formulations for nanoherbicides use a variety of nanoparticle kinds, including inorganic and polymeric nanoparticles. This nanocapsule showed significant control over the diversity of species, minimal levels of genotoxicity, and a significant reduction in atrazine mobility in soil.

Nanomaterials for disease management: Each year, microbiological (virus, fungus, and bacteria) illnesses cause significant losses in the agriculture sector. Specific antibacterial nanomaterials aid in preventing microbial infestations. Many nanoparticles, including copper and nickel ferrite nanoparticles, have potent antifungal properties and are utilized to manage infection.

Chitosan nanoparticles, zinc oxide nanoparticles, and silica nanoparticles are beneficial in the treatment of viral infections, including the mosaic virus for tobacco, potatoes, and alfalfa.

Nano fertilizers: To address nutrient insufficiency in plants, scientists used nanotechnology to create a smart delivery system that would transfer nutrients to the targeted site slowly and under regulated conditions. By improving the availability of vital nutrients to the plant, nano fertilizers boost agricultural output. A notable boost in crop yields following the use of nano phosphors fertilizers in dry conditions has been seen.

Nanotechnology in seed development: The crop yield depends on several crucial factors, including seed quality. Carbon nanotubes have been seen to penetrate the tough seed coat of tomatoes, greatly enhancing germination and plant growth. Similar to this, spraying multiwall carbon nanotubes on soybean and maize seedlings boosted the seedling growth.

Nano biosensors: Modern nano biosensors are much more sensitive and precise. Through the use of a microprocessor, these devices transform biological reactions into electrical reactions. For the explicit or implicit detection of infectious microorganisms, drug resistance, insecticides, toxins, and heavy metal pollutants, nano biosensors provide real-time signal monitoring. Additionally, crop distress, soil quality, plant development, nutritional composition, and quality of the product are all monitored using this technology.

Futuristic strategies for sustainable farming using agricultural nanotechnology:

1. Nanoparticles made using a regulated green synthesis.
2. Knowledge of the significance of root endophytes and mycorrhizal fungi's production of nanoparticles in plant growth and infection control.
3. How nanoparticles interface with agroecosystems, such as the way they move inside of crops.
4. A critical assessment of the detrimental effects of nanoparticles on various environmental circumstances.
5. Creation of lightweight, user-friendly nano biosensors for quick analyses and decision-making on water, soil, plants, and pesticides.

17.4 Conclusion

Smart, precision farming using IoT is the way of the future of agriculture. Of all, if they do not deal with global problems, none of these advances will be of any benefit. Future generations will benefit from wise farming techniques. The idea that a network of intelligent sensors, actuators, cameras, robots, drones, and other connected devices will provide agriculture with an incredible measure of influence and analytical decision-making has been propelled by the Internet of Things. This will pave the way for a bold ecosystem of innovation in the most established companies. Agricultural IoT and smart farming are establishing the foundation for a Third Green Revolution. The plant hybridization and genomics innovations were followed by the Third Green Revolution, which is now transforming agriculture. The Internet of Things (IoT), big data analytics, robots, unmanned aerial vehicles (UAVs or drones), and other data-driven analytics technologies are all utilized in this renaissance in precision agricultural machinery. This smart farming boom predicts a future with less agrochemical use and greater overall performance. IoT technologies will lead to an improvement in food security by improving food monitoring. Additionally, it will improve the environment by, for instance, enabling more efficient irrigation use, purification, and supply optimization. Therefore, precision farming provides a chance of bringing about a more resilient and profitable kind of crop production built on a more accurate and resource-efficient strategy. The long-held ambition of humanity will, at last, be realized thanks to new farms. It will feed the next generation, which is predicted to number 9.6 billion by 2050.

Acknowledgement

"Team work is the secret that makes common people achieve uncommon results."

We are thankful to the almighty for his abundance of blessing showered on us in all walks of our life.

Our profound gratitude goes out to the president of the Vels Institute of Science, Technology, and Advanced Studies for supporting us and giving us the chance to learn about and implement contemporary scientific technologies.

We would like to extend our deep appreciation to the Vels Institute of Science, Technology, and Advanced Studies' administration and personnel for their unwavering support and inspiration throughout the writing of this chapter.

It gives us great pleasure to express our sincere appreciation to everyone who has supported our efforts to finish the task.

References

[1] Muhammad Ayaz, Mohammad Ammad-Uddin, Zubair Sharif, Ali Mansour and El-Hadi M. Aggoune," Internet-of-Things (IoT)-Based Smart Agriculture: Toward Making the Fields Talk", IEEE Access2019, vol 7, pp 129551–129583

[2] Heena Panchasara, Nahidul Hoque Samrat and Nahina Islam, "Greenhouse Gas Emissions Trends and Mitigation Measures in Australian AgricultureSector—A Review", Agriculture 2021, Vol 11, no 85

[3] AndrésVilla-Henriksen, Gareth T.C.Edwards, Liisa A. Pesonen, OleGreen, Claus Aage Grøn Sørensen." Internet of Things in arable farming: Implementation, applications, challenges and potential", Biosystems Engineering 2020,Vol 191, pp 60–84.

[4] Jeongeun Kim, Seungwon Kim, Chanyoung Ju, Hyoung II Son, "Unmanned aerial vehicles in agriculture: A review of perspective of platform, control, and applications". IEEE Access 2019, Vol 7, pp 105100–105115

[5] UM RaoMogili, B B V LDeepak, "Review on application of drone systems in precision agriculture.", Procedia Computer Science 2018, Vol 133, pp 502–509.

[6] Pankaj Agarwal, Vijander Singh, G. L. Saini and Deepak Panwar," SustainableSmart-farmingframework: smartfarming",.InSmartFarming Technologies for Sustainable Agricultural Development; IGI Global: Hershey, PA, USA, 2019; pp. 147–173. Available online: https://www.igi-global.com/chapter/sustainable-smart-farming-framework/209549

[7] Manlio Bacco, Andrea Berton, Erina Ferro, Claudio Gennaro, Alberto Gotta, Stefania Matteoli," Smartfarming: Opportunities, challenges and technology enablers", In Proceedings of the 2018 IoT Vertical and Topical Summit on Agriculture-Tuscany (IOT Tuscany), Tuscany, Italy 2018, pp. 1–6

[8] Anandhavalli Muniasamy," Machine Learning for Smart Farming: A Focus on Desert Agriculture", Published in: 2020 International Conference on Computing and Information Technology (ICCIT-1441), IEEE, doi : 10.1109/ICCIT-144147971.2020.9213759

[9] Ahmad F. Subahi , Kheir Eddine Bouazza," An Intelligent IoT-Based System Design for Controlling and Monitoring Greenhouse Temperature "IEEE 2020, Vol 8.

[10] Zongyou Lv, Rui Jiang, Junfeng Chen,Wansheng Chen, "Nanoparticle-mediated gene transformation strategies for plant genetic engineering", The Plant Journal 2020 , Vol 104, pp 880–891

[11] Madhu Kamle, Dipendra Kumar Mahato, Sheetal Devi, Ramendra Soni, Vijay Tripathi, Awdhesh Kumar Mishra & Pradeep Kumar," Nanotechnological interventions for plant health improvement and sustainable agriculture", 3Biotech 2020, Vol 10.

[12] Henry I. Miller, "Genetic engineering applied to agriculture has a long row to hoe", Biotechnology in Agriculture and the Food Chain 2018, Vol 9, no 1.

[13] K. Thea, C. Martin, M. Jeffrey, E. Gerhard, Z. Dimitrios, M. Edward, and P. Jeremy, "Food safety for food security: Relationship between globalmegatrends and developments in food safety," Trends Food Sci. Technol. Oct 2017, Vol. 68, pp. 160–175.

[14] Code of Conduct on Agricultural Data Sharing Signing. Accessed: Apr. 13, 2019. [Online]. Available: https://www.ecpa.eu/news/code-conduct-agricultural-data-sharing-signing

[15] Xiaojie Shi, Xingshuang An, Qingxue Zhao, Huimin Liu, Lianming Xia, Xia Sun and Yemin Guo, "State-of-the-art Internet of Things in protected agriculture", Sensors 2019, Vol. 19, no. 8, pp 1833.

Index

S

T

U

V

About the Editors

Dr. K. Kalaiselvi presently working as Associate Professor in the Department of Computer Applications, Saveetha College of Liberal Arts and Sciences (SIMATS), Chennai, India. She has more than 20 years of teaching experience at both UG and PG level. Her research interests include knowledge management, data mining, embedded systems, wireless sensor networks, big data analytics and knowledge mining. She has published more than 90 research papers in various international journals, including Scopus Indexing, Web of Science and UGC care list Journals. She has also published chapters for various publishers like Springer, Wiley, and Emerald Publishing. She has acted as editor and reviewer for various reputed journals. She acted as editor for the book "Handbook on Intelligent Healthcare Analytics – Knowledge Engineering with Bigdata" in 2022. She has also published books including "Protocol to learn C Programming Language", "Tryambaka Programming Code {C, C++, Java}", and "A Handbook on Python Programming for Naive Seekers to Expertise", and "Programming in Java Beginners Guide". She has received the Best Researcher award, Young Educator and Scholar award, and Best Faculty award from various reputed organizations.

Dr. A. Jose Anand received his Diploma from the State Board of Technical Education, Tamil Nadu in 1995, Bachelor of Engineering Degree from Institution of Engineers (INDIA), Calcutta in 2003, Master of Engineering in Embedded System Technologies from Anna University, Chennai in 2006, Master of Arts in Public Administration from Annamalai University in 2000 Master of Business Administration from Alagappa University in 2007, and Ph.D. (Information and Communication Engineering) from Anna University in 2017. He received State 3rd Rank in Bachelor of Engineering. He is a Member of CSI, IEI, IET, IETE, ISTE, INS, QCFI and EWB. He has one year of industrial experience and 23 years of teaching experience. He has presented several papers at National Conferences and International Conferences. He has published several papers in national and international journals, and also books for polytechnic and engineering subjects. He is recognized as a research supervisor by Anna University.

Dr. Poonam Tanwar received his B.Tech and M.Tech in Computer Science & Engineering from Maharishi Dayanand University, India in 2001 and 2009 respectively, and his Ph.D. in computer science and engineering from the Uttarakhand Technical University, India in 2015. She has 18 years of teaching experience working as associate professor in Manav Rachna International Institute of Research & Studies, Faridabad, India. She has published more than 50 research papers in various International Journals and Conferences. She has one copyright and filled 6 patents. She edited the book "Big Data Analytics and Intelligence: A Perspective for Health Care" in 2019 published by Emerald. She was Guest Editor for a special issue "Advancement in Machine Learning (ML) and Knowledge Mining (KM)" for International Journal of Recent Patents in Engineering (UAE). She has been awarded the woman researcher award by VDGOODS Academy Chennai. She has organized various Science and Technology awareness programs for rural development. She is a technical program committee member for various international conferences like ICIC 2018, ICFNN, Rome (Italy), European Conference on Natural Language Processing and Information Retrieval, Berlin (Europe), etc.

Dr. Haider Raza, School of Computer Science and Electronic Engineering (CSEE), University of Ulster, UK. Haider Raza received a bachelor's degree in computer science and engineering from the Integral University, India in 2008, a master's degree in computer engineering from the Manav Rachna International University, India in 2011, and a PhD in computer science from University of Ulster, UK in 2016. During his PhD, he won the best-literature review award sponsored by McGraw Hill. He is currently a Lecturer in AI for decision making at the School of Computer Science and Electronics Engineering, University of Essex, UK. Dr. Raza has a strong academic track record and a developing international profile. Highlights of his work include research funding (£620k) as PI for a range of research projects in the domains of healthcare analytics, business, and environmental science by Innovate UK, ESRC and NHS Trusts. He has published in peer-reviewed journals such as IEEE TNSRE, IEEE JBHI, and IEEE TCDS, and 16 conference papers in major international conferences. He is currently leading on two knowledge transfer partnership projects (KTP) and supervising two post-doc KTP associates funded by Innovate UK and has led on a project in healthcare with East Suffolk North Essex Foundation Trust on AI-based cancer pathways.